Recent Developments in Poultry Nutrition

Editors
D.J.A. Cole, PhD
W. Haresign, PhD
University of Nottingham School of Agriculture

Butterworths
London Boston Singapore Sydney Toronto Wellington

First published 1989

© The several contributors named in the list of contents, 1989

British Library Cataloguing in Publication Data

Recent developments in poultry nutrition.
   1. Poultry. Nutrition
   I. Haresign, William II. Cole, D. J. A.
   (Desmond James Augustus), *1935–*
   636.5′0852

   ISBN 0-407-01513-2

Library of Congress Cataloging-in-Publication Data applied for

Printed and bound in Great Britain by Anchor Press Ltd, Tiptree, Essex

Recent Developments in Poultry Nutrition

CONTENTS

INTRODUCTION

The success of previous volumes in this series, *Recent Developments in Ruminant Nutrition, Recent Developments in Pig Nutrition* and *Recent Developments in Ruminant Nutrition – 2,* has stimulated the production of a sister text based on poultry nutrition. Like the others in the series, *Recent Developments in Poultry Nutrition,* draws together, in one volume, important chapters from recent past proceedings of the University of Nottingham Feed Manufacturers Conference. It is inevitable therefore that each chapter will be based on sound scientific principles, while at the same time paying particular attention to the practical application of the information contained in them. The importance of nutrition on the profitability of animal production is well recognized, and there is continued need to seek to improve the efficiency with which feedstuffs are used for productive purposes. The chapters contained in this volume have been chosen to examine important developments which have taken place in poultry nutrition in recent years. Since the original proceedings were published it is recognized that some further progress has been made in certain areas of poultry nutrition research. However, in the limited instances where this applies, the chapters are still considered to be landmarks in our understanding of poultry nutrition, and it is hoped that the book will be of value to students, teachers, advisory staff, research workers and many others.

Of particular significance in the application of research to practical nutrition is the correct interpretation of response data. It is pertinent therefore that the first chapter should seek to discuss this very important issue. Recent years have seen the introduction of the metabolizable energy (ME) system for poultry as well as the compulsory declaration of the ME value of poultry feeds. A number of chapters are therefore devoted to this topic, and seek to provide information on how the ME content of feeds can be measured directly or predicted from values determined by simple laboratory techniques for legislative and formulation purposes. In addition, attention is focused on the influence of fibre on the digestibility of poultry feeds and the importance of dietary anion-cation balance on the efficiency of production.

It is important that any animal production system pays particular attention to consumer perceived aspects of product quality and responds to changes in them. A series of chapters is devoted to consideration of the various factors which influence quality, and how these might be manipulated to maximize it as economically as possible. For example, with the current consumer trend towards more 'natural' foods, it is timely that a chapter is devoted to the use of 'natural' rather than 'artificial' products for egg yolk pigmentation. The downgrading of eggs is primarily due to cracking of the shell, and attention is therefore given to the various factors influencing eggshell formation and quality, particularly the effect of dietary phorphorus.

Advances in the nutrition and production of animals have often had their origins in poultry. For many reasons scientific and technological developments have been most highly developed in the poultry industry. Clearly the way ahead will involve much greater precision in both our understanding and application of knowledge of poultry nutrition.

1

THE INTERPRETATION OF RESPONSE DATA FROM ANIMAL FEEDING TRIALS

T.R. MORRIS
Department of Agriculture and Horticulture, University of Reading, UK

Many animal feeding trials are conducted every year in universities and research institutes and on feed manufacturers' research farms, both in the UK and elsewhere in the world. Some of the data eventually get published. In view of the large investment in this activity it is surprising how little has been written about the methodology of interpreting such experiments. There are a few recent papers (e.g. Lerman and Bie, 1975; Robbins, Norton and Baker, 1979; Heady, Guinan and Balloun, 1980; Ware *et al.*, 1980) but most of those who carry responsibility for nutritional trials either use their own preferred method to interpret data (which, though lacking a logical foundation, often produce wise judgements) or hand the results over to a professional statistician for analysis. The statistician applies rigorous logic but does not always arrive at appropriate conclusions. The worst conclusions of all are apt to be drawn by graduate students, who write most of the papers that eventually get published and who lack both the wise judgement of the experienced feed formulator and the extensive skills of the professional statistician. Thus it comes about that many dose/response trials are interpreted with the aid of nothing more elaborate than a Student's t-test or a multiple range test, which is rather like trying to peel an apple with an axe.

Before proceeding to the main discussion of how response data should be interpreted, there are some important assumptions to be made. Firstly, we will assume that the data come from trials which are properly designed and adequately replicated. These matters are discussed in the other two chapters in this section and will not be reviewed again here. Nevertheless, it is important to understand that many nutritional trials have too few treatments or too few replicates to give any hope of answering the question for which they were designed. Secondly, we will assume that thought has been given to the scales used for measuring inputs and responses. For example, trials investigating amino acid responses in *ad-libitum* fed animals may make more sense if the intake of the limiting amino acid is plotted on the abscissa, rather than the dietary concentration of that amino acid; trials with dairy cows may make more sense if energy output (calculated from

First published in *Recent Advances in Animal Nutrition – 1983*

milk yield and composition and liveweight changes) is taken as the response criterion, rather than the volume of milk. The choice of appropriate scales is an important matter, but further discussion of it is beyond the scope of this chapter. Thirdly, we will assume that nobody is trying to estimate 'the requirement' for a nutrient. This is not because requirements are known, but because the term 'requirement' is unhelpful. What the practical nutritionist needs to know is the rate at which animals in a given class, in a reasonably well-defined nutritional and environmental context, will respond to incremental inputs of a given nutrient. Armed with this information, and a knowledge of his marginal costs and the value of extra output, he can calculate an optimum dose.

To illustrate the point that the method of analysing results can make a large difference to the conclusions drawn, let us consider some data from a

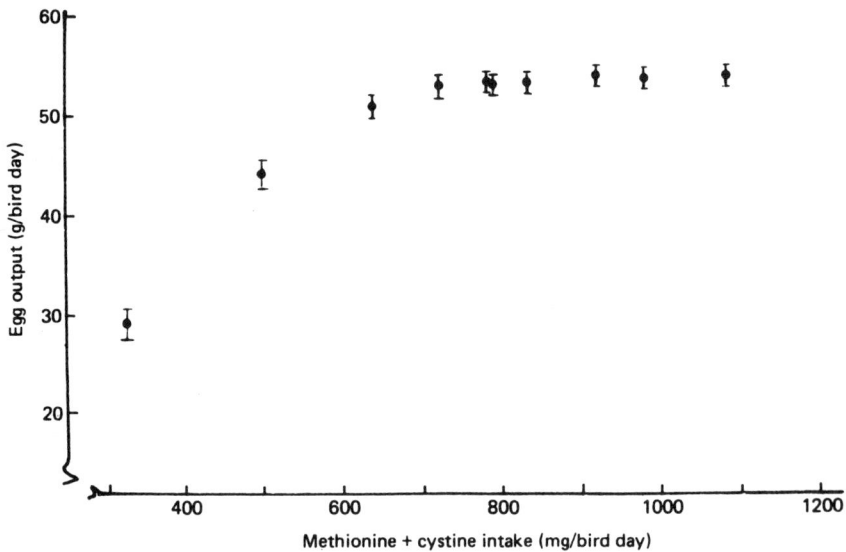

Figure 1.1 Data from an experiment reported by Morris and Blackburn (1982), in which laying pullets were fed from 30 to 40 weeks of age on diets of varying protein content (methionine being the first limiting amino acid in the protein mixture used). The data plotted are means with standard errors, taken from the last four weeks of the trial

recently published chicken experiment, as reproduced in *Figure 1.1*. This was a well-replicated trial (54 groups of 72 laying pullets) with ten dietary treatments. The data are therefore more precise and more wide-ranging than the results of most animal feeding trials and, whatever problems may arise in interpretation, they cannot be blamed on the use of inadequate resources. We will consider in turn a number of procedures which are commonly used to interpret results such as those in *Figure 1.1*.

Duncan's multiple range test

If one compares treatment means with the aid of a standard error calculated from replicate groups (there were three replicate groups of 72 pullets allocated to the two lowest treatments and six replicates for each of the remaining treatments), diets 1, 2 and 3 all differ significantly, but there are no significant differences ($P = 0.05$) between diet 3 and any of the remaining treatments. This conclusion is reached whether one uses a variance ratio (F test) or a multiple range test (such as that due to Duncan, 1955). This method of handling data has been used more commonly than any other in the literature (and especially so in the USA) and it is wrong for two reasons. Firstly, the comparison of treatments by means of a multiple range test is inappropriate when there is a logical structure to the set of treatments. Secondly, the use of a conventional 5% probability value is inappropriate when trying to obtain the best estimate of some end point, as opposed to requiring a high degree of confidence that we have not gone too far along some input scale. For example, if we can show that the odds are 5:1 that by spending another £1 on *dl*-methionine we can get back another £2 worth of eggs, it would be foolish to conclude that we should not spend the £1, because the odds in our favour are less than 20:1. The conventional 5% probability level is akin to a judgement made in a criminal trial when an issue has to be proved 'beyond reasonable doubt'. In making judgements about the optimum input of a nutrient we need something more akin to the 'balance of probability' argument which is used when apportioning liability in a civil suit. Since these are well worn arguments to the professional statistician, and he would immediately counsel some form of regression analysis for an input/output trial such as *Figure 1.2*, it is surprising that so many papers have been published in which Duncan's multiple range test has been used to support false conclusions.

The bent stick

Numerous authors have used simple regression analysis to interpret their data. The commonest procedure is to assume that response is a linear

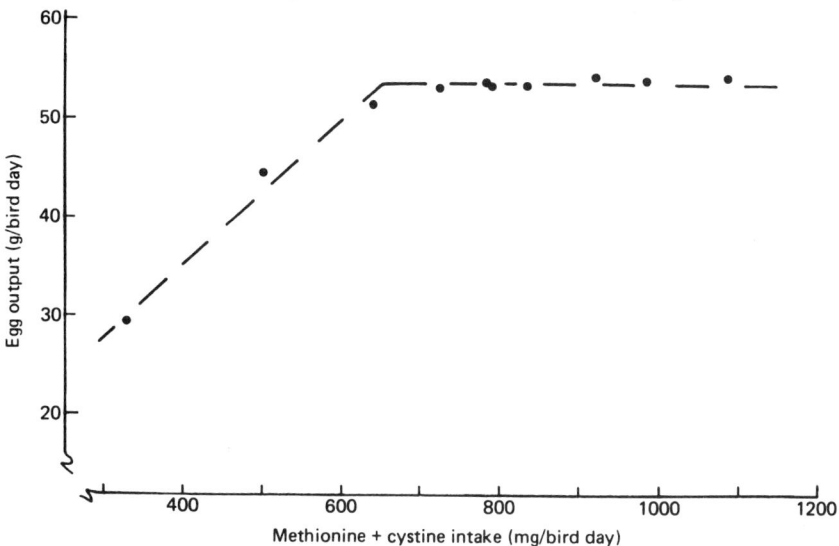

Figure 1.2 The data from *Figure 1.1*, with a 'bent-stick' model fitted by minimizing the residual sum of squares

function of input up to some threshold value, at which the response abruptly ceases. This 'bent-stick' model is fitted to the data of Morris and Blackburn (1982) in *Figure 1.2*. The model is a good fit to the data, the deviation's mean square being slightly less than the error mean square calculated from replicate groups. However, it is a simple matter to show that this particular model always leads to false deductions about the optimum input.

Suppose, for the sake of argument, that an individual animal shows an input/output response exactly corresponding to the bent stick model. This is not an unreasonable proposition and is exactly the assumption made when nutrient requirements are calculated in the familiar factorial way:

$$y = aW + bP \tag{1.1}$$

where y is a nutrient requirement, aW is a maintenance allowance proportional to body weight (W) and bP is a production allowance proportional to the output (P) of the individual animal. Note that the model cannot be tested, since one cannot measure the output of milk or eggs or weight gain from the same animal under the same physiological conditions at enough different input levels to allow precise definition of the shape of the individual response relationship. However, if we suppose that

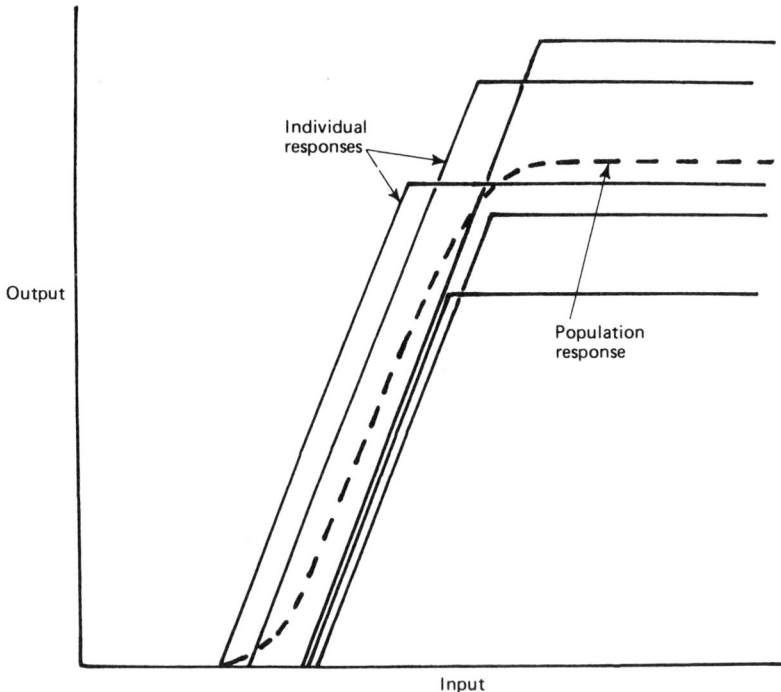

Figure 1.3   The input–output relationship for a population of animals, where individual responses conform to a bent-stick model, but individuals vary in their productive potential and in their maintenance requirement. In the case illustrated, individuals do not vary in their net efficiency of nutrient utilization for production, as indicated by the slope of the response lines

the model is reasonable and applies to all individuals in a population, but that those individuals vary (as they must) in W and P, then we have the situation shown in *Figure 1.3*. The integration of a set of bent stick responses, where there is variation in individual maximum response levels, necessarily produces a curve showing diminishing returns to increasing inputs and reaching a plateau at the point where the individual with the highest requirement ceases to respond. Notice that *Figure 1.3* is drawn without supposing that individuals vary in their net efficiencies of nutrient utilization for production. There may be some variation amongst individuals in net efficiency (it is very hard to tell) but the proposition of curvilinearity in the population response rests solely on the incontrovertible statement that individuals vary in size (and therefore in maintenance requirements) and in productive potential. There is a special set of conditions under which all the bent sticks in a population would turn at the same input value (when the correlation of P with W is -1.0 and the standard deviations of P and W are in the ratio $b:a$) but these conditions will not occur in practice. Therefore, in any feeding trial in which more than one animal has been used, the real response function must be curvilinear and will show diminishing returns.

The consequences of fitting a bent stick model to a set of experimental data is almost always to underestimate the optimum dose. In *Figure 1.2*, the model suggests an optimum input of 650 mg methionine plus cystine (M + C) per day, but profitable responses are continuing well beyond that point. Although a bent stick model may, in a particular case, be a good fit in the statistical sense it is always a bad fit philosophically and for practical purposes.

The parabola

By fitting the model

$$y = a + bx + cx^2 \tag{1.2}$$

where x is input and y is response, one can often obtain a curve which fits the data well. *Figure 1.4* shows the same results as *Figures 1.1* and *1.2*, but this time with two parabolic curves superimposed. Curve 1 is fitted to the data for all ten diets and reaches a maximum at 900 mg M + C per day. Some users would adopt this as their estimate of requirement. Others would calculate the cost of methionine and the value of eggs and, if these were £2/kg and 50p/kg respectively, would derive an optimum input of 886 mg M + C per day.

One difficulty with a parabolic curve is that it predicts a reduction of output beyond the optimum dose. This may not be of concern to particular users in particular contexts, but it is not in accord with the evidence for most nutrients and it can lead to serious trouble when incorporating the equations into computer prediction models. Many nutrients do cause a reduction in output when fed to excess, but the response curve is seldom symmetrical. More typically, there is a substantial range within which a

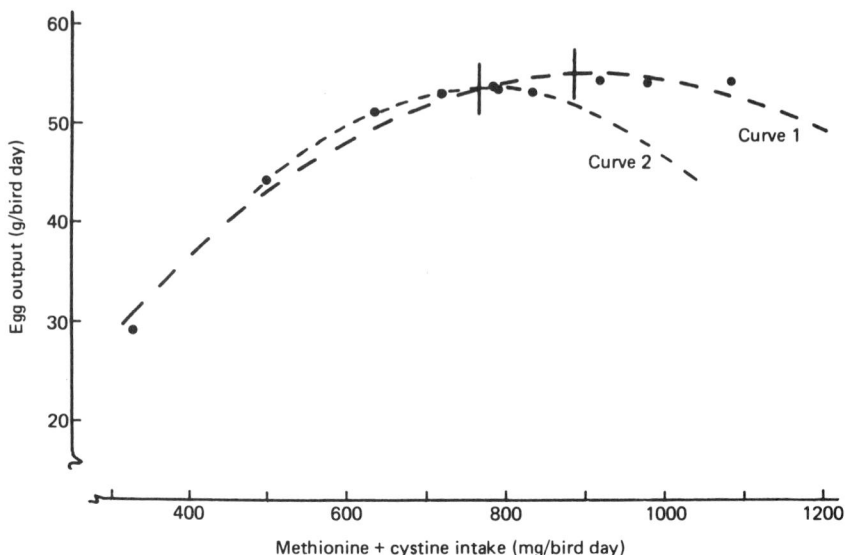

Figure 1.4 Parabolic models fitted to the data of *Figure 1.1*. Curve 1 uses all the experimental values and reaches a maximum at 900 mg input. The optimum input of methionine plus cystine is 886 mg/day, assuming a cost of £2/kg for methionine and a value of 50p/kg for egg output. Curve 2 is fitted to six of the ten treatment means and leads to an estimated optimum input of 759 mg methionine plus cystine per bird day

nutrient can be in surplus without causing any adverse effects on performance. This is illustrated by the experiment shown in *Figure 1.1* in which M + C was increased by varying the protein content of the diet, *not* by adding free methionine.

By choosing a restricted range of diets, the experimenter can usually obtain data which fit a parabolic curve rather well and avoid the problems of an extended plateau. This is often used as an argument to justify the model, but it is a dangerous argument. It presupposes that the experimenter has a good idea where the optimum input will lie (he usually does) and that the slope of the curve in this region can be best estimated by spacing the treatments closely about this assumed optimum input. Unfortunately, the shape of a fitted quadratic curve is very sensitive to the range of input values selected. This is illustrated by curve 2 in *Figure 1.4*, which is fitted to six of the ten treatments (still a larger experiment than most) and gives an estimated maximum output at 776 mg M + C per day and an optimum input (for the prices given above) of 759 mg M + C per day. Curve 2 is an excellent fit to the restricted set of data, but the estimate of optimum input is only 85% of the estimate obtained with the full set of data.

The parabolic model is thus philosophically wrong, in that it presumes symmetrical responses to deficiency and excess, with no intervening plateau. It is practically dangerous because it is apt to fit a limited set of data rather well and, in so doing, fails to give the experimenter any warning that his conclusions could be quickly falsified by another experiment. If he has taken the precaution of conducting a number of experiments before drawing any conclusions, he will be faced with the problem that curves fitted independently to each trial cannot be reconciled by pooling the coefficients of the quadratic equations; nor can the data be pooled to produce a single equation if the maximum output levels vary appreciably in the several trials.

The hyperbola (exponential and inverse polynomial models)

These models are asymptotic in form and so incorporate the notion that, within the range of interest, the output response rises towards a plateau and does not diminish again at high input values.

Various exponential equations have been used, a simple form being:

$$y = a - bC^{-x} \tag{1.3}$$

where y is the output, x is the input and a is the maximum value of output towards which the curve is proceeding.

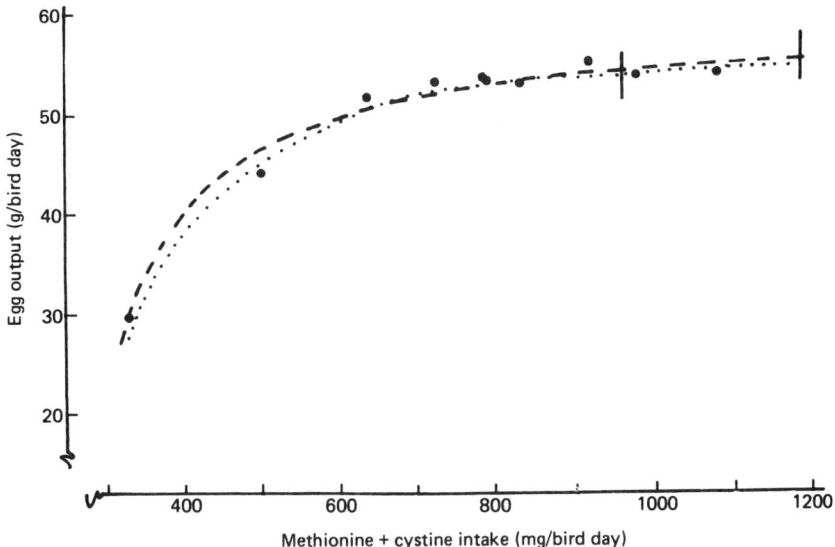

Figure 1.5 Exponential ($\cdots$) and inverse polynomial ($---$) models fitted to the Morris and Blackburn data. Taking prices of £2/kg for methionine and 50p/kg for eggs, the estimates of optimum input are 1184 mg/day for the inverse polynomial curve and 964 mg/day for the exponential

The inverse polynomial models were introduced by Nelder (1966) and a particular form is described by Morris and Blackburn (1982) as applicable to the data of their experiment. Curves derived from these two models are shown in *Figure 1.5*, again using the same set of data as in *Figure 1.1, 1.2* and *1.4*.

Both curves are a satisfactory fit, as judged by the residual mean square, but they round off the 'corner' of the response in a rather unsatisfactory way, which makes the user rightly suspicious. The curves also predict continued small responses in a region where it is very doubtful whether any response is occurring and thus they both regularly overestimate the optimum dose. This is particularly true when dealing with an input, such as methionine or an antibiotic, whose marginal cost is rather small in relation to the value of the output, so that the economic optimum is close to the true maximum yield. With more expensive inputs, such as biotin or tryptophan, a hyperbolic function will sometimes lead to a satisfactory estimate of optimum dose.

An advantage of these models, in comparison with the parabola, is that, in both cases, the coefficients of the equations are estimates of meaningful biological parameters, such as maximum output or rate of decline in output, and it is thus possible to calculate response curves for populations with output characteristics differing from those observed in the experiment. However, the tendency of these functions to overestimate the optimum dose is a serious flaw.

One temptation which must be resisted at all costs, is to fit an asymptotic model to experimental data and then to choose some arbitrary proportion of maximum output (e.g. 95%) as the 'requirement' (e.g. D'Mello and Lewis, 1970). If the model is appropriate, then the user should be given the equation of the model so that he can determine his own optimum for a local set of prices.

The Reading model

Fisher, Morris and Jennings (1973) have described a model which starts with the assumption that an individual animal responds in a bent-stick fashion (as in *Figure 1.2*). From this postulate, a population curve is constructed (as in *Figure 1.3*), based on information about the standard deviations of body weight and output. The equations needed to fit the model to a set of experimental data have been given by Curnow (1973) and examples of its application to laying hens have been published for lysine (Pilbrow and Morris, 1974) and tryptophan (Morris and Wethli, 1978). Fisher (1981) and Clark, Gous and Morris (1982) have used the model to describe responses of growing birds to amino acid intake.

Figure 1.6 shows the Reading model fitted to the same set of data as was given in *Figures 1.1, 1.2, 1.4* and *1.5*. The model is a good fit (though not significantly better than any of the alternatives discussed previously). One advantage of the Reading model is that its curvature depends upon the variability of the experimental animals, which does not change much from trial to trial, and is independent of the choice of dietary treatments. Thus, a Reading model fitted to the data from treatments 2 to 7 only, gives essentially the same curve and so leads to the same conclusion about optimum dose as the full data. This is in marked contrast to the results of fitting quadratic equations to sub-sets of the data.

Another advantage of the Reading model is that the coefficients of the response equation are meaningful numbers, being the net efficiencies of nutrient utilization for maintenance and for production. Three consequences flow from this. Firstly, given a set of experimental data for animals with particular productive characteristics, one can reasonably extrapolate to make an estimate of the response curve for another group of animals of different mean body size or with different potential output. Secondly, given the results of a number of experiments one can pool them to obtain best estimates of the response coefficients for future use. Thirdly, one can estimate net efficiencies by procedures other than a simple feeding trial and so obtain independent confirmation of the response coefficients.

The major disadvantages of the Reading model are that it assumes that outputs of individual animals are normally distributed about the mean and it requires a meaningful estimate of mean body size. These conditions are usually satisfied in short-term trials, but in long-term egg production or milk production experiments individual yields are not normally distributed

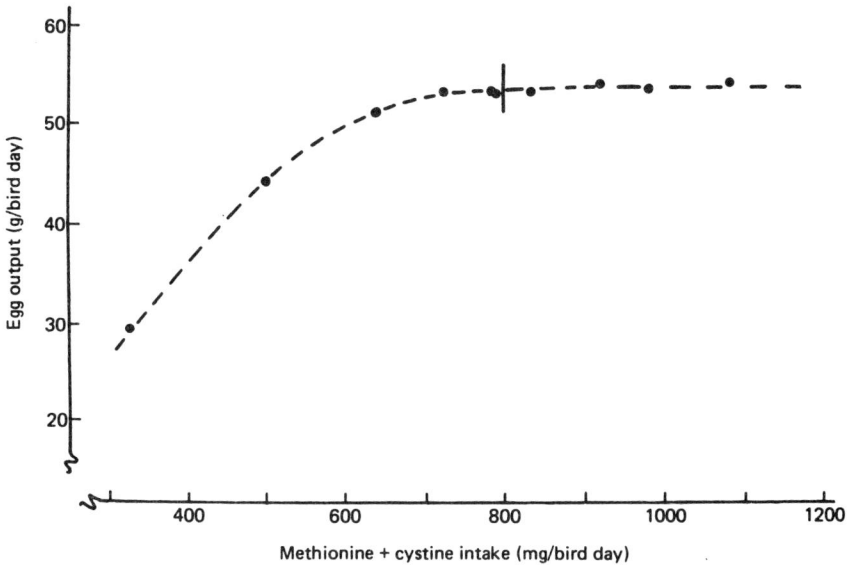

Figure 1.6 The Reading model fitted to the Morris and Blackburn data. The underlying assumption is that individual birds have a methionine + cystine requirement (M + C) defined by M + C = 13.04W + 10.15E, where W = body weight (kg) and E = egg output (g/bird day). The population response curve is then derived, by assuming normal variation in E and W and integrating individual response lines, as illustrated in *Figure 1.3*. Using the model, the estimated optimum input is 794 mg/day

about the mean; and in long-term growth trials mean body size changes during the course of the trial. A modification of the Reading model to cope with non-normal distributions is theoretically possible, but no suitable computer program is yet available for this.

Some animal feeding trials cannot be interpreted with the aid of a Reading model, either because the input scale is complex or ill-defined (e.g. a comparison of feeding programmes) or because the measured output does not depend directly on the input variable (e.g. measuring liveweight changes where body composition is variable and the real response parameters are protein deposition and fat deposition). Empirical solutions can sometimes be found for these problems by fitting one of the curvilinear models described above. However, a real understanding of dose–response relationships and ability to predict responses in future

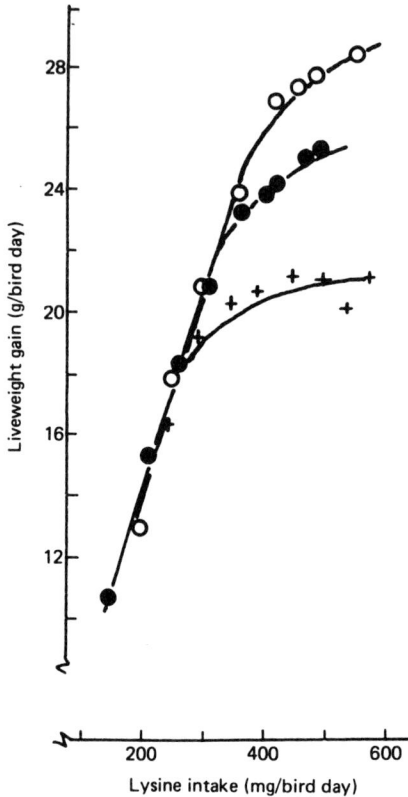

Figure 1.7 Results from Clark, Gous and Morris (1982) showing the application of the Reading model to chick growth data. The curves represent three separate experiments, but all are derived from the simple model that the lysine intake (L, mg/bird day) required by an individual chicken is given by $L = 0.03W + 12.94\Delta W$, where W is mean live weight (g) and ΔW is the rate of liveweight gain (g/bird day)

situations depends upon reformulating the problem in more fundamental biological terms and then building a simulation model which incorporates the necessary information.

The fact that a large computer is needed to fit a Reading model has not so far proved a barrier to its use to interpret suitable sets of data. The user of the output from the model does not need a computer to calculate optimum doses for his local prices, since a set of tables can be prepared giving optimum inputs for various cost ratios (*see*, for example, Morris and Wethli, 1978).

The Reading model was originally developed to help in the interpretation of input–output experiments involving laying hens fed on diets limiting in particular essential amino acids. However, the underlying concept, that organisms are variable and therefore the essential problem is to determine what proportion of a population shall be supplied with non-limiting levels of a nutrient, is applicable to nutrients other than amino acids and to species other than chickens. It seems equally applicable to the problems of optimizing fertilizer application to fields of wheat. *Figure 1.7* shows an example of the use of a Reading model to reconcile data from three growth trials. The curves fit the data quite well even though the maximum growth rates differed in the three trials and, since they derive from a single equation, it seems reasonable to suppose that one could predict the response curve for a flock of chickens with a much higher growth rate, if that were required.

Conclusions

It is always wrong to use either a multiple range test or linear regression to interpret the results of a dose–response trial, where the ultimate objective is to arrive at an estimate of optimum dose. Some form of curvilinear analysis is required. A parabolic curve will often give a good fit to experimental data, but this may give a false sense of security, since the shape of a fitted parabolic curve is unduly sensitive to the range of treatments selected. Inverse polynomial and exponential functions give asymptotic curves which sometimes fit well, but are apt to predict continuing economic responses at high inputs where the real response has ceased. The Reading model has a curvature which is largely independent of the choice of treatments and it therefore gives realistic estimates of optimum dose even when the data are scanty. It also readily allows the combination of evidence from trials with disparate levels of performance and is suitable for extrapolation to levels of performance which lie outside the range of experimental data.

References

CLARK, F.A., GOUS, R.M. and MORRIS, T.R. (1982). *Br. Poult. Sci.*, **23**, 433

CURNOW, R.N. (1973). *Biometrics*, **29**, 1

D'MELLO, J.P.F. and LEWIS, D. (1970). *Br. Poult. Sci.*, **11**, 367

DUNCAN, D.B. (1955). *Biometrics*, **11**, 1

FISHER, C. (1981). In *Protein Deposition in Animals*. Ed. Buttery, P.J. and Lindsay, D.B. Butterworths, London

FISHER, C., MORRIS, T.R. and JENNINGS, R.C. (1973). *Br. Poult. Sci.*, **14**, 469

HEADY, E.O., GUINAN, J.F. and BALLOUN, S.L. (1980). *Poult. Sci.*, **59**, 224

LERMAN, P.M. and BIE, S.W. (1975). *J. Agr. Sci., Camb.*, **84**, 459

MORRIS, T.R. and BLACKBURN, H.A. (1982). *Br. Poult. Sci.*, **23**, 405

MORRIS, T.R. and WETHLI, E. (1978). *Br. Poult. Sci.*, **19**, 455

NELDER, J.A. (1966). *Biometrics*, **22**, 128

PILBROW, P.J. and MORRIS, T.R. (1974). *Br. Poult. Sci.*, **15**, 51

ROBBINS, K.R., NORTON, H.W. and BAKER, D.H. (1979). *J. Nutr.*, **109**, 1710

WARE, G.O., PHILLIPS, R.D., PARRISH, R.S. and MOON, L.C. (1980). *J. Nutr.*, **110**, 765

2

METABOLIZABLE ENERGY EVALUATION OF POULTRY DIETS

I.R. SIBBALD
Animal Research Institute, Ottawa, Ontario, Canada

Introduction

Knowledge of dietary energy levels is important today and may assume even greater importance in the future. In North America many feed manufacturers use metabolizable energy (ME) as the common denominator in poultry diet formulation (Dansky, 1978). The level of dietary ME determines the minimum levels of inclusion of amino acids and, to a lesser extent, minerals and vitamins. Work is in progress in Canada to assess the feasibility of ensuring that manufacturers guarantee the ME levels of poultry diets. If appropriate legislation is promulgated it is probable that regulations establishing acceptable ME:drug ratios will follow.

The term ME, as used in poultry nutrition, tends to be generic rather than specific. There are four types of ME value and, unfortunately, it is not always apparent which type is being reported. Additional confusion occurs because values of a particular type may vary according to the assay procedures used in their derivation. Poultry nutrition has reached a point where it is desirable to adopt a single type of ME value and a standardized procedure for its measurement.

The purpose of this chapter is to describe the current state of energy evaluation of poultry feedstuffs. Metabolizable energy values are defined and discussed, assay procedures are described and compared, and attention is directed to the problems of ME measurement and application. This review is not exhaustive, but perhaps it can form the basis for further discussion of an important topic.

Metabolizable energy values

Apparent metabolizable energy (AME), sometimes referred to as classic ME, is the difference between the energy of the feed and the energy of the faeces + urine which, in poultry, are combined as a single excreta. The energy lost as gaseous products of digestion is insignificant and therefore can be ignored (Harris, 1966).

First published in *Recent Advances in Animal Nutrition – 1979*

$$\text{AME/g of feed} = \frac{(F_i \times GE_f) - (E \times GE_e)}{F_i} \tag{2.1}$$

where F_i is the feed intake (g); E is the excreta output (g); GE_f is the gross energy/g of feed; and GE_e is the gross energy/g of excreta.

Nitrogen corrected apparent metabolizable energy (AME$_n$) is the most commonly used estimate of metabolizable energy. It differs from AME because a correction is made for nitrogen retention (NR) which may be either positive or negative. Proponents of the correction argue that body nitrogen, when catabolized, is excreted as energy containing products and that it is desirable to bring AME data to a basis of nitrogen equilibrium. This is not a new development and it was used by Armsby and Fries (1918) in their work with cattle. Hill and Anderson (1958), working with chicks, introduced a correction factor of 34.4 kJ/g of retained nitrogen. This is the GE value of uric acid which is the principle nitrogen excretion product of poultry. Subsequently, Titus *et al.* (1959) proposed a correction factor of 36.5 kJ/g of nitrogen which describes more accurately the GE value of the nitrogen constituents of chicken urine. Unfortunately, both factors are being used and this contributes to some of the variation among AME$_n$ data.

$$\text{AME}_n\text{/g of feed} = \frac{\left[(F_i \times GE_f) - (E \times GE_e)\right] - (NR \times K)}{F_i} \tag{2.2}$$

where $NR = (F_i \times N_f) - (E \times N_e)$; N_f is the nitrogen/g of feed (g); N_e is the nitrogen/g of excreta (g); and K is a constant, usually either 34.4 or 36.5 kJ.

Although AME$_n$ data are used by many poultry nutritionists the need for the nitrogen correction has been questioned (Swift and French, 1954; Baldini, 1961). It is claimed that protein storage is characteristic of growth and egg production, and that it is difficult to justify the exactment of a penalty from a diet which permits nitrogen retention. While this may be true when comparisons are being made between complete diets, it is not a valid reason for disposing of the correction because, in the assay of individual ingredients, it is often necessary to feed imbalanced diets.

Sibbald and Slinger (1963a) made a regression analysis of the AME and AME$_n$ values of 1375 feedstuffs and obtained a correlation coefficient of 0.995 with an equation of the form:

$$\text{AME}_n = 0.009 + 0.948 \text{ AME} \tag{2.3}$$

which suggests that the correction was proportional to the AME value. It was also demonstrated that the small increase in precision associated with the correction can be a function of the method of calculation and, therefore, more apparent than real. These findings suggest that the additional work involved in measuring nitrogen retention is of questionable value. However, in a more recent study of the effect of the nitrogen correction it was found that AME values were influenced by the protein content of the reference diet and that correction for NR decreased the differences (Leeson *et al.*, 1977). This is logical because a change in NR must affect the AME value. Thus, if the assay diet (reference diet

+ test material) permits greater NR than does the reference diet alone then the AME value of the test material is inflated. Conversely, if NR is decreased the AME value of the test material is penalised for the reduction. Data published by Sibbald and Slinger (1963c) provide some support for this explanation. The differences between the AME and AME_n values of high protein feedstuffs were less when they were added singly to a basal diet than when added as mixtures.

The controversy surrounding the use of the nitrogen correction will probably continue. In most situations it has only a small effect and the additional work involved in its determinations is difficult to justify. However, because assay diets are frequently imbalanced, situations such as that described by Leeson *et al.* (1977) can occur. It is apparent that AME data are less likely to be additive than are AME_n values. Therefore, it is recommended that the correction be retained for the evaluation of poultry feedstuffs.

True metabolizable energy (TME) is a term used by Harris (1966) to describe an estimate of ME in which correction is made for metabolic faecal (FE_m) and endogenous urinary (UE_e) energy. The FE_m is the energy of that portion of the faeces other than the feed residues and is present as abraded intestinal mucosa, bile and digestive fluids. The UE_e is that portion of the urinary energy not of direct feed origin. Combined, the FE_m + UE_e represent a maintenance cost which should not be charged against the feed.

$$\text{TME/g of feed} = \frac{\left[(F_i \times GE_f) - (E \times GE_e)\right] + (FE_m + UE_e)}{F_i} \qquad (2.4)$$

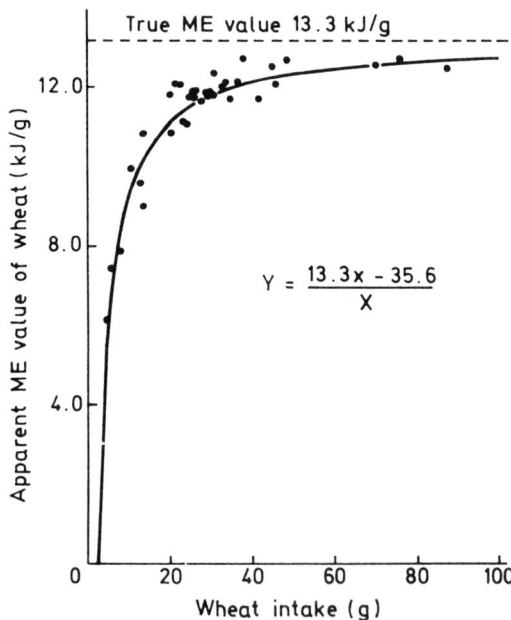

Figure 2.1 The effect of level of intake on the AME value of wheat (After Sibbald, 1975a)

The correction for FE_m + UE_e has several important effects. Using theoretical data Guillaume and Summers (1970) showed that AME and AME_n values would vary with feed intake whereas TME values would be independent of this variable. This was confirmed experimentally by Sibbald (1975a). Under standardised conditions the excretion of FE_m + UE_e is constant. When feed intake is large the energy loss as FE_m + UE_e is relatively small, but as feed intake is reduced this energy loss becomes increasingly significant and depresses the AME value; this is illustrated in *Figure 2.1*. The problem of variation in feed intake is important when feedstuffs of low palatability are being assayed, because they are usually included in assay diets at high levels to reduce experimental variation.

The maintenance cost of a bird varies according to its metabolic size, its physiological state and the environment in which it is kept. It is reasonable to assume that these factors also affect the FE_m + UE_e production and, therefore, AME and AME_n values. The measurement of TME controls, at least partially, the effects of these variables and causes the data to be more accurate and reproducible.

Nitrogen corrected true metabolizable energy (TME$_n$) bears the same relationship to TME as does AME_n to AME. The correction to nitrogen equilibrium is made in the same manner and its value is subject to the same debate.

TME_n/g of feed
$$= \frac{[(F_i \times GE_f) - (E \times GE_e) - (NR \times K)] + [(FE_m + UE_e) + NR_O \times K)]}{F_i} \quad (2.5)$$

where NR and NR_O are estimates of nitrogen retention for fed or fasted birds, respectively.

AME Assay procedures

Assays for AME_n differ from those for AME only because NR is measured. As this only involves measuring the nitrogen content of feed and excreta there is no need to describe the two procedures separately.

There are numerous assays for AME but many of the differences are minor and beyond the scope of this chapter. Four assays are outlined below, following which some of the important variables are discussed.

SINGLE INGREDIENT ASSAY

The simplest assay involves feeding the test material alone or in combination with an inert indicator, such as chromic oxide. The birds are acclimatised to the diet following which feed consumption and excreta production are measured for several days. The GE of the feed and excreta are measured and the AME value is calculated. When an indicator is used it is not necessary to measure feed intake and excreta production. Instead, the ratios of GE to indicator in pooled samples of feed and excreta are measured.

McIntosh *et al.* (1962) measured the AME_n values of cereal grains fed alone and in combination with a basal diet. The overall mean values revealed no major differences between the two methods, but individual treatment comparisons showed a degree of variation which cast doubt on the merits of the single ingredient assay. Pryor and Connor (1966) found the single ingredient approach satisfactory for some grains but not for others, while the data of Lockhart, Bryant and Bolin (1967) suggest that feeding wheat alone is satisfactory. Sibbald, Slinger and Ashton (1962) obtained AME_n data by feeding grain in combination with a small quantity of a mineral-vitamin supplement.

While the single ingredient assay is satisfactory for a few feedstuffs it is not recommended because many feedstuffs are unpalatable when fed alone. In addition, most feedstuffs are very imbalanced with respect to nutrient content and may have adverse effects on body functions if fed alone for several days.

GLUCOSE REPLACEMENT ASSAY

Hill *et al.* (1960) described an assay which involves feeding a reference diet containing 44.1% of glucose (Hill and Anderson, 1958) and a similar diet in which a portion of the glucose is replaced by the test material. The AME values of the diets are measured, as in the single ingredient assay, and the AME value of the test material is calculated using a value of 15.2 kJ/g for glucose. The latter value was established by Anderson, Hill and Renner (1958). A similar assay in which α-cellulose is the standard was described by Potter *et al.* (1960).

This assay is used in many laboratories and has yielded much useful information. Nevertheless, it is open to criticism. The glucose standard is not assayed in each experiment but is assumed to be constant under all conditions. Total reliance on the value of Anderson, Hill and Renner (1958), which is based on rather variable data, is a weakness. The reference diet is semipurified yet the values obtained from the assay are used to formulate practical diets. The AME values of fats have been shown to vary with the nature of the reference diet (Kalmbach and Potter, 1959; Cullen, Rasmussen and Wilder, 1962). Rao and Clandinin (1970) assayed rapeseed meal both by glucose and by practical diet replacement and obtained significantly different values. On the other hand Lockhart, Bryant and Bolin (1967) found diet composition to have no effect on the AME_n value of wheat.

PRACTICAL DIET REPLACEMENT ASSAY

Sibbald, Summers and Slinger (1960) used reference diets composed of practical ingredients and prepared test diets by replacement of a portion of the reference diet. The assay was refined in the light of subsequent experiments and is described in detail by Sibbald and Slinger (1963a). An important modification was the inclusion of the test material at more than one level. This makes it possible to determine whether the AME value is constant irrespective of the level of inclusion and permits the estimation of values by regression analysis (Miller, 1974).

A major advantage of this assay is that the reference diet serves as a standard and is assayed in every experiment. A disadvantage is that the AME value of a feedstuff may vary with the composition of the reference diet (Sibbald, Summers and Slinger, 1960), but such differences are usually small.

NEW RAPID ASSAY

Farrell (1978) recently described a very rapid assay for AME which involves the use of adult cockerels trained to consume their daily feed allowance within one hour. The test material may be fed alone or as partial replacement for a reference diet. The total amount of feed consumed is recorded and the excreta voided during the subsequent 24 hours is collected quantitatively. Preliminary results indicate that this could be a valuable technique because of the reproducibility of the data and because of the short time required to obtain results. Additional information is required, however, before the assay can be fully appraised.

ASSAY VARIABLES

A variety of birds have been used in AME assays and the resulting data have been used indiscriminately in diet formulation. However, there is evidence that the AME value obtained may be influenced by species (Slinger, Sibbald and Pepper, 1964; Bayley, Summers and Slinger, 1968; Fisher and Shannon, 1973; Leeson *et al.*, 1974; Sugden, 1974), strain (Sibbald and Slinger, 1963b; Slinger, Sibbald and Pepper, 1964; Foster, 1968a, b; Proudman, Mellen and Anderson, 1970; March and Biely, 1971) and age (Renner and Hill, 1960; Lockhart, Bryant and Bolin, 1963; Bayley, Summers and Slinger, 1968; Zelenka, 1968; Lodhi, Renner and Clandinin, 1969, 1970; Rao and Clandinin, 1970). It should be noted, however, that many of the bird differences were small and that some experiments have failed to demonstrate such differences. Nevertheless, this variation in AME assays can contribute to data variability.

Some assays are based on the total collection system while others use an indicator. Total collection necessitates accurate measurement of feed consumption and quantitative collection of excreta over a period of several days. Feed spillage makes consumption measurements difficult and can result in contamination of excreta. However, the use of an indicator introduces a new set of problems. It must be uniformly dispersed through the feed, totally inert, excreted in proportion to the feed residues and be amenable to physical or chemical assay. The most popular indicator is undoubtedly chromic oxide. The total collection and chromic oxide indicator techniques were compared by Sibbald, Summers and Slinger (1960) and Pryor and Connor (1966); both decided in favour of the indicator. However, use of chromic oxide introduces many problems (Halloran, 1972; Carew, 1978), and suggested alternatives have included polyethylene (Roudybush, Anthony and Vohra, 1974), naturally occurring fibre (Almquist and Halloran, 1971) and acid insoluble ash (Vogtmann, Pfirter and Prabucki, 1975).

The method of drying excreta samples can introduce variation in AME values. The preferred method is lyophilization but this is slow and expensive. However, there is evidence that oven drying can cause substantial energy losses (Manoukas, Colovos and Davis, 1964; Shannon and Brown, 1969). This subject requires further attention as it is of importance in many assay procedures.

There are many other variables which can have minor effects on AME values, but the foregoing examples illustrate that the assay procedure influences the data obtained. If AME or AME_n is to remain the estimate of choice then a serious attempt should be made to standardize the assay conditions.

True metabolizable energy bioassay

The assay for TME is a relatively recent development and will be described in detail. The concept of a single rapid bioassay arose during an investigation of the effects of feed intake on AME values. When starved adult cockerels were fed various amounts of wheat there was a linear relationship between GE excreted and wheat intake (*Figure 2.2*). The intercept of the regression line (8.5 kcal $\equiv$ 35.6 kJ) was an estimate of the $FE_m + UE_e$ excretion of the birds, and its slope

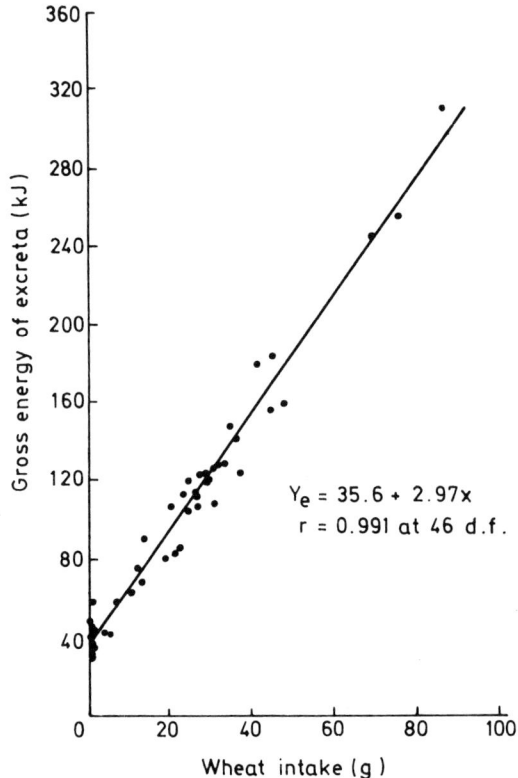

$$Y_e = 35.6 + 2.97x$$
$$r = 0.991 \text{ at } 46 \text{ d.f.}$$

Figure 2.2 The relationship between wheat consumption and gross energy voided as excreta (After Sibbald, 1975a)

(0.709 kcal $\equiv$ 2.97 kJ/g) was an estimate of the GE excreted for each gram of wheat consumed. By subtracting the slope from the GE value of the wheat (3.88 kcal $\equiv$ 16.2 kJ/g) a TME value (3.17 kcal $\equiv$ 13.3 kJ/g) was obtained. The obvious advantage of the TME value was that it was independent of $FE_m + UE_e$ excretion and therefore did not vary with feed intake (*Figure 2.1*).

Subsequent experiments demonstrated that for each feedstuff tested there was a linear relationship between GE excreted and feed input (Sibbald, 1975a; 1976a). By assuming that linearity holds for all feedstuffs it was possible to develop a simple bioassay based on a single level of feed input: negative control birds, receiving no feed, are required to measure $FE_m + UE_e$ output. The assay, which has been modified slightly since it was first published, is outlined below, a detailed description of which follows:

1. Birds are starved for 24 hours to empty their alimentary canals of feed residues
2. A bird is selected and force-fed a known quantity of test material
3. The bird is placed in a wire cage over an excreta collection tray and the time is recorded
4. A similar bird is selected and placed in a cage over a tray at a known time but is not fed
5. Exactly 24 hours after putting the birds in the cages their excreta is collected quantitatively, frozen, freeze-dried and weighed
6. Samples of the test material and excreta are ground and then assayed for GE content
7. $$\text{TME (kJ/g)} = \frac{(F_i \times GE_f) - (Y_f - Y_e)}{F_i} \qquad (2.6)$$
    where: Y_f and Y_e are $(E \times GE_e)$ for the fed and control birds, respectively.

The assay has been performed with adult cockerels, laying hens, meat-type hens and turkeys (Sibbald, 1976b), and also with egg-type and meat-type chicks of different ages (Sibbald, 1978a). However, for routine assay work the adult, single comb, White Leghorn cockerel is preferred. It tends to maintain a steady state, does not become obese, and has good liveability. Of the alternatives, meat-type birds tend to become obese and 24 hours of starvation followed by a suboptimal feed intake causes many laying birds to produce soft-shelled eggs which break and make quantitative excreta collection impossible. Chicks and growing birds are satisfactory once they are feathered, but they must be replaced after each assay if several experiments are to be made with birds in a uniform physiological condition.

The birds are housed in a windowless room where they receive 12 hours of light daily (06.00–18.00 hours). Between assays the birds are fed a maintenance diet *ad libitum* and fresh water is available at all times, including the starvation and excreta collection periods. Birds have been used for as many as 30 assays which are spaced 14 days apart without any adverse effects. After an assay the birds are returned to the maintenance regime for a minimum of one day but a longer rest period is preferred (Sibbald, 1978b). Recent work on the rate of passage of feed residues through the alimentary canal has confirmed the adequacy of a 24 hours starvation period (Sibbald, 1979a), but, if more convenient, a starvation period of up to 96 hours may be used without altering the data obtained (Sibbald, 1976c).

Birds selected for the assay must be healthy and should not be in a heavy moult. Feathers and scale contaminate excreta and make quantitative excreta collection difficult. It is desirable to use birds having bodyweights within a narrow range because the $FE_m + UE_e$ excretion of negative control birds is assumed to be the same as that of the fed birds. However, in an examination of a large number of $FE_m + UE_e$ values, only a small amount of the variation could be explained by differences in bodyweight (Sibbald and Price, 1978).

Force feeding is accomplished by use of a stainless steel funnel and plunger. The funnel, which has a stem 40 cm long and is 1.3 cm in external diameter, is pushed down the oesophagus until the end is in the crop. The feed is poured into the funnel and pushed into the crop with the plunger. It is very important that the feed enters the crop and not the oesophagus which forces the funnel out. When feed is put in the oesophagus the incidence of regurgitation increases.

After feeding the funnel is removed with a rotary motion, pressure being applied to the wall of the oesophagus to remove any adhering feed particles. Feeding can normally be completed in less than a minute.

The feed is prepared in advance of the assay and the amounts to be fed are stored in sealed plastic containers. Gross energy and dry matter measurements are made at the time that the feed containers are prepared. Feeding is made easier if the feed is pelleted but this is not essential. Ground feeds necessitate additional care in avoiding losses due to adherence to the funnel. Very finely ground feed may form a lump in the crop which is slow to disintegrate and which causes delayed passage through the alimentary canal.

Most feedstuffs are assayed as single ingredients but a reference diet and a blend of the reference diet with the test material may be used, although this introduces an extra treatment (Sibbald, 1977a). Fats are usually assayed in conjunction with a reference diet (Sibbald and Price, 1977a; Sibbald and Kramer, 1977, 1978; Sibbald, 1978c), although corn oil was administered as a single ingredient (Sibbald, 1975a).

The optimal amount of feed input depends upon the size of the bird and the form and nature of the feed. The greater the input the smaller the effect of experimental errors; however, as feed intake rises the incidence of regurgitation increases (Sibbald, 1977b). For adult White Leghorn cockerels the optimum input is 30–40 g of pellets or 25–30 g of ground feed.

The excreta trays must have smooth surfaces to facilitate quantitative collection and should extend beyond the cage in all directions. If the test material causes diarrhoea the excreta may travel several centimetres beyond the walls of the cage.

In most experiments several feedstuffs are assayed simultaneously. This reduces the amount of work because there need be only one negative control bird per replication. Thus in planning an experiment the required number of birds is:

(number of test materials + 1) X the number of replications

The usual excreta collection period is 24 hours. It is essential that the period selected be the same for all birds and it must be sufficient to permit the voiding of all feed residues. Residues of rapeseed meal were found to be voided more than 24 hours after feeding (Sibbald, 1978b) and in a subsequent experiment, using both 24 and 48 hours excreta collections (Jones and Sibbald, 1979), the TME value of a sample of ground seed was reduced by the extended collection period, indicating that all residues were not voided in 24 hours. However, the values of other products, including hulls, were not changed significantly. These findings led to a study of the rate of passage of feed residues under the conditions of the TME assay in which it was found that the residues of meat meal, fish meal and dehydrated alfalfa required more than 24 hours to be completely voided (Sibbald, 1979a). A subsequent experiment confirmed that the TME assay of the three feedstuffs, plus peanut skins, requires an excreta collection period of more than 24 hours (Sibbald, 1979b). Work is in progress to find methods of overcoming problems of delayed passage which appear to be associated with only a small group of feeds. In the meantime, if assay values obtained with a 24 hours collection are erratic, it is possible that an extension of the collection period will provide more uniform data.

Excreta collection and processing is relatively simple but it demands care if experimental errors are to be minimized. When birds are force-fed feathers and scale are loosened and fall on to the excreta trays. By blowing feathers and scale off the trays about one hour after feeding a considerable amount of contamination is avoided. When collecting the excreta any feathers must be washed clean of adhering excreta before they are discarded. Each tray must also be examined for regurgitated feed. When regurgitation occurs data relating to that bird are useless and should be discarded. Freeze-drying is the preferred method of removing water from the excreta for reasons mentioned earlier. When excreta is dry it may be weighed, ground and assayed, but the last two steps require care because dry excreta picks up moisture from the atmosphere. If time permits, the dry excreta should be equilibrated with atmosphere moisture for two or three days before weighing and further processing. Grinding of excreta is usually done in a mill but a mortar and pestle are usually adequate and easier to clean between samples.

In calculating TME values a common procedure is to use a mean value for the $FE_m + UE_e$. This tends to reduce data variability. If a TME value is obviously high, relative to replicate determinations, it is probable that there was an incomplete collection of feed residues. This could be due to excreta missing the collection tray, incomplete passage or regurgitation. An extremely low TME value is usually due to regurgitated feed being mixed with the excreta. Regurgitation is not a major problem once the art of force-feeding has been mastered; the incidence should not exceed 1%.

Advantages and disadvantages of true metabolizable energy

The TME bioassay is relatively new but it has been adopted by some feed manufacturers and is being tested by others. Those who have adopted it claim to be better able to predict bird performance than they could with AME_n data.

In attempting to evaluate the assay it is necessary to examine the quality of the data it produces, as well as the logistics of the technique. The correction for $FE_m + UE_e$ is an important part of the assay but it may be argued that the $FE_m + UE_e$ excretion of a negative control bird differs from that of a fed bird. There is validity in such an argument and indeed it was shown that the $FE_m + UE_e$ excretion decreases with the duration of starvation (Sibbald *et al.*, 1976c). But while the decrease introduces a small error it is outweighed by the benefits resulting from the correction.

By making correction for $FE_m + UE_e$, TME values are independent of variations in feed intake. In AME assays a high level of test material is included in the assay diet to minimize the effects of experimental errors. If the test material is unpalatable (e.g. feather meal) feed intake decreases and a low incorrect AME value is obtained. MacAuliffe and McGinnis (1971) found that the AME_n value of rye fell from 13.2 to 10.5 kJ/g when the level of dietary inclusion increased from 20% to 40%. The palatability problem can be overcome by feeding the test material at a practical level where the lack of palatability is masked by the other diet components, but this is self-defeating because the resulting data are very variable.

Reference was made to variations in AME values associated with the type of assay bird. Work on TME is not as extensive but it appears that values obtained with adult cockerels can be used in the formulation of diets for other birds (Sibbald, 1976b, 1978a). This requires confirmation with a wider array of feedstuffs. It is probable that differences in feed intake relative to $FE_m + UE_e$

output contribute to between bird type variation in AME data. The TME assay controls this source of variation.

Mention was made that AME values sometimes vary according to composition of the diet with which the test material is combined. In one experiment the TME values of five feedstuffs and of 10 diets prepared therefrom were measured (Sibbald, 1977c). The observed values of the diets were not significantly different from those calculated using the values for their component parts. This is strong evidence of additivity. Fats are an exception because of interactions with other diet components (Sibbald and Price, 1977a; Sibbald and Kramer, 1977, 1978; Sibbald, 1978c), but this problem also plagues AME measurements (Kalmbach and Potter, 1959; Cullen, Rasmussen and Wilder, 1962) and there is no apparent solution.

A strong argument in favour of retention of AME or AME_n as the standard measure of available energy in poultry feeds is that most of the existing energy requirement data are expressed in these terms. This is not an insurmountable problem because changes have been made before, as evidenced by the change from productive to metabolizable energy around 1960. An interim solution based on the relative TME and AME_n values of feedstuffs which are usually palatable, and for which reliable data are available, was outlined by Sibbald (1977d). If TME data are to be used as a standard then intensive work on energy requirements will be required.

For a while there was a shortage of TME values for feedstuffs, but this problem is being overcome and data for many ingredients are published (Sibbald, 1977a,e; Sibbald and Price, 1977b; Sibbald and Kramer, 1977). Several laboratories are measuring TME values and it is reasonable to expect a large body of data to become available in the near future.

The TME bioassay is working well, the only apparent problem being that some feed residues take more than 24 hours to clear the alimentary canal. This is being worked on and a solution may be possible. It is not a major problem because the collection period can be extended by several hours but this is not desirable.

The TME assay is simple and can be initiated at short notice if a flock of birds is maintained for this purpose. This is a major advantage over the conventional chick assays for AME. The amount of test material required for four replicated measurements plus associated dry matter and GE determinations is about 200 g compared to about 10 kg for a chick assay. This is important because it becomes feasible to mail samples to a central quality control or assay laboratory. The time required to complete the assay depends upon the drying procedures employed, but if notice of sample arrival is received 24 hours in advance a TME value can be available within 36 hours of receipt. This is substantially less than the AME assays currently in use, although the assay of Farrell (1978) has a similar time requirement. The overall cost of a TME assay is less than conventional AME assays.

An attractive feature of the TME assay is the reproducibility of data between laboratories (Sibbald, 1977f; Kessler and Thomas, 1978). In a collaborative study recently completed by the Animal Nutrition Research Council eight laboratories measured the TME value of a sample of corn and the range of their mean values was 16.6–17.4 kJ/g whereas AME_n values, measured in nine laboratories, ranged from 12.9 to 16.9 kJ/g (Sibbald, 1978d). It was noted that five of the eight laboratories that reported TME values had no prior experience with the assay.

While not directly relevant, it is interesting to note that Likuski and Dorrell (1978) have applied the basic methodology of the TME bioassay to measure amino acid availability. Sibbald (1979c) has confirmed that there are linear relationships between amino acid excretion and feed intake which are not affected by a supplementary energy input. Thus, it appears possible to measure TME and available amino acids as part of a single assay. This is still under development but it lends support to the adoption of the TME bioassay.

The feed industry and regulatory bodies will make the final decision regarding which measurement of ME will be adopted. The TME assay has several distinct advantages, but there is need for TME requirement data before it can be fully recommended. In the meantime efforts should be directed to the adoption of a single measurement by a standardized procedure to overcome the existing confusion in the energy field.

Indirect measurement of metabolizable energy

No review of methods of energy measurement would be complete without mention of the indirect procedures. Most of these are based on physical or chemical measurements and were developed to permit rapid and inexpensive estimates of ME in feedstuffs. With the development of rapid assays for AME (Farrell, 1978) and TME (Sibbald, 1976a), the need for indirect methods has diminished to some degree but they remain interesting because they do not require the maintenance of birds and the predictive data may be obtained in most quality control laboratories.

A review of indirect procedures was published by Sibbald (1975b). One paper not mentioned was that of Watts and Davenport (1971) who described a method for predicting the ME value of cottonseed meals. More recent publications are those of Guirguis (1975), Sibbald and Price (1976a, b; 1977b, c), Lodhi, Singh and Ichhponani (1976), Coates *et al.* (1977a, b), Hartel *et al.* (1977) and Moir and Connor (1977).

While prediction equations can be used to obtain ME values from chemical data the results are usually quite variable and therefore of limited practical value. Most equations use proximate analysis data and some use values for carbohydrates and other nutrients. Unfortunately there are other constituents of feedstuffs which may affect energy utilisation; for example trypsin inhibitors (Sambeth, Nesheim and Serafin, 1967), gossypol (Rojas and Scott, 1969), and tannins (Yapar and Clandinin, 1972); and if they are not taken into account it is unlikely that accurate predictions can be made. Processing, such as steam treatment and pelleting, can also affect ME values (Summers, 1975) but is not taken into account in indirect ME assays. Another uncontrolled variable is the interaction between constituents in mixed diets.

While the search for indirect assays continues, it seems unlikely that one will be found which is an adequate replacement for biological assay. This is not to say that existing indirect assays are without value but rather to stress that the quality of the data which they yield is inadequate for many purposes and that occasionally they will produce very erroneous values.

Conclusion

There are several methods for assessing the available energy content of poultry feedstuffs. There are advantages and disadvantages associated with each procedure. Because of the variation in methodology the available ME data should be treated with caution. Indiscriminate use in feed formulation may lead to imbalanced diets and inefficiencies.

There is need for the adoption of a single type of ME value and for a standardised method for its measurement. Until this occurs our information about energy values of feedstuffs and requirements of birds will contain variability which is unnecessary and confusing.

References

ALMQUIST, H.J. and HALLORAN, H.R. (1971). *Poult. Sci.*, **50**, 1233

ANDERSON, D.L., HILL, F.W. and RENNER, R. (1958). *J. Nutr.*, **65**, 561

ARMSBY, H.P. and FRIES, J.A. (1918). *J. Agr. Res.*, **15**, 269

BALDINI, J.T. (1961). *Poult. Sci.*, **40**, 1177

BAYLEY, H.S., SUMMERS, J.D. and SLINGER, S.J. (1968). *Cereal Chem.*, **45**, 557

CAREW Jr., L.B. (1978). *Feedstuffs*, **50** (20), 22

COATES, B.J., SLINGER, S.J., SUMMERS, J.D. and BAYLEY, H.S. (1977a). *Can. J. Anim. Sci.*, **57**, 195

COATES, B.J., SLINGER, S.J., ASHTON, G.C. and BAYLEY, H.S. (1977b). *Can. J. Anim. Sci.*, **57**, 209

CULLEN, M.P., RASMUSSEN, O.G. and WILDER, O.H.M. (1962). *Poult. Sci.*, **41**, 360

DANSKY, L.M. (1978). *Feedstuffs*, **50** (28), 23

FARRELL, D.J. (1978). *Br. Poult. Sci.*, **19**, 303

FISHER, C. and SHANNON, D.W.F. (1973). *Br. Poult. Sci.*, **14**, 609

FOSTER, W.H. (1968a). *Rec. Agric. Res.*, **17**(1), 13

FOSTER, W.H. (1968b). *J. Agric. Sci. Camb.*, **71**, 153

GUILLAUME, J. and SUMMERS, J.D. (1970). *Can. J. Anim. Sci.*, **50**, 363

GUIRGUIS, N. (1975). *Aust. J. Exp. Agric. Anim. Husb.*, **15**, 733

HALLORAN, H.R. (1972). *Feedstuffs*, **44**(7), 38

HARRIS, L.E. (1966). *Biological Energy Interrelationships and Glossary of Energy Terms*. Washington, D.C.: National Academy of Sciences

HARTEL, H., SCHNEIDER, W., SEIBOLD, R. and LANTZSH, H.J. (1977). *Arch. Geflugelk.*, **41**, 152

HILL, F.W. and ANDERSON, D.L. (1958). *J. Nutr.*, **64**, 587

HILL, F.W., ANDERSON, D.L., RENNER, R. and CAREW Jr., L.B. (1960). *Poult. Sci.*, **39**, 573

JONES, J.D. and SIBBALD, I.R. (1978). *Poult. Sci.*, **8**, 385

KALMBACH, M.P. and POTTER, L.M. (1959). *Poult. Sci.*, **38**, 1217

KESSLER, J.W. and THOMAS, O.P. (1978). In *Proceedings of the Maryland Nutrition Conference*, p. 18. University of Maryland

LEESON, S., BOORMAN, K.N., LEWIS, D. and SHRIMPTON, D.H. (1974). *Br. Poult. Sci.*, **15**, 183

LEESON, S., BOORMAN, K.N., LEWIS, D. and SHRIMPTON, D.H. (1977). *Br. Poult. Sci.*, **18**, 373

LIKUSKI, H.J.A. and DORRELL, H.G. (1978). *Poult. Sci.*, **57**, 1658

LOCKHART, W.C., BRYANT, R.L. and BOLIN, D.W. (1963). *Poult. Sci.*, **42**, 1285

LOCKHART, W.C., BRYANT, R.L. and BOLIN, D.W. (1967). *Poult. Sci.*, **46**, 805

LODHI, G.N., RENNER, R. and CLANDININ, D.R. (1969). *Poult. Sci.*, **48**, 964

LODHI, G.N., RENNER, R. and CLANDININ, D.R. (1970). *Poult. Sci.,* **49**, 991
LODHI, G.N., SINGH, D. and ICHHPONANI, J.S. (1976). *J. Agric. Sci. Camb.,* **86**, 293
MACAULIFFE, T. and MCGINNIS, J. (1971). *Poult. Sci.,* **50**, 1130
MANOUKAS, A.G., COLOVOS, N.F. and DAVIS, H.A. (1964). *Poult. Sci.,* **43**, 547
MARCH, B.E. and BIELY, J. (1971). *Poult. Sci.,* **50**, 1036
MCINTOSH, J.I., SLINGER, S.J., SIBBALD, I.R. and ASHTON, G.C. (1962). *Poult. Sci.,* **41**, 445
MILLER, W.S. (1974). In *Energy Requirements of Poultry,* p. 91. Eds T.R. Morris and B.M. Freeman. Edinburgh: British Poultry Science
MOIR, K.W. and CONNOR, J.K. (1977). *Anim. Feed Sci. Tech.,* **2**, 197
POTTER, L.M., MATTERSON, L.D., ARNOLD, A.W., PUDELKIEWICZ, W.J. and SINGSEN, E.P. (1960). *Poult. Sci.,* **39**, 1166
PROUDMAN, J.A., MELLEN, W.J. and ANDERSON, D.L. (1970). *Poult. Sci.,* **49**, 961
PRYOR, W.J. and CONNOR, J.K. (1966). *Aust. Vet. J.,* **42**, 141
RAO, P.V. and CLANDININ, D.R. (1970). *Poult. Sci.,* **49**, 1069
RENNER, R. and HILL, F.W. (1960). *Poult. Sci.,* **39**, 849
ROJAS, S.W. and SCOTT, M.L. (1969). *Poult. Sci.,* **48**, 819
ROUDYBUSH, T., ANTHONY, D.L. and VOHRA, P. (1974). *Poult. Sci.,* **53**, 1894
SAMBETH, W., NESHEIM, M.C. and SERAFIN, J.A. (1967). *J. Nutr.,* **92**, 479
SHANNON, D.W.F. and BROWN, W.O. (1969). *Poult. Sci.,* **48**, 41
SIBBALD, I.R. (1975a). *Poult. Sci.,* **54**, 1990
SIBBALD, I.R. (1975b). *Feedstuffs,* **47**(7), 22
SIBBALD, I.R. (1976a). *Poult. Sci.,* **55**, 303
SIBBALD, I.R. (1976b). *Poult. Sci.,* **55**, 1459
SIBBALD, I.R. (1976c). *Poult. Sci.,* **55**, 1578
SIBBALD, I.R. (1977a). *Poult. Sci.,* **56**, 1652
SIBBALD, I.R. (1977b). *Poult. Sci.,* **56**, 1662
SIBBALD, I.R. (1977c). *Poult. Sci.,* **56**, 363
SIBBALD, I.R. (1977d). *Feedstuffs,* **49** (43), 23
SIBBALD, I.R. (1977e). *Poult. Sci.,* **56**, 380
SIBBALD, I.R. (1977f). In *Proceedings of the Maryland Nutrition Conference,* p. 32. University of Maryland
SIBBALD, I.R. (1978a). *Poult. Sci.,* **57**, 1008
SIBBALD, I.R. (1978b). *Poult. Sci.,* **57**, 455
SIBBALD, I.R. (1978c). *Poult. Sci.,* **57**, 473
SIBBALD, I.R. (1978d). *Feedstuffs,* **50**(48), 20
SIBBALD, I.R. (1979a). *Poult. Sci.,* **58**, 446
SIBBALD, I.R. (1979b). *Poult. Sci.,* **58**, 896
SIBBALD, I.R. (1979c). *Poult. Sci.,* **58**, 668
SIBBALD, I.R. and KRAMER, J.K.G. (1977). *Poult. Sci.,* **56**, 2079
SIBBALD, I.R. and KRAMER, J.K.G. (1978). *Poult. Sci.,* **57**, 685
SIBBALD, I.R. and PRICE, K. (1976a). *Can. J. Anim. Sci.,* **56**, 255
SIBBALD, I.R. and PRICE, K. (1976b). *Can. J. Anim. Sci.,* **56**, 775
SIBBALD, I.R. and PRICE, K. (1977a). *Poult. Sci.,* **56**, 2070
SIBBALD, I.R. and PRICE, K. (1977b). *Poult. Sci.,* **56**, 1329
SIBBALD, I.R. and PRICE, K. (1977c). *Can. J. Anim. Sci.,* **57**, 365
SIBBALD, I.R. and PRICE, K. (1978). *Poult. Sci.,* **57**, 556
SIBBALD, I.R. and SLINGER, S.J. (1963a). *Poult. Sci.,* **42**, 313
SIBBALD, I.R. and SLINGER, S.J. (1963b). *Poult. Sci.,* **42**, 1325

SIBBALD, I.R. and SLINGER, S.J. (1963c). *Poult. Sci.,* **42**, 707

SIBBALD, I.R., SUMMERS, J.D. and SLINGER, S.J. (1960). *Poult. Sci.,* **39**, 544

SIBBALD, I.R., SLINGER, S.J. and ASHTON, G.C. (1962). *Poult. Sci.,* **41**, 107

SLINGER, S.J., SIBBALD, I.R. and PEPPER, W.F. (1964). *Poult. Sci.,* **43**, 329

SUGDEN, L.G. (1974). *Poult. Sci.,* **53**, 2227

SUMMERS, J.D. (1975). In *Proceedings of the Georgia Nutrition Conference,* p. 113, University of Georgia

SWIFT, R.W. and FRENCH, C.E. (1954). In *Energy Metabolism and Nutrition,* p. 95. Washington, D.C.: The Scarecrow Press

TITUS, H.L., MEHRING Jr., A.L., JOHNSON Jr., D, NESBITT, L. and TOMAS, T. (1959). *Poult. Sci.,* **38**, 1114

VOGTMANN, H., PFIRTER, H.P. and PRABUCKI, A.L. (1975). *Br. Poult. Sci.,* **16**, 531

WATTS, A.B. and DAVENPORT, R.F. (1971). *Poult. Sci.,* **50**, 1643

YAPAR, Z. and CLANDININ, D.R. (1972). *Poult. Sci.,* **51**, 222

ZELENKA, J. (1968). *Br. Poult. Sci.,* **9**, 135

3

ENERGY EVALUATION OF POULTRY RATIONS

C. FISHER
Agricultural Research Council, Poultry Research Centre, Roslin, Midlothian, Scotland

Introduction

If a requirement to declare energy levels is to be included in the regulations which govern the sale of poultry feeds, then important questions arise as to how the values used are to be defined and how they can be monitored and verified. This chapter is largely concerned with methods of verification based on chemical prediction equations. It seems to be generally assumed that this is the method of control most likely to be used, although alternatives, such as rapid bioassays and the simulation of digestion *in vitro*, do exist and will be mentioned briefly. Other aspects of energy evaluation, such as the description of feedstuffs and of animal responses, which will not be discussed, have been reviewed recently (Janssen, 1976; Härtel, 1979; Sibbald, 1980a; Farrell, 1981).

At the Poultry Research Centre a series of experiments has been carried out to test existing equations for the prediction of energy levels for poultry in mixed feeds and to establish new equations. This work will be described elsewhere (Fisher, 1982) but some of the results will be used here to illustrate points of principle. Extensive reference will also be made to the work of Härtel *et al.* (1977), who suggested an equation which is being strongly canvassed as the basis of EEC regulations on energy declarations.

There is, of course, nothing new in the idea that energy values can be calculated in terms of the chemically defined constituents of a feed. Certainly by the turn of the century Atwater had defined his 'factors', stating that protein, fat and carbohydrate, when digested and absorbed, yielded 4, 9 and 4 kcal/g 'available' energy, respectively. With this starting point a whole variety of chemical predictors have been suggested for poultry feeds and for feed ingredients. These have been reviewed by Sibbald (1975) and Janssen (1976) and incorporated widely into tables of energy values from Titus (1961) to Janssen *et al.* (1979). All of this is familiar and such work has undoubtedly contributed greatly to the efficiency of feed formulation for poultry. However, it appeared that existing data and equations, although illustrating very well the general features of a solution, failed when faced with the problem of verification in

First published in *Recent Advances in Animal Nutrition – 1982*

two main respects. First, no single set of data brought together a wide range of analytical variables and the potential pay-off between the complexity of equations and their accuracy could therefore not be fully explored. Second, the results could not be related directly to recent developments in energy evaluation systems and were not available in sufficient detail for recalculations to be made. It therefore appeared that further experimentation was required, although it was recognized from the outset that in many respects this could only repeat work that had been done well in the past.

Abbreviations and conventions

The following abbreviations and conventions are followed in this chapter.

ME   Metabolizable energy as a general concept. Units for ME are kJ/g DM.
AME, TME   Apparent and true ME, without N correction—i.e. classical ME.
AME_n, TME_n Apparent and true ME corrected to zero N retention.
GE   Gross energy, kJ/g DM.
DM   Dry matter. Unless explicitly stated, all concentrations and equations are expressed on a dry matter basis.
FAT   Fat content of feeds, method not specified.
CP   Crude protein, N × 6.25.
CF   Crude fibre.
ASH   Ash.
NFE   Nitrogen-free extractives, 1 − (FAT+CP+CF+ASH).
STC   Starch, method unspecified.
SUG   Free sugars, method unspecified.
NDF, ADF, ADL   Neutral and acid detergent fibre, acid detergent lignin.
OM, InM   Organic, inorganic matter.
D-, -D   Digestible nutrient (e.g. DOM) and digestibility (e.g. DMD).
rsd   Residual standard deviation after fitting an equation.
r^2 Multiple correlation coefficient; indicates the proportion of the total variability in the dependent variable accounted for by an equation.

All equations are expressed so that the coefficients are in units of kJ/g. Dietary compositions in equations are therefore expressed as proportions or g/g. Otherwise dietary compositions are in g/kg.

The experimental approach

The basic experimental approach is to determine the ME content of a series of mixed feeds and to relate the variation observed to differences in chemical composition, usually by regression methods. Although this is simple, it is not very efficient and it is important to recognize that the results obtained can be influenced by many details of the methods used.

An initial problem is to relate the sample of feeds studied to the population of mixed feeds to which the resulting equation(s) will be applied; assuming for the time being that such a population is homogeneous with respect to the relationships under investigation. An empirical solution to this problem is to use practical ingredients and to cover thoroughly the range of compositions which is likely to be found in practice. In our experiments 28 feeds were used, made from 29 practical feed ingredients which were constrained to levels typical of poultry feeds. The test feeds were drawn from a factorial matrix of 32 defined by the following variables; crude protein, 120 and 250 g/kg; total fat, 20, 40, 80 and 160 g/kg; and calculated AME values, 9, 11, 13 and 15 kJ/g. The last of these ensured a wide range in the carbohydrate composition of the feeds. This design encloses most practical feeds, although higher protein levels are occasionally used for turkeys, and emphasizes variations in fat level and carbohydrate composition. It also breaks the correlation between fat level and ME which is a feature of practical feeds and offers an opportunity to study the non-linear effects of fat.

Figure 3.1 illustrates some of the chemical dimensions of the test feeds in three-factor diet spaces (Parks, 1973). *Figure 3.1(a)* shows the partition of the dry matter between InM, DOM and non-DOM. It can be seen that there is little variation in the proportions of inorganic and organic matter, and the design therefore assumes that both water and ash act simply as diluents and do not influence the utilization of energy from other sources. Some recent evidence on the effect of acid:base ratios on ME values (McNab and Dewar, 1981; Nelson *et al.*, 1981) suggests that this assumption might require reconsideration.

Figure 3.1(b) shows the diet space for digestible fat, protein and carbohydrate (NFE), the major determinants of ME. It can be seen that only a relatively small part of this space is being explored and that there is some correlation between the proportions of DFAT and DNFE. Clearly if, as assumed here, these three factors account for all the DOM, there must be a correlation between at least two of the dimensions in such a space.

The correlations among the chemical characteristics of feeds made from practical ingredients are in fact extensive and the complete correlation matrix for the feeds used in these studies is shown in *Table 3.1*. These correlations are largely as expected, the single surprising feature being that ash, in spite of the small amount of variation involved (*Figure 3.1a*), is highly correlated with fibre content and behaves much as fibre does in relation to the other components.

Such correlations among the chemical characteristics of the feeds mean that the assumptions of the least-squares regression methods are strictly not met and the estimates of the coefficients obtained for each predictive variable are not independent. The detailed consequences of this have not been worked out, although it has been found that the solutions are fairly 'stable', in accordance with theoretical expectation and in good agreement with other studies. This theoretical weakness might therefore not be too serious a problem.

The prediction equations resulting from such experiments are also influenced by the relative importance of each chemical dimension or group of dimensions, and these characteristics are inescapably 'built in' to the

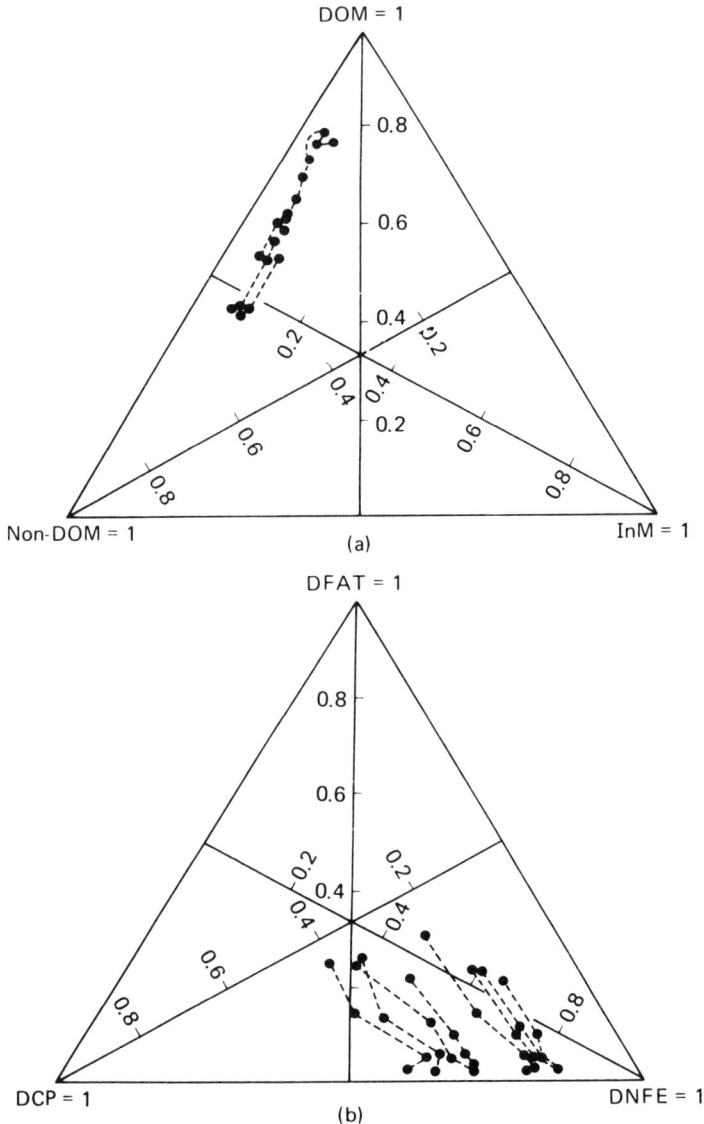

Figure 3.1   The distributions in two diet spaces of test feeds used to determine prediction equations in two diet spaces. The lines join feeds with the same nominal CP and AME content but differing in FAT. (a) Eighteen test feeds in a space partitioning DM into DOM, non-DOM and InM. (The remaining ten feeds could not be shown separately in this diagram.) (b) Twenty-eight test feeds in a space partitioning DOM into DFAT, DCP and DNFE

design. For example, Janssen *et al.* (1979) suggest a prediction equation for the ME content of wheat and wheat products based only on fibre content. However, the sample of wheat products from which this equation was derived (Janssen, 1976) had a 14-fold range in fibre contents, but much narrower ranges for other variables (ash, 3.8-; protein, 1.5-; fat, 2.3-;

Table 3.1   THE CORRELATIONS BETWEEN CHEMICAL COMPONENTS OF 56
TEST FEEDS MADE FROM PRACTICAL INGREDIENTS

	FAT	CP	CF	ASH	STC	SUG	NDF	ADF	ADL	GE
FAT	–	–0.06	0.32	0.31	0.54	0.10	0.35	0.35	0.41	0.95
CP		–	–0.05	0.13	–0.37	0.29	–0.15	–0.02	–0.14	0.22
CF			–	0.91	–0.80	0.48	0.97	0.99	0.95	0.22
ASH				–	–0.89	0.66	0.90	0.91	0.80	0.22
STC					–	–0.71	–0.79	–0.83	–0.71	–0.54
SUG						–	0.52	0.50	0.32	0.06
NDF							–	0.97	0.92	0.22
ADF								–	0.95	0.26
ADL									–	0.31
GE										–

starch, 6.8-fold). Since the correlation between ME and fibre level was
very high, $r = -0.998$, there was little opportunity for the effect of these
other variables to be estimated. The conclusion that such an equation is
suitable, for example, for distinguishing between samples of wheat grain,
which will differ little in fibre content, is unjustified. While this is an
extreme example of the danger of allowing a single chemical dimension to
predominate, it is also a reminder that in any set of data based on practical
ingredients there will be effects of this sort.

Types of interpretative model

To some extent these limitations of experimental method may be overcome
by the judicious selection of interpretative models. Obviously, when an
equation is finally selected for practical use, consideration of cost and ease
of application will be of great importance. However, the selected equation
may also be judged to be more 'robust' for general use if, in addition to
being an effective empirical description of the data from which it was
derived, it is also consistent with external evidence and expectations about
the nature of the underlying relationships involved.

This can be illustrated by considering three aspects of chemical prediction equations for energy values. Should they contain a constant value?
Should they contain negative predictors? Can the coefficients be compared
with theoretical values for the energy content of each chemical component?

Models which are a summation of the energy-yielding components of a
feed are attractive. If they include all such components, a constant term
should not be required, and if each component can reasonably be
represented by a single coefficient across all feeds, then other dietary
characteristics and interactions should not be required. Such arguments led
Härtel (1979) to propose an equation with FAT, CP, STC and SUG as
energy-yielding variables and with no constant.

Obviously this argument has limitations. If the energy value of a digested
nutrient is constant, then the use of single coefficients is equivalent to
assuming constant digestibility for all feeds. This is clearly untrue, and
factors such as fibre level may be required in a prediction equation as an
empirical index of digestibility rather than as an energy source *per se*.

Table 3.2   EQUATIONS DERIVED FROM THE DATA OF HÄRTEL *ET AL.* (1977)
TO ILLUSTRATE DIFFERENT TYPES OF MODEL. IN ALL CASES THE
EQUATIONS PREDICT AME_n (kJ/kg DM)

Equation No.		r^2	rsd
1	$-2.664 + 34.87x_1 + 17.72x_2 - 15.23x_3 + 17.42x_4$	0.85	0.93
2	$31.97x_1 + 14.14x_2 - 18.37x_3 + 14.48x_4$	0.84	0.94
3	$-3.064 + 34.82x_1 + 17.21x_2 + x_4(18.52 - 31.2x_3)$	0.85	0.92
4	$31.36x_1 + 12.75x_2 + x_4(15.30 - 38.1x_3)$	0.85	0.93
5	$0.909 + 31.02x_1 + 13.65x_2 + 15.29x_5 + 11.27x_6$	0.92	0.70
6	$32.01x_1 + 15.48x_2 + 16.40x_5 + 12.39x_6$	0.91	0.71

x_1, FAT (g/g); x_2, CP (g/g); x_3, CF (g/g); x_4, NFE (g/g); x_5, STC (g/g); x_6, SUG (g/g).

Some of these points are illustrated by the equations in *Table 3.2*, which
were calculated from the data of Härtel *et al.* (1977). Before doing this it
was shown that recalculation could reproduce the published equations with
only minor rounding errors.

Equations 1 and 2 both involve the components of the proximate
analysis and differ only by the constant term in equation 1. Since this
constant term is aliased with ASH, by the proximate sum, and since it is
not significantly different from zero ($P > 0.05$), it seems reasonable to omit
it. The calculated energy values of FAT, CP and NFE are 32.2, 15.1 and
14.8 kJ/g in equation 1, and 32.0, 14.1 and 14.5 kJ/g in equation 2. If these
are compared with the energy yields of the digested nutrients (*Table 3.3*),
then digestibilities of about 0.845, 0.81 and 0.85 for fat, protein and
carbohydrate are implied by the coefficients. The observed mean values
were 0.78, 0.74 and 0.71, respectively (Härtel *et al.*, 1977).

Table 3.3   EQUATIONS FOR THE CALCULATION OF AME_n OF MIXED FEEDS
FROM DIGESTIBLE NUTRIENTS (kJ/g DM)

Reference[a]		r^2	rsd
1	$39.0x_1{}^b + 18.5x_2 + 17.1(x_3 + x_4)$	–	1%
2	$35.2x_1 + 17.8x_2 + 17.4(x_3 + x_4)$	–	0.30
3	$38.8x_1 + 17.5x_2 + 17.5(x_3 + x_4)$	–	0.26
4	$38.5x_1 + 18.5x_2 + 17.2x_4$	0.99	0.23
	$38.4x_1 + 18.6x_2 + 17.4x_5 + 15.6x_6 + 20.5x_7$	0.99	0.20

[a]References: 1, Chudy and Schiemann (1971); 2, Hoffman and Schiemann (1973); 3, Janssen
et al. (1976); 4, Härtel *et al.* (1977).
[b]A value of 39.8 was reported originally for this coefficient, but was subsequently corrected
(*see* Härtel *et al.*, 1977, footnote 8).
x_1, digestible FAT (g/g); x_2, digestible CP (g/g); x_3, digestible CF (g/g); x_4, digestible NFE
(g/g); x_5, digestible STC (g/g); x_6, digestible SUG (g/g); x_7, digestible residual NFE
$(x_4 - x_5 - x_6)$ (g/g).

The estimated coefficients are therefore broadly in line with expectation,
but close comparisons are invalidated by the large, and highly significant,
effect of fibre level. Clearly, the variable 'fibre' is acting here as an index of
factors associated with lower energy values, since it is expected that the
crude fibre fraction *per se* would make a zero or small positive contribution
to digested energy. The observed mean digestibility of the crude fibre was

−0.023. It can therefore be argued that a preferable model would not necessitate a large negative effect for fibre. One possible approach is to consider that fibre exerts its effect by reducing the digestibility of other components; this can be represented by introducing interaction terms into the model. When this is done with these data, the interactions of fibre with oil and protein are not significant but equations 3 and 4 in *Table 3.2* include a significant interaction between CF and NFE. When this is included in the model, the direct effect of CF becomes non-significant. The coefficients indicate that the energy yield of NFE is reduced by 0.3–0.4 kJ/g for each 1 per cent increase in CF. For equation 4 the average digestibilities of the components, calculated as above, are FAT 0.83, CP 0.71 and NFE (for average CF = 0.053 g/g) 0.77, which compare well with the observed values. However, although this model might be preferable to equations 1 and 2, it does not provide a better description of the data.

The replacement of NFE by STC and SUG does achieve a considerable improvement in goodness of fit and also has the effect of making the fibre effect non-significant (see equations 5 and 6 in *Table 3.2*). Again the constant term is non-significant and the coefficients realistic, and it was this combination of characteristics that led Härtel (1979) to suggest this form of equation for practical uses. (Although equation 6 is of the same form as that recommended by Härtel, it is not identical, because he removed some data points before reaching a final equation. This has not been done here.)

It seems inevitable that issues of the sort discussed here will arise in the analysis of any similar data set. There are no general answers to the problems of interpretation that are raised, but they should be borne in mind when different equations are being compared. It is inescapable that in selecting an equation for practical use a pragmatic balance will have to be struck between purely empirical analysis of the data and the introduction of theoretical constraints.

Accuracy of prediction

Any discussion of the accuracy with which ME values may be predicted by an equation must start with the assumption that all feeds, including those used to generate the equation, are drawn at random from a population which is defined by the model elements. A major source of potential error is that this might not be true but it is difficult to see how to avoid this possibility. No single equation, of conceivable complexity, could deal with variations in the toxic, anti-nutritive or exceptional factors which the commercial nutritionist is considering all the time. It is easy to list the candidates—trypsin inhibitors, tannins, toxic substances, lactose, and so on—but not easy to incorporate them into a prediction system for ME. For any reasonable equation it will always be possible to devise a feed mixture which makes a nonsense of its predictive properties. Whether such mixtures could ever be sold against all the other constraints acting in the market-place is a matter for conjecture. However, in the context of the verification of energy values, the only way to guard against these possibilities is to use a biological test either as the primary evaluation or as a back-up to a chemical prediction equation.

An approximate expression for the variance of an ME value for a single feed predicted by a given equation has been suggested by Wainman *et al.* (1981). This takes account of not only the error with which the equation is determined, but also the additional variation involved in the measurement of the chemical parameters. The data of Wainman *et al.* show that the rank order of different equations is vitally affected by the information used to assess the variance of prediction, and this is also a feature of the poultry data (Fisher, 1982). However, the discussion here is restricted only to a consideration of the accuracy with which an equation might describe the data from which it is derived.

We can identify three probable upper limits to accuracy in this restricted sense. These are: the accuracy with which ME values can be predicted with knowledge of nutrient digestibility; the accuracy with which GE values can be predicted; and the accuracy with which ME values can be determined experimentally. It seems unlikely that the ability to account for measured ME values in terms of total nutrients will exceed any of these.

Four series of experiments in which ME values were related to measured differences in digestible nutrients are summarized by the equations in *Table 3.3*. The residual standard deviation in these studies averages about 0.25 kJ/g, about 2 per cent of the mean value. There is also a striking consistency in the coefficients obtained in these different experiments; the values agree closely with the theoretical values, and the means of 37.9, 18.0 and 17.3 kJ/g for DFAT, DCP and DNFE can be used to judge the suitability of the coefficients in other equations.

The GE values of mixed feeds can be predicted from their chemical composition with considerable accuracy. Widely used values are 39, 24 and 17 kJ/g for FAT, CP and carbohydrate (NFE + CF), respectively, and these agree very closely with the following equation derived from data from the Poultry Research Centre for 56 feeds:

$$GE, kJ/g = 39.4 \, FAT + 23.5 \, CP + 17.5 \, (CFE + NFE) \qquad (3.1)$$

This equation had a residual standard deviation of 0.19 kJ/g, 0.96 per cent of the mean, and the standard deviation of the measurement of GE values was about 0.5 per cent of the mean.

The accuracy with which ME values can be determined experimentally is also high. In experiments at the Poultry Research Centre, using six replicates, the standard error of the mean AME_n value for a single diet, based on a pooled error, was 0.15 kJ/g, or 1.15 per cent of the overall mean.

These various indicators of the upper limits to the accuracy which might be achieved in the determination of equations contain common elements and cannot be combined in any way. However, they suggest that about 98 per cent of the observed variation in ME might be explicable, giving a residual standard deviation (of determination only) of 0.2–0.3 kJ/g. In addition to this, there will be errors associated with the chemical analyses. In a later section it will be shown that these possible upper limits to accuracy have in fact been achieved.

The other aspect of accuracy that will be mentioned is how a given statement of standard error should be interpreted; with what probability should the limits of accuracy be calculated? It is suggested that this will vary for different purposes. For setting legal limits it will probably be appropriate to set a range which includes a 90 or 95 per cent level of confidence. This gives a *t* value of about 2. Thus, a standard deviation—for example, of 0.4 kJ/g—will give an 'accuracy' of ± 0.8 kJ/g. For assessing the value of the information, however, an average accuracy might be more suitable—for example, as a basis of decision making by the producer. This gives a *t* value of about 0.7; a standard deviation of 0.4 kJ/g, indicating an average 'accuracy' of ± 0.28 kJ/g.

Metabolizable energy systems

Agreement on which definition of energy levels should be used is an obvious prerequisite for a system of declarations. In principle, any of the alternatives can be used—for example AME corrected to zero or some other level of N retention, TME or even net energy values—but there has to be prior agreement and uniform practice. There are two aspects of this question which are of importance. First, what effect will agreement and legislation based on a single system at one time have on the prospects for further innovation and improvement in the future? Second, on what criteria should the choice be made at any one time?

It is not intended to discuss the first of these questions here, although it is a matter of importance which could usefully be debated. The second question emphasizes the fact that this is a difficult time to settle on one system, since new alternatives have recently been suggested and are currently being extensively tested and compared. The most important of these developments is the suggestion by Sibbald (1976a) that determined ME values should be corrected for energy loss of endogenous origin to yield a TME value. The detailed justification for using TME values has been extensively presented by Sibbald (1980a) and debated by Farrell (1981) and McNab and Fisher (1981).

It is far from clear at this time to what extent Sibbald's proposals will be accepted and whether in a few years' time energy declarations on the basis of TME will seem more appropriate than AME. It was therefore decided that in our experiments data should be collected on both AME and TME values so that equations could be calculated in terms of either.

The second important development is the use of so-called 'rapid' assays which require the identity of excreta with a single input of feed (Sibbald, 1976a; Farrell, 1978). This is done by starting and finishing the assay with the digestive tract empty and to this end we used a 40 h pre-experimental fast and a 48 h post-feeding collection period. The experimental programme involved a total of some six hundred energy balances and it was therefore essential to use one of the rapid methods. The tube feeding method of Sibbald was found to be both convenient and accurate and therefore was selected. For the routine determination of energy balance a food intake of 30 g was used.

THE RELATIONSHIP BETWEEN TME AND AME

In the present context, this relationship is important for two reasons. First of all, in our experimental work use was made of the theoretical rela-

tionship between TME and AME in order to determine both simultaneously on the same animals. Second, the theory shows how variations in
endogenous energy loss (EEL) are important in the determination of ME,
and this needs to be taken into account in the development of equations.

By definition the relationship can be derived as follows:

AME/g food = energy intake/g food − energy excreted/g food

TME/g food = energy intake/g food − energy excreted/g food + EEL/g
food

By rearrangement

TME/g food = AME/g food + EEL/g food

If it is further assumed that the amount of energy excreted is a linear
function of intake, then TME is constant and the relationship between the
two measures of energy takes the familiar form shown in *Figure 3.2*.

This leads to two important conclusions. First of all, for a given TME
value, AME depends on the ratio between EEL and food intake (FI).
Second, given knowledge of energy balance at a known food intake and of
EEL, then the energy content of the feed can be transformed from one
scale to the other simply by calculation.

If TME and EEL are constant, then AME values vary with food intake
(*Figure 3.2*). Thus, AME determinations made under free feeding conditions will contain an element of variation associated with rate of feed
intake. Such variation might turn out to be related to the chemical
composition of the feeds (for example, if fibrous feeds reduce intake), but
such a relationship is spurious and tends to conceal the true cause and
effect. The feeds used in our studies varied greatly in palatability and
therefore energy balances were made at one intake only, 30 g, and AME
values at other intakes were obtained by calculation:

$$TME = AME_{30} + EEL/30$$

$$AME_{80} = TME - EEL/80$$

There is one further complication. Experimental energy balance and
EEL were measured over 48 h. For practical purposes both require to be
expressed on a 24 h basis and for conversion it was assumed that EEL per
24 h is $0.55 \times$ EEL per 48 h. The factor 0.55 was derived from a number of
experiments in which EEL was measured concurrently over 0–24 and 0–48 h
(Sibbald, 1980c, 1981a; Sibbald and Price, 1980), the mean of fifteen
independent observations being 0.544 ± 0.016. Finally, all estimates of ME
were corrected to zero N retention by use of a factor of 34.4 kJ/g N. For
this purpose estimates of EEL were also corrected, a procedure about
which there is a lack of agreement in the literature. Sibbald (1980a) does
not recommend N correction in the negative control birds for the calculation of TME_n. Mutzar and Slinger (1981) followed this recommendation
but the Poultry Research Centre group and Shires *et al.* (1980) believe this
to be incorrect.

Although these theoretical relationships are derived from the definitions

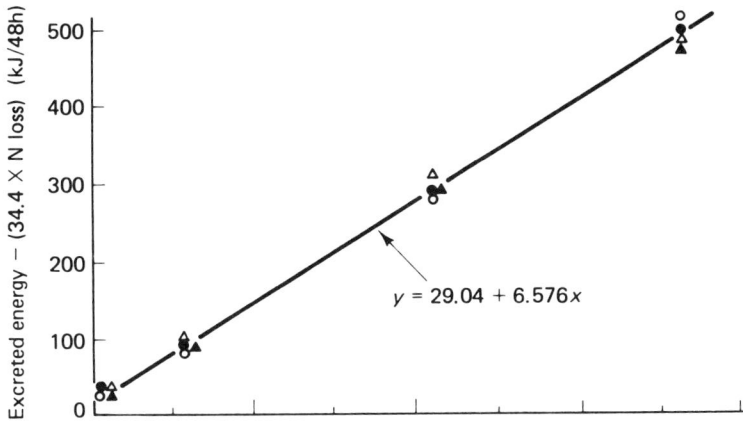

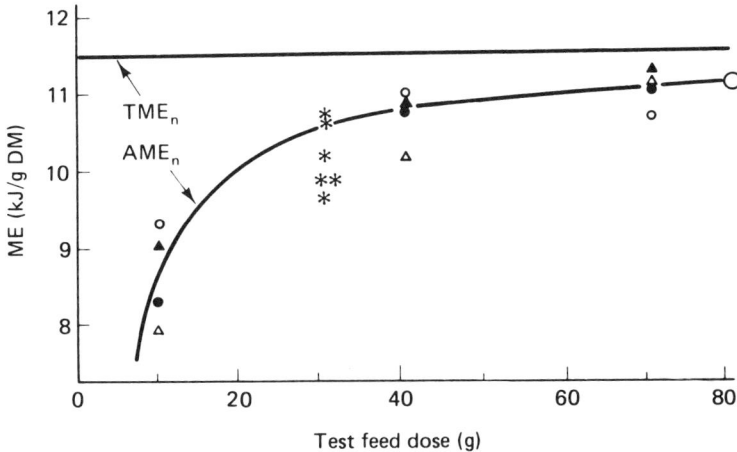

Figure 3.2 The relationship between TME and AME; an experimental demonstration for one feed. (a) Excreted energy output (corrected for N loss) for four birds given glucose/test feed in the ratios (g/g) 25/0, 15/10, 0/40 and 0/70 in a change-over design. (b) The same data plotted as AME_n, kJ/g feed and the derived TME value. Also shown (★) are six independent observations at 30 g intake and the calculated AME_n at 80 g intake (0) derived from these six points using $EEL_n = 18$ kJ per bird per 24 h

of AME and TME, they have also been tested empirically with some of the feeds used in our experiments. There were two main reasons for doing this. First of all, although it is a necessary condition for constant TME values that the relationship between excreted energy and food intake should be linear, it is not a sufficient condition. If there are linear increases in EEL with feeding level (or decreases), the theory will not apply. It was therefore decided to carry out energy balances at different levels of intake. Second, Farrell (1981) has presented evidence that EEL, when estimated by regression methods, varies among feeds and is related to neutral detergent fibre content. Since a common value of EEL for all feeds was used, it was obviously important to check this point.

The results of these supplementary experiments are described by Fisher (1982), but those for one feed are illustrated in *Figure 3.2* and show that, for all feeds tested, the theoretical calculations were justified empirically. The results did not support Farrell's (1981) observation on the relationship between EEL and NDF content.

The pursuit of this approach to the determination of ME values depends, of course, on the proper definition of endogenous energy losses in the experimental birds. During the course of this work the approach to this problem has changed, and in the final analysis a single estimate of endogenous loss of both energy and nitrogen has been used in all calculations. The data used were derived from 133 birds which received 25 g glucose at the start of a 48 h balance period with the mean results shown

Table 3.4   MEAN VALUES FOR ENDOGENOUS ENERGY LOSSES IN MATURE WHITE LEGHORN ROOSTERS (BODY WEIGHT ABOUT 1.8 kg)

		Over 48 h	*Over* 24 h[a]
Total or uncorrected loss (kJ)	EEL	71	39
N-corrected energy loss (kJ)[b]	EEL$_n$	32.5	18

[a]The values for 24 h losses are 0.55 times those observed over 48 h.
[b]N correction assumes 34.4 kJ/g N.

in *Table 3.4*. The use of a single value is justified by the fact that the six observed AME values (30 g intake) for each feed fall within a very narrow range. This would not be so if there were large variations in the endogenous fraction. This topic is discussed in detail by Fisher (1982).

VARIATIONS IN ME VALUES

A difficult, and at present unresolved, question is whether different equations are required to predict ME values for different classes of stock or whether a single one will suffice. In the work described adult roosters have been used and it is not possible to say with certainty, one way or the other, whether the results are applicable to growing broilers, turkeys and adult hens. Certainly, any differences will be small but they may not be negligible. It is equally uncertain whether such differences, if established, would be associated differentially with the energy yield of chemical components or whether they could be described by additional constants in the equations. The importance of this topic has been heightened by the fact that the ME values observed in these experiments are generally higher, by about 0.5 kJ/g, than those from other studies (see below).

There is an extensive literature on the effect of fixed factors such as breed, strain, age and sex on ME values. However, it does not lead to any clear conclusions and the reader is referred to Sibbald (1980a) and Farrell (1981) for recent bibliographies. Sibbald (1980a and elsewhere) has argued that the effects of such factors on TME are smaller, or even non-existent, when compared with AME, and this aspect alone will be briefly discussed here.

The theoretical relationships between AME and TME, which have been discussed above, show that if different AME values are observed in two circumstances, three explanatory hypotheses can be advanced. Either digestibility, and, hence, TME, differs in the two cases or else TME is constant and EEL/FI ratios differ. Both effects may of course be involved. For adult roosters in these experiments a ratio between EEL and FI of 39 kJ/80 g = 0.49 kJ/g for uncorrected ME has been used. The question is whether this ratio might vary sufficiently to reconcile our results with others and to explain differences in AME values that have been reported.

The ratio can be affected by differences in food intake. Insofar as low-energy feed ingredients are less palatable than those of higher energy content, there is an expectation of bias in existing AME data, with low-energy materials being relatively underestimated. This could explain the observation that the differences between our ME values and those from other laboratories were greater at the lower end of the range studied. The suggestion by Farrell (1981) that EEL is increased on feeds of high fibre content, although not supported by our data, would also produce a bias in this direction. In the experiments of Härtel *et al.* (1977) such bias was avoided by using constant intakes, but the overall level of observed AME values was depressed by restricting intake to 70 g per day. Thus, in comparison with the values predicted by Härtel's (1979) equation our values are high (*Table 3.7*).

From the data collected it would be possible to calculate AME values and equations for any EEL/food intake ratio. At present, however, there is relatively little information about what these ratios should be and whether they differ widely from the figures used for the adult rooster.

Miski and Quazi (1981) have reported a suitable experiment for the study of endogenous losses in broiler chicks of both sexes at different stages of growth. A possible interpretation of their data, which requires some assumptions to be made, is shown in *Figure 3.3*. In spite of the assumptions that have to be made, these data are clearly in accord with the hypothesis that age and sex have, at most, only minor effects on endogenous losses and that the ratios with food intake are similar to the value of 0.49 kJ/g used for the rooster. They also give a very limited amount of data on N loss which can be used to calculate EEL_n/FI ratios of 0.22, 0.26 and 0.25 kJ/g at 32, 77 and 92 days, respectively; again in reasonable agreement with the figure of 0.225 kJ/g in our data.

No data have been found which permit a similar calculation for mature laying and non-laying birds. However, Sibbald (1981b) has recently published some data on total losses. Just prior to maturity, at 136 days, White Leghorn hens of four different strains lost 62.7 kJ/48 h and just after maturity, at 157 days, 72.2 kJ/48 h. As food intake would have increased over the same period, it seems unlikely that this represents a change in ratio, and in any case these values are very similar to those for adult roosters. In broiler breeders mature males lost 101.7 kJ/48 h and females (of different strain and age) 100.8 kJ/48 h. It is not clear how these figures should be scaled so as to compare them with those for White Leghorns, but

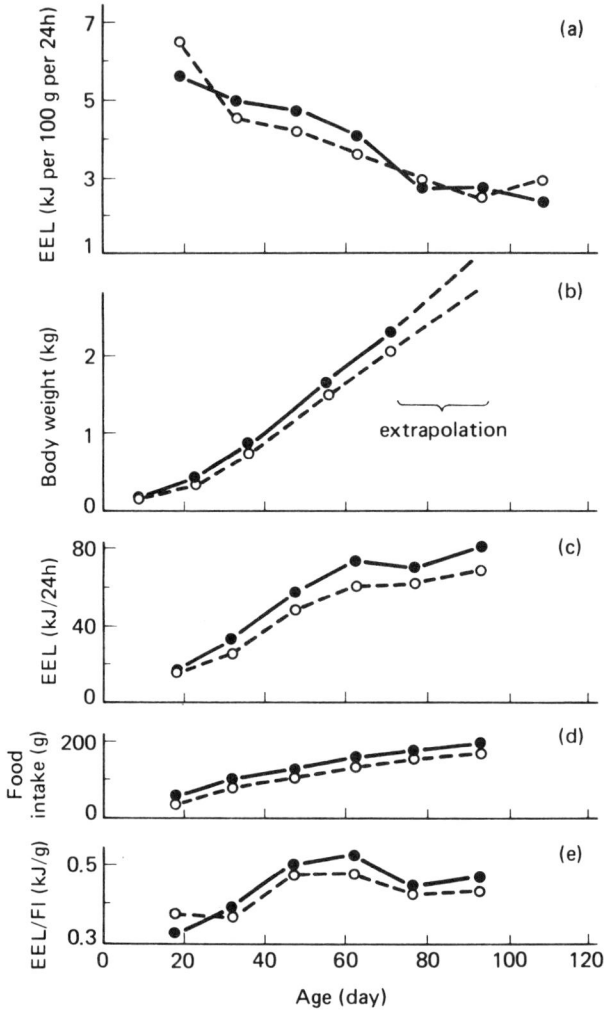

Figure 3.3 A possible interpretation of the results of Miski and Quazi (1981) on endogenous energy losses of growing broiler chickens. Losses per 100 g body weight (as published) are shown in (a). Body weight at each age can be estimated by interpolation and extrapolation (to 92 day) of data given (b) and total energy losses calculated (c). Food intakes (d) are assumed to give estimated EEL/FI ratios (e). ● and ○ refer to male and female, respectively

they clearly imply that there is little difference between the sexes in EEL/FI ratios.

The alternative hypothesis must therefore be tested; namely, that birds of different types have different TME values or alternatively that there is an artefactual difference between procedures. There are some comparisons of TME values determined with different types and ages of bird (Sibbald, 1976b, 1978; Mutzar *et al.*, 1977; Kussaibati, 1979; Dale and Fuller, 1980; Shires *et al.*, 1980; Boldaji *et al.*, 1981). However, a review of these led to many problems of interpretation and will not be presented in detail. At the present time the question of whether different ME data and prediction equations need to be used with different types of stock remains to be unambiguously resolved.

Existing prediction equations

In reviewing existing prediction equations in the light of the issues discussed above, a number of limitations quickly become apparent. First, it is necessary to recalculate the equations into comparable forms and to derive common statistical parameters. This inevitably produces rounding errors and a need for some approximation. Second, the chemical characteristics have to be grouped into a few broad categories—for example, fat, protein, starch—without paying attention to the details of the methods used. Third, the equations have been derived, variously, from data on individual ingredients, groups of ingredients and mixed feeds, and these distinctions have also to be ignored. Of course, if ME values are additive and all the experiments involve a common population of feedstuffs, then in theory data from different sources can be compared directly. Whether

Table 3.5 SOURCES OF EQUATIONS FOR THE PREDICTION OF THE METABOLIZABLE ENERGY CONTENTS OF POULTRY FEEDS AND FEED INGREDIENTS

Reference	Feeds/ingredients	Chemical predictors
1	All	FAT, CP, STC, SUG
2	All	FAT, CP, STC, SUG
3	All	FAT, CP, STC, SUG
4	All	FAT, CP, STC, SUG
5	All (protein concentrates)	FAT, CP, CF, STC, SUG
6	All	FAT, CP, CF, NFE
7	All ingredients	Various; mainly proximate components
8	Wheat	FAT, CP, CF, ASH, NFE, selenium
8	Barley	CP, CF, ASH, SUG
9	Wheat, oats, barley	FAT, CP, CF, ASH, STC, SUG, NFE
10	Oats	CF
11	Oats	Bushel weight; % groats
12	Maize/maize products	CP, CF, STC, SUG, GE
12	Barley	CF
13	Sorghum	Tannin
14	Sorghum	'Fibre'
15	Fats	Melting point; iodine, saponification and acid values; fatty acid composition
16	Tallow	Free fatty acids
17	Cottonseed meal	FAT, CP, CF
18	Blood, fish and meat meals	(?)

References: 1, Carpenter and Clegg (1956); 2, Bolton (1960); 3, Sibbald *et al.* (1963); 4, Härtel *et al.* (1977); 5, Lodhi *et al.* (1976); 6, Moir *et al.* (1980); 7, Janssen *et al.* (1979); 8, Coates *et al.* (1977a); 9, Sibbald and Price (1976a,b, 1977); 10, Thomke (1960); 11, Lockhart *et al.* (1961); 12, Guillaume (1979); 13, Nelson *et al.* (1975); 14, Moir and Connor (1977); 15, Terpstra (1976); 16, Shannon (1971); 17, Watts and Davenport (1971); 18, Cave and Slinger (1970).

Table 3.6 PREDICTION EQUATIONS FOR ME CONTENT OF POULTRY FEEDS, WHEAT AND SOYABEAN MEALS (CONVERTED TO COMMON FORMAT)

Equation No.[a]	kJ/g DM		r^2	rsd	Notes
COMPLETE FEEDS					
1	AME	$0.247 + 35.8x_1 + 15.9x_2 + 17.5x_6 + 15.9x_7$	–	0.80	Adult hens
2	AME	$0.836 + 33.4x_1 + 14.9x_2 + 17.1x_6 + 15.5x_7$	0.90	–	Adults
2a	AME	$0.427 + 33.4x_1 + 14.9x_2 + 17.1x_6 + 15.5x_7$	–	–	Chicks
3	AME_n	$32.8x_1 + 14.7x_2 + 17.2x_6 + 14.9x_7$	–	–	'Final' equation
3a	AME_n	$-0.159 \quad 31.5x_1 + 14.1x_2 + 16.5x_6 + 14.2x_7$	0.77	0.64	'Fitted' equation
4	AME_n	$32.2x_1 + 15.1x_2 + 17.0x_6 + 10.9x_7$	0.97	0.31	Adult hens
5	AME_n	$1.549 + 27.5x_1 + 10.2x_2 - 3.4x_3 + 18.4x_6 + 18.4x_7$	0.73	1.97	Chicks
6	AME_n	$-2.13 + 35.6x_1 + 21.2x_2 + 15.8\,(x_3+x_5) - 27.9\,(x_3/(x_3+x_5))$	0.92	0.31	Chicks
WHEAT					
7	AME_n	$16.673 - 85.8x_3$	1.00	–	Meals
7a	AME_n	$16.426 - 75.7x_3$	–	0.25	Pellets
8	AME_n	$15.171 + 9.6x_2 - 41.4x_3 - 37.2x_4$	0.53	0.87	Chicks
8a	AME_n	$-1.874 + 156.5x_1 + 18.8x_2 - 25.5x_3 - 45.2x_4 + 15.5x_5$	0.84	–	Turkeys
8b	AME_n	$-5.104 + 92.5x_1 + 28.9x_2 - 13.4x_3 - 51.0x_4 + 20.5x_5$	0.82	–	Roosters
9	AME	$-23.702 + 78.6x_1 + 44.6x_2 + 25.4x_3 + 37.8x_5$	0.19	0.52	Roosters
9a	TME	$-14.464 + 37.1x_1 + 39.5x_2 - 0.6x_3 + 31.1x_5$	0.42	0.32	Roosters
9b	AME	$5.699 + 35.3x_1 + 13.8x_2 + 10.7x_6 - 30.8x_7$	0.42	0.44	Roosters
9c	TME	$8.410 + 3.4x_1 + 14.7x_2 + 7.3x_6 + 19.1x_7$	0.44	0.31	Roosters
SOYABEANS					
7b	AME_n	$11.585 + 25.7x_1 - 24.7x_3$	–	0.39	Wholebeans, meals
7c	AME_n	$11.029 + 34.5x_1 - 23.3x_3$	–	0.49	Wholebeans, pellets
7d	AME_n	$11.305 + 30.1x_1 - 24.0x_3$	–	–	Extracted meals

[a]The numbers correspond to the references in Table 5.

x_1, FAT (g/g); x_2, CP (g/g); x_3, CF (g/g); x_4, ASH (g/g); x_5, NFE (g/g); x_6, STC (g/g); x_7, SUG (g/g).

these conditions are fulfilled remains a matter for conjecture. Notwithstanding these limitations, it seemed worth while to pull together the various pieces of evidence, a task also done recently by Sibbald (1975) and Janssen (1976). It is also useful to re-examine the origins of some of these equations.

Table 3.5 lists sources of the prediction equations that have been found, many of them for use only with single ingredients. In *Table 3.6* the six general equations and those suggested specifically for two ingredients, wheat and soyabean meal, are expressed as far as possible in a common format.

The three most widely quoted general equations are those of Carpenter and Clegg (1956), Bolton (1962) and Sibbald *et al.* (1963). They have a lot in common, which is not surprising, since all three embody similar assumptions about the theoretical values of nutrients and the relationships between them.

Carpenter and Clegg (1956) assumed that protein, fat, starch and sugar yield ME in the proportions 1:2.25:1.1:1. This combined function was then regressed onto AME values determined with adult hens for nine cereals, four cereal by-products and four mixed feeds. The regression coefficient was 3.8 kcal/g (15.9 kJ/g) and the intercept 0.053 kcal/g (0.222 kJ/g). The original equation was on a 0.9 DM basis but is corrected to 1.0 DM in *Table 3.6*. This widely quoted equation had a residual standard deviation of 0.79 kJ/g.

Bolton (1962) used 'available' carbohydrate (STC plus 0.91 SUG) in his equation and showed (1960) that digestible carbohydrate was equal to 'available' carbohydrate + k, where k = 49 and 25 g/kg for adults and chicks, respectively. He further assumed that ME is derived from digestible protein, fat and carbohydrate in the ratio 1:2.25:1, that protein and fat have digestibilities of 0.87 and, from Hill (1962), that digestible carbohydrate (starch) has an ME value of 4.08 kcal/g (17.07 kJ/g). Transformation of these assumptions leads to the equation in *Table 3.6*, which therefore combines experimental determinations of digestible carbohydrate, of the digestibility of protein and fat and of the energy value of digested starch. When Bolton (1960) evaluated this equation using the data of Carpenter and Clegg (1956), it explained 0.903 of the observed variation in AME values.

Sibbald *et al.* (1963) used a large body of experimental data (22 ingredients and 81 feeds) to test the effectiveness of the Carpenter and Clegg equation and to derive an alternative. For both purposes they found it necessary to remove data on 16 feeds with more than 120 g/kg crude fibre or which contained maize gluten feeds or certain fats. They then deduced from the 'available' data AME_n values of 4.10, 3.55, 7.85 and 3.52 kcal/g for starch, glucose, fat and protein, respectively. By regressing the resulting calculated AME_n values on the observed data they obtained the 'fitted' equation shown in *Table 5.6*. It was then argued, since the intercept value in this equation was not significantly different from zero and the slope (observed ME on calculated ME) not different from unity, that the assumed values could be used. The 'final' equation in *Table 3.6* was therefore recommended by the authors.

The three remaining general equations in *Table 3.6* are the result of more recent work, although it may be incorrect to include Lodhi *et al.* (1976) in this list, since their equation was derived from data for protein concentrates only. However, they recommend it for general use in the tropics.

Moir *et al.* (1980) analysed the variation in AME_n values among 24 feeds based on Australian ingredients and fed to young chicks. The equation, which, unlike the others in *Table 3.6,* is non-linear, was obtained by regressing the observed AME_n values on crude fibre and calculated GE. For calculating GE the heats of combustion used were 39.5, 23.5 and 17.5 kJ/g for fat, protein and carbohydrate, respectively.

Lodhi *et al.* (1976) developed their equation from AME_n values for 5 fish meals, 1 blood meal and 23 oilseed cakes as determined with young chicks. The equation had a high residual standard error and the significance of the individual coefficients was not given.

The equation proposed by Härtel *et al.* (1977) is of great importance, because it has been proposed as the basis of legislation in the EEC. Forty-five experimental feeds were used in its derivation, five being independent assessments of a basal ration and the remainder 0.33:0.67 mixtures of test ingredients and a basal ration. The basal feed contained (g/kg) soyabean meal, 360; maize, 420; oat hulls, 55; soyabean oil, 20; vitamins/minerals, 145. Each test feed, in meal form, was given to six White Leghorn hens, each bird receiving about 70 g food per day. The standard error of the mean AME_n for a single feed was 0.156 kJ/g, or 1.3 per cent.

Many equations were reported but none had a lower standard deviation than the one in *Table 3.6*. However, this was only obtained after the results for six feeds were removed; these had residuals greater than 3 rsd. The feeds removed contained sugar, milk powder, sugar beet chips, potato starch, soyabean oil and tallow. When these were retained in the analysis, the coefficients were not greatly altered but the statistics became: $r^2 = 0.911$, rsd = 0.71 kJ/g. Equations can, of course, always be 'improved' by removing data, but in this case all the feeds involved contained high levels of single nutrients and might therefore be expected to deviate widely when single coefficients are used for all feeds.

The equations for predicting the ME value of wheat in *Table 3.6* are less consistent, both in form and in the individual coefficients, than the general equations. This observation led Coates *et al.* (1977a) to argue the merit of developing separate equations for each ingredient. However, individual feed ingredients may only provide a very inefficient way of sampling the general population, giving equations which are rather unstable and of limited value, even when applied to just the one ingredient. A rather narrow range of variability will also be found for many ingredients.

In the past 10 years there has been a huge investment of effort in Canada aimed at elucidating the variations in ME values of cereals in chemical terms. The failure of all this work to establish widely applicable methods of prediction serves as a stark reminder of the limitations of the experimental approach.

March and Biely (1973) studied 33 samples of wheat. Although the AME_n values ranged from 12.6 to 15.9 kJ/g, the authors were unable to demonstrate significant relationships either with proximate components or with a series of physical measurements.

Coates *et al.* (1977a,b) present data on 16 samples each of wheat and barley and on a wide range of chemical and physical characteristics. They also report AME_n values determined with chicks, turkeys and roosters. Equations were derived empirically by stepwise regression, and include all factors which were significant at the 5 per cent level. Those obtained for wheat are shown in *Table 3.6*. In fact, better equations were obtained if selenium content was included as a variable, r^2 values being increased from 0.53 (*Table 3.6*) to 0.79 for chicks, from 0.84 to 0.87 for turkeys and from 0.82 to 0.96 for roosters. While such an effect, whether direct or indirect, of a single element is of considerable interest, it lies outside the scope of the present discussion.

The equations in *Table 3.6* show some idiosyncratic variations between the types of bird used. The chick equation, which is a poor description of the data, is little more than a constant value corrected for fibre and ash level. In the turkey and rooster the constant is replaced by a reasonable coefficient for NFE and a very high one for FAT.

A further extensive series of experiments with AME values determined on roosters is reported by Sibbald and Price (1976a). Thirty-five samples of wheat, 28 of oats and 40 of barley were used. These results were compared with various existing equations, and equations estimated with the same variables as each existing equation. The 'best' equations were not reported. Two of the equations for wheat are shown in *Table 3.6*. When proximate components were used, there is a large negative constant, aliased with ASH, and a high coefficient for FAT. The correlation coefficient is very low, and, in fact, most of the coefficients are not significantly different from zero. When STC and SUG are used as prediction variables, the constant is smaller and the coefficients for FAT, CP and STC are realistic. There is a large negative value for SUG. The results for the other cereals, not given here, showed that the predictions for oats were much better than for wheat, and for barley much worse.

Sibbald and Price (1976b, 1977) argued that differences in voluntary food intake, especially between barley samples, would have affected the AME values and concealed the relationships under investigation. They therefore studied variations in TME values, and the resulting equations for wheat are also given in *Table 3.6*. These may reasonably be compared with those for AME, since many of the samples were common to both experiments. The important finding is that the TME values were much more closely correlated with the chemical components and indeed the standard deviations are impressively low. This is a practical example of the importance of avoiding bias in the ME data, due to food intake, when studying chemical prediction equations.

For soyabean meals only the equations suggested by Janssen *et al.* (1979) have been found.

EVALUATION OF EXISTING PREDICTION EQUATIONS

The general prediction equations shown in *Table 3.6* were each evaluated against the data for the test feeds used in our experiments. The observed values were AME_n for an intake of 80 g/day; the predicted values varied between equations and are variously AME and AME_n. Equations of similar forms were also derived from our data, and these, together with the results of the evaluations, are given in *Table 3.7*.

There is a high degree of correlation between the observed values and those predicted by the equations of Carpenter and Clegg (1956), Bolton (1962) and Sibbald *et al.* (1963), which all contain terms for FAT, CP, STC and SUG. On statistical grounds these three equations are indistinguishable, both among themselves and from similar equations calculated from the data.

Conversely, the equation of Lodhi *et al.* (1976), which contains the same variables plus CF, is much less effective, although an equation of this form is satisfactory when calculated from the data. The coefficients suggested by these authors are clearly not applicable to our results. The equation of Moir *et al.* (1980) is also less effective, although this is based only on the proximate components. There seems to be no advantage in this particular configuration, since more effective combinations of the same chemical components have been found (Fisher, 1982).

The equation suggested by Härtel (1979) is an effective predictor of our data, although no more so than the other, similar equations. However, slightly, but significantly, better equations could be derived from the data.

When the residuals (observed minus fitted ME values) obtained with each equation were regressed on the chemical components, a number of significant relationships were found in all cases. A consistent observation was that consideration of the non-linear effect of fat level would give an improvement in prediction.

Although the correlation coefficients between the observed and predicted ME values are high for several of these equations, there are variable, and in some cases considerable, differences between the means. In general, the observed values are higher than those predicted, to a small extent when compared with Carpenter and Clegg (1956) and Bolton (1962) and to a large extent when compared with Sibbald *et al.* (1963) and Härtel *et al.* (1977). This grouping may reflect the fact that N correction was made by the latter two authors but not by the former two.

These differences in mean values raise the same issues that have already been discussed in connection with variations in ME values. For the present, the results in *Table 3.7* only underline the uncertainties that persist about this question.

Discussion of prospects

From the work carried out at the Poultry Research Centre, and from published work, it is clear that the metabolizable energy values of mixed poultry feeds can be predicted effectively from their chemical composition. Several thousand equations have been computed in the search for the best description of our results, and the report by Fisher (1982) should be consulted for a summary of these. The 'best' equation, that is the one with the lowest residual standard deviation in which all the coefficients are significant, had an rsd of 0.24 kJ/g, less than twice the error of determination of 0.15 kJ/g. This equation contained five variables—FAT, CP, STC, NDF and a measure of fatty acid saturation. It is therefore fairly complex,

Table 3.7 THE RELATIONSHIP BETWEEN OBSERVED AME_n VALUES FOR 56 FEEDS (kJ/g DM) AND THOSE PREDICTED BY DIFFERENT EQUATIONS SHOWN IN TABLE 5.6, AND DERIVED EQUATIONS OF SIMILAR FORMS

Prediction equation[a]	Regression of observed values (y) on predicted values (x) (± S.E.[b])	r^2	rsd (kJ/g)	Mean values observed (kJ/g)	Mean values predicted (kJ/g)	Additional variables[b]
1	y = 0.5588 + 0.9627x (±0.0219)	0.973	0.318	14.19	14.15	FAT, FAT²
2	y = 0.0882 + 0.9975x (±0.0216)	0.975	0.304	14.19	14.14	FAT, FAT², CF
3	y = 1.081 + 0.9913x (±0.0216)	0.975	0.305	14.19	13.22	FAT, FAT², CP
4	y = 1.482 + 0.9761x (±0.0216)	0.974	0.310	14.19	13.01	FAT, FAT², STC, SUG
5	y = 2.545 + 0.8449x (±0.0409)	0.887	0.648	14.19	13.78	FAT, FAT², CP, ASH, CF, STC
6	y = 2.043 + 0.9281x (±0.0437)	0.893	0.631	14.19	13.08	FAT, FAT², CP, ASH, STC, SUG, ADL

Derived equations

		FAT	CP	CF	NFE+CF	STC	SUG	r^2	rsd
(1,2)[c]	y = 0.8662	+ 32.78	+ 15.78					0.977	0.304
(3,4)	y =	34.25	+ 16.66	− 12.86[d]				0.975	0.311
(6)	y = 56.64	+ 94.04	+ 78.62	− 25.37[d]	+ 75.22	+ 16.94	+ 14.02	0.943	0.476
(6)	y =	35.15	+ 17.62		+ 15.10	+ 17.86	+ 18.47	0.904	0.609
(5)	y = 5.443	+ 27.79	+ 11.18	− 14.11		+ 11.72	+ 2.67	0.981	0.278
(5)	y =	34.00	+ 16.65	+ 1.28		+ 17.89	+ 17.37	0.975	0.312

[a] Prediction equations as indicated in Table 5.5.
[b] These variables accounted for a significant, $p = 0.05$, or nearly significant amount of the residual variation in observed values about predicted values.
[c] Numbers of corresponding published equations; see Table 5.6.
[d] Variable is CF/NFE + CF in these equations.

and costly, from an analytical point of view and will have to be compared with other more simple but less accurate equations. As Wainman *et al.* (1981) pointed out, such a cost–benefit analysis of different equations will have to take into account the variability of the analytical methods if the accuracy of prediction, as opposed to the accuracy of determination, is to be properly calculated.

In our past data it was found that the regression of AME_n values on FAT, CP and CF, as currently declared, gave an rsd of 0.60 kJ/g. The combination of all the proximate components, including a quadratic effect of fat level, reduced this to 0.42 kJ/g, but the most successful equation based on proximate components alone had an rsd of only 0.29 kJ/g. This contained terms for the interactions between crude fibre and other nutrients. In all equations based on proximate components the substitution of NDF for CF improved the predictions. The additional components of the van Soest detergent fibre analysis scheme did not produce further gains. There are a series of highly effective four-factor equations with FAT, CP, STC and a measure of 'fibre' with rsd values between 0.26 and 0.28 kJ/g. The quadratic effect of fat level was not significant when added to these equations. Finally, the model suggested by Härtel (1979) with FAT, CP, STC and SUG was highly effective (rsd = 0.31 kJ/g), and gave coefficients very similar to Härtel's (*see Tables 3.6, 3.7*). However, it could be improved by the addition of a non-linear fat effect (rsd = 0.29), and, in these data, was slightly inferior to equations with significant constant terms and negative 'fibre' effects.

Whether the levels of accuracy obtained are sufficient for the purpose of legal declarations is a matter for discussion among the interested parties. In making this judgement, however, it is important to remember that the residual errors are only slightly greater than the errors involved in well-replicated and carefully conducted bioassays. As already noted, there will inevitably be some feeds which are not well described by the sort of equations discussed here. Variations in anti-nutritive factors are the most likely source of systematic error, and it is only by the direct use of a bioassay that such possible errors can be completely avoided. Whether a bioassay should replace chemical equations or be used in certain conditions as a final arbiter cannot be discussed here. It should be noted, however, that recent developments in technique make the definition of a standard assay procedure more feasible than it has been hitherto.

These overall conclusions are slightly more optimistic than those reached by Sibbald *et al.* (1980b), who determined the TME values of 419 commercial poultry feeds and regressed these on a range of chemical variables. A series of 42 feeds made in the laboratory from maize, wheat, barley, oats, soyabean meal, rapeseed meal and meat meal was also examined (Sibbald *et al.*, 1980a). Analytical work was done under 'routine quality control conditions', which the authors imply are less precise than 'in a research laboratory'. Only limited details of the work have been published, but in both data sets 80–85 per cent of the variation in observed TME values was explained by the components of the proximate analysis. This is lower than what we have found ($r^2 = 0.945$) and, in contrast to our results, there was no further gain when starch and sugar were included in the equations.

Equations with terms for FAT, CP and NFE had rsd values of 0.305 and

0.609 kJ/g (2.0 and 4.1 per cent of the mean TME values) when determined with the laboratory and commercial feeds, respectively. The reason for this difference is not clear. The standard error of a mean TME value for one of the laboratory feeds (four replicates) was 0.117 kJ/g (0.9 per cent), according to Sibbald *et al.* (1980a), but the corresponding value for the commercial feeds was not given.

A more detailed assessment of this work is not possible from the available data, but Sibbald (private communication) has expressed the view that they show that chemical prediction equations cannot be used for regulatory purposes.

Finally, the results of Sibbald *et al.* (1980b) have to be qualified for three reasons. First, excreta collection was carried out for 24 h only and this is now known to be insufficient; second, N correction was not carried out; and third, while it is correct to consider the errors of routine control quality analysis in assessing the accuracy of an equation, this source of variation should be eliminated as far as possible in the development of equations.

The final alternative to chemical prediction equations and bioassays is that an *in vitro* simulation of the digestive process can be used to predict energy values. Such methods are, of course, well established in ruminant nutrition (Osbourn and Terry, 1977) and feature widely in the prediction equations reported by Wainman *et al.* (1981). Recently Furuya *et al.* (1979) have suggested a similar method using pepsin/HCl in the first stage and a duodenal extract from the pig in the second. Sakamoto *et al.* (1980) subsequently showed that this *in vitro* method gave results which correlated well with *in vivo* digestibility determined with hens.

The potential value of knowing organic matter digestibility can be examined by further re-analysis of the data of Härtel *et al.* (1979). They report determined values for each feed and these can be used in several ways to predict ME, as shown in *Table 3.8*.

Dr G.R. Tanner (personal communication) at the Poultry Research Centre has carried out some preliminary investigations using the feeds from the ME studies. A two-stage assay was used, similar to that of Furuya *et al.* (1979), except that the second stage used a porcine pancreatin preparation (Sigma Chemicals) instead of a duodenal extract and the assay was assessed both by dry matter loss, to give DMD, and by energy loss, to give an *in vitro* DE (IVDE) value.

The method was applied to 28 of the experimental feeds (meal form only). The mean DMD value was 0.822, with a standard deviation, based on duplicate determinations, of 0.024, or 2.9 per cent of the mean. The mean IVDE was 14.7 kJ/g, compared with 14.2 kJ/g for the mean AME_n value, with a standard deviation of 0.32 kJ/g, or 2.2 per cent of the mean. This average disappearance or solubilization of energy in the system appears to be very realistic.

When the observed AME_n values were regressed on IVDE, the correlation was 0.87 ($P < 0.001$); however, the residual standard deviation was 1.0 kJ/g, which is high in comparison with many of the chemical prediction equations.

The values for DMD were combined with GE and chemical nutrients to give estimates of digestible nutrients and these evaluated as predictors of ME, as in *Table 3.8*. Good predictions were obtained but in no case was

Table 3.8 THE PREDICTIVE VALUE OF ORGANIC MATTER DIGESTIBILITY (OMD). CALCULATIONS ON THE DATA OF HÄRTEL $et\ al.$ (1977). THE PREDICTED VARIABLE (y) IS AME_n, kJ/g DM AND ALL DIETARY CONCENTRATIONS ARE EXPRESSED AS g/g (EXCEPT GE)

Equation No.		r^2	rsd
1	11.59	0	2.26
2	$-0.476 + 0.9918\text{GE}\star\text{OMD}$	0.97	0.41
3	$-0.265 + 0.9767\text{GEC}\star\text{OMD}$	0.96	0.46
4	$0.925 + 38.7x_1\star\text{OMD} + 17.3x_2\star\text{OMD} + 15.8x_4\star\text{OMD}$	0.98	0.37
(5	$-2.664 + 34.9x_1 + 17.7x_2 + 17.4x_4 - 15.2x_3$	0.85	0.93)
6	$5.198 + 33.4x_1\star\text{OMD} + 11.1x_2\star\text{OMD} + 12.2x_5\star\text{OMD} + 9.0x_6\star\text{OMD}$	0.96	0.46

[a]GEC = calculated GE = $39x_1 + 24x_2 + 17x_3 + 17x_4$ (kJ/g).
x_1, FAT (g/g); x_2, CP (g/g); x_3, CF (g/g); x_4, NFE (g/g); x_5, STC (g/g); x_6, SUG (g/g).

the improvement over using chemical components alone significant. For example, an equation containing FAT, CP and STC had a residual standard deviation of 0.32 kJ/g whether nutrient levels were defined in total or in 'digestible' terms.

Thus, in its present state of development, this *in vitro* method provides a reasonable estimate of the availability of dry matter or energy to the animal but, as judged from the present results, provides no advantage over chemical data for the prediction of ME values. The assay could undoubtedly be improved, although both the error of determination and simple correlations with energy values are very similar to those found by Wainman *et al.* (1981) with sheep. To date a sample size of 0.5 g has been used and this leaves a very small residue for determination of anything other than dry matter disappearance. By doing the bomb calorimetry in gelatin capsules the energy content of the total residue could also be determined. Organic matter or nutrient solubilization in the system could be determined with a larger sample size, and the prediction of energy values might thereby be improved.

References

BOLDAJI, F., ROUSH, W.B., NAKAUE, H.S. and ARSCOTT, G.H. (1981). *Poult. Sci.*, **60**, 225

BOLTON, W. (1960). *Analyst*, **85**, 189

BOLTON, W. (1962). *Proc. XIIth World's Poultry Congress*, **2**, 38

CAVE, N.A.G. and SLINGER, S.J. (1970). *Proc. Can. Fed. Biol. Soc.*, **13**, 124 (not seen in original; quoted by Sibbald, 1975)

CARPENTER, K.J. and CLEGG, K.M. (1956). *J. Sci. Fd Agric.*, **7**, 45

CHUDY, A. and SCHIEMANN, R. (1971). In *Energetische Futterbewertung und Energienormen*, p.168. Eds R. Schiemann, K. Nehring, L. Hoffman, W. Jentsch and A. Chudy. Berlin; VEB Deutscher Landwirtsch.-Verlag

COATES, B.J., SLINGER, S.J., ASHTON, G.C. and BAYLEY, H.S. (1977a). *Can. J. Anim. Sci.*, **57**, 209

COATES, B.J., SLINGER, S.J., SUMMERS, J.D. and BAYLEY, H.S. (1977b). *Can. J. Anim. Sci.*, **57**, 195

DALE, N.M. and FULLER, H.L. (1980). *Poult. Sci.*, **59**, 1941

DAVIDSON, J. and GRAHAM, S. (1981). *J. agric. Sci., Camb.*, **96**, 221

FARRELL, D.J. (1978). *Br. Poult. Sci.*, **19**, 303

FARRELL, D.J. (1981). *Wld's Poult. Sci. J.*, **37**, 72

FISHER, C. (1982). *Energy Values of Compound Poultry Feeds*. PRC Occasional Publication No. 2. Edinburgh; Poultry Research Centre

FURUYA, S., SAKAMOTO, K. and TAKAHASHI, S. (1979). *Br. J. Nutr.*, **41**, 511

GUILLAUME, J. (1979). *Proc. 2nd European Symposium on Poultry Nutrition*, p.1

HÄRTEL, H. (1979). *Proc. 2nd European Symposium on Poultry Nutrition*, p.6

HÄRTEL, H., SCHNEIDER, W., SEIBOLD, R. and LANTZSCH, H.J. (1979). *Arch. Geflugelk.*, **41**, 152

HILL, F.W. (1962). In *Nutrition of Pigs and Poultry*, p.3. Eds J.T. Morgan and D. Lewis. London; Butterworths

HOFFMAN, L. and SCHIEMANN, R. (1973). *Arch. Tierernähr.*, **23**, 105

JANSSEN, W.M.M. (1976). *Proc. Int. Symp. on Computer Use in Feed Formulation*, p.93. Brussels; National Renderers Association

JANSSEN, W.M.M., TERPSTRA, K., BEEKING, F.F.E. and BISALSKY, A.J.N. (1979). *Feeding Values for Poultry*, 2nd edn. Beekbergen, The Netherlands; Spelderholt Institute for Poultry Research

JANSSEN, W.M.M., WAANDERS, J. and TERPSTRA, K. (1976). *Proc. 7th Symp. Energy Metabolism*, EAAP Publ. No. 19, p. 273. Ed. M. Vermorel. Clermont-Ferrand; G. de Bussac

KUSSAIBATI, R. (1979). *Proc. 2nd European Symposium on Poultry Nutrition*, p.14

LOCKHART, W.C., BOLIN, D.W., OLSON, G. and BRYANT, R.L. (1961). *Poult. Sci.*, **40**, 327

LODHI, G.N., SINGH, D. and ICHHPONANI, J.S. (1976). *J. agric. Sci., Camb.*, **86**, 293

McNAB, J.M., BLAIR, J.C., DEWAR, W.A. and DOWNIE, J.N. (1981). *Proc. 3rd European Symposium on Poultry Nutrition*, p. 132

McNAB, J.M. and FISHER, C. (1981). *Proc. 3rd European Symposium on Poultry Nutrition*, p. 45

MARCH, B.E. and BIELY, J. (1973). *Can. J. Anim. Sci.*, **53**, 569

MISKI, A.M.A. and QUAZI, S. (1981). *Poult. Sci.*, **60**, 781

MOIR, K.W. and CONNOR, J.K. (1977). *Anim. Feed Sci. Technol.*, **2**, 197

MOIR, K.W., YULE, W.J. and CONNOR, J.K. (1980). *Aust. J. exp. Agric. Anim. Husb.*, **20**, 151

MUTZAR, A.J. and SLINGER, S.J. (1981). *Poult. Sci.*, **60**, 835

MUTZAR, A.J., SLINGER, S.J. and BURTON, J.H. (1977). *Poult. Sci.*, **56**, 1893

NELSON, T.S., KIRBY, L.K. and JOHNSON, Z.B. (1981). *Poult. Sci.*, **60**, 786

NELSON, T.S., STEPHENSON, E.L., BURGOS, A., FLOYD, J. and YORK, J.O. (1975). *Poult. Sci.*, **54**, 1620

OSBOURN, D.F. and TERRY, R.A. (1977). *Proc. Nutr. Soc.*, **36**, 219

PARKS, J.R. (1973). *J. theor. Biol.*, **42**, 349

SAKAMOTO, K., ASANO, T., FURUYA, S. and TAKAHASHI, S. (1980). *Br. J. Nutr.*, **43**, 389

SHANNON, D.W.F. (1971). *J. agric. Sci., Camb.*, **76**, 217

SHIRES, A., ROBBLEE, A.R., HARDIN, R.T. and CLANDININ, D.R. (1980). *Poult. Sci.*, **59**, 396

SIBBALD, I.R. (1975). *Feedstuffs, Minneapolis*, **47**, 22

SIBBALD, I.R. (1976a). *Poult. Sci.*, **55**, 303

SIBBALD, I.R. (1976b). *Poult. Sci.*, **55**, 1459

SIBBALD, I.R. (1978). *Poult. Sci.*, **57**, 1008

SIBBALD, I.R. (1980a). In *Recent Advances in Animal Nutrition—1979*, p.35. Eds W. Haresign and D. Lewis. London; Butterworths

SIBBALD, I.R. (1980b). *BioScience*, **30**, 736

SIBBALD, I.R. (1980c). *Poult. Sci.*, **59**, 836

SIBBALD, I.R. (1981a). *Poult. Sci.*, **60**, 805

SIBBALD, I.R. (1981b). *Poult. Sci.*, **60**, 2672

SIBBALD, I.R., BARRETTE, J.P. and PRICE, K. (1980a). *Poult. Sci.*, **59**, 805

SIBBALD, I.R., CZARNOCKI, J., SLINGER, S.J. and ASHTON, G.C. (1963). *Poult. Sci.*, **42**, 486

SIBBALD, I.R. and PRICE, K. (1976a). *Can. J. Anim. Sci.*, **56**, 255

SIBBALD, I.R. and PRICE, K. (1976b). *Can. J. Anim. Sci.*, **56**, 775

SIBBALD, I.R. and PRICE, K. (1977). *Can. J Anim. Sci.*, **57**, 365

SIBBALD, I.R. and PRICE, K. (1980). *Poult. Sci.*, **59**, 1275

SIBBALD, I.R., PRICE, K. and BARRETTE, J.P. (1980b). *Poult. Sci.*, **59**, 808

TERPSTRA, K. (1976). *Proc. Int. Symp. on Computer Use in Feed Formulation*, p. 85. Brussels; National Renderers Association

THOMKE, W. (1960). *Arch. Geflügelk.*, **24**, 557

TITUS, H.W. (1961). *The Scientific Feeding of Chickens,* 4th edn. Danville, Illinois; The Interstate

WAINMAN, F.W., DEWEY, P.J. and BOYNE, A.W. (1981). Third Report, Feedingstuffs Evaluation Unit, Rowett Research Institute. Edinburgh; Dept Agriculture for Scotland

WATTS, A.B. and DAVENPORT, R.F. (1971). *Poult. Sci.*, **50**, 1643

4

TECHNIQUES FOR DETERMINING THE METABOLIZABLE ENERGY CONTENT OF POULTRY FEEDS

C. FISHER AND J.M. McNAB
Institute for Grassland and Animal Production, Poultry Division, Roslin, Midlothian, UK

Introduction

Apart from the economic importance of energy in feed formulation the present sustained level of interest in metabolizable energy (ME) determination stems from two main events. The first was the introduction of rapid bioassays for ME in the mid-1970s and in particular the work of Dr I.R. Sibbald in Canada. The second was the adoption of energy declarations, and the associated chemical control equation, into the feed trade in Europe. This latter development has focused attention on the accuracy, repeatability and suitability of different methods of measuring ME.

The subject has been widely reviewed, both here and elsewhere. Sibbald (1979) described the development of his methods and in 1982 produced a further very detailed review. Attention is also drawn to the review by Farrell (1981). Sibbald (1986) provided an up-to-date description of his bioassay and a complete list of references, although this was not a review. The development and application of prediction equations for ME of poultry feeds was discussed by Fisher (1983) and the derivation of the EEC equation has been described (Fisher, 1986). Since 1975 a huge literature has developed on the topic; Sibbald (1986) lists 561 references concerned directly or indirectly with this field and only five of these predate 1975. Many of these papers are concerned with methodology but the present review is concerned mainly with those which have appeared since Sibbald's substantive work in 1982.

Recent developments have tended to highlight the questions of reproducibility in ME bioassays, across laboratories and across time, and of variation in ME data. The introduction of energy declarations and of a control equation does this, because it is presumed that the whole system is based on a defined and reproducible biological characteristic — the ME of a feed. Attempts to test or verify equations obviously founder if this cannot be observed consistently. Further development of ME values for feed ingredients also requires prediction equations to be established relating ME to chemical composition or to other quality control factors. Progress in this field is facilitated if results from different laboratories can be combined and again variations in technique are brought into focus, especially if they lead to different biasses. It might have been hoped that the introduction of effective rapid assays would help to achieve the laudable aim of standardization of technique but, in fact, this has not happened, apparently for two main reasons. First, the rapid

First published in *Recent Advances in Animal Nutrition – 1987*

assays require the use of starved birds which has proved controversial; second, it is clear that the adoption of published techniques led to problems in some laboratories and a whole series of minor and major variations have been introduced. Thus, we now have a plethora of 'methods' for ME determination described in the literature and standardization is probably further away now than it was ten years ago.

TERMINOLOGY AND DEFINITIONS

These topics are relatively free from disagreement and a widely used convention is followed here, mainly in agreement with Sibbald (1982, 1986). The term ME is used in a general sense rather than bioavailable energy (BE) as proposed by Sibbald (1982 and elsewhere). We also continue to use the term endogenous energy loss (EEL) even though it is usually defined, not as a biological entity, but in empirical experimental terms, e.g. energy loss from a starved bird. This is convenient and not necessarily confusing. The convention of ignoring energy lost as gases produced in fermentation is also followed. Pesti and Edwards (1983) propose that the nomenclature used in this field should be completely changed to reflect the methods used in experiments but we find this unhelpful. A preferable approach is to modify the methodology until it measures and reflects well defined biological entities. The point made by Pesti and Edwards however, that more care is needed in relating experimental observations to supposedly well defined biological elements, cannot be repeated too often.

   Finally, in the introduction to this chapter it is relevant to comment critically on the standard of reporting work in this field in journals. ME values are not observed or measured but are derived by calculation from a whole series of measurements and far too little basic information is normally reported. Thus it is frequently impossible to make critical comparisons between different experiments. Furthermore, as the background of the topic develops, far more use could be made of existing data, and the results from different studies could be combined, if more detailed tabulation of results was introduced.

Methods for determining the ME of feeds and ingredients

Metabolizable energy is erroneously considered to be a characteristic of a feed; it is really a characteristic of an animal to whom the feed is given. ME measurement relates to the complete feed given and values for feed components or ingredients must, in most cases, be obtained by comparing the results for two or more suitable feeds (substitution methods). In a few cases (e.g. cereals) this distinction between feed and ingredients can be eliminated. In either case the assumption of additivity of ME values amongst feedstuffs is essential and very little progress can be made if this is not upheld. Energy is of course a useful currency for describing mass conversion of food elements in the bird. There is a set of problems, analogous to those discussed here for ME, in determining the 'metabolizability' of any nutrient; lipid, protein, carbohydrate. For many purposes, and especially for prediction, it would be preferable if both ME values and digestibility coefficients for the main components were measured concurrently, but this is rarely done.

McNab and Fisher (1981) suggested that the observations required for an ME bioassay were threefold: (i) a knowledge of energy balance (EB) at (ii) a known food intake (FI) and (iii) an appropriate measure of EEL. For correction to zero N-retention (NR_0) then N-balance (NB) must also be measured. It is useful when discussing methods to bear in mind the relationships shown in *Figure 4.1*, which, in particular, have been discussed by Wolynetz and Sibbald (1984).

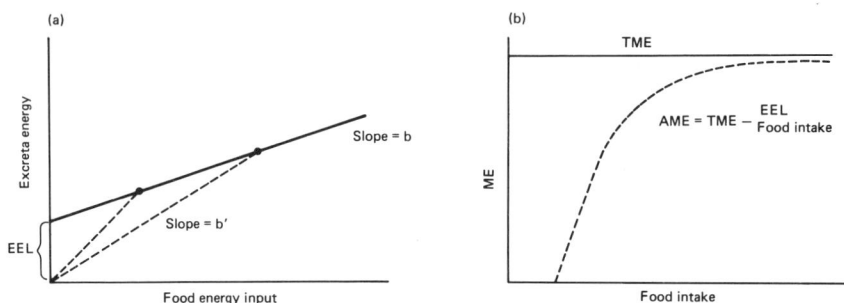

Figure 4.1   (a) Regression of excreta energy on food energy input. (b) Relationship between AME, TME and food intake as derived from (a), assuming that EEL $\neq 0$

Figure 4.1a shows the regression of excreta energy (EXE or EXE_n if corrected for N loss) and food (gross) energy input (GEI). The intercept on the y axis provides an operational definition of EEL (or EEL_n); that is, energy excretion at zero energy input, and the slope of the line yields the true metabolizable energy (TME or TME_n) of the feed as TME = GE(1 − b) where GE is the gross energy of the feed. Estimates of apparent metabolizable energy (AME or AME_n) correspond in a similar way to the slope of lines joining a given energy balance with the origin of the graph; thus for the example in *Figure 4.1a*, AME = GE(1 − b′). The derivation of *Figure 4.1b* is obvious if a range of intake is envisaged. Notice also that if the intercept is zero then AME = TME and AME is independent of intake. Negative intercepts suggest an artefact of measurement.

Three general types of energy balance experiment can be identified. (Squibb (1971) and others have suggested that ME could be assessed by growth but this is not pursued here.) The three types are as follows:

(1) Traditional assays which involve preliminary feeding periods to establish 'equilibrium' conditions. Differences in carryover in the digestive tract between the beginning and end of the assay period ('end-effects') are controlled by trying to ensure that they are the same. Complete diets must be used in most cases and substitution methods used for ingredients.
(2) Rapid assays, using starvation before and after giving a known aliquot of test feed to control 'end-effects' but which permit the birds free access to the feed. Again, complete diets and substitution methods must be used in most cases.
(3) Rapid assays, as above, but using tube-feeding (or force-feeding or precision feeding) to place the test feed directly in the bird's crop. These methods usually avoid the need to substitute feed ingredients into a basal diet.

Whilst many variations are found in these general groups, this classification provides a convenient framework within which the many details of procedure can be discussed. This is done here under a number of headings although this arbitrary separation oi ᴛechniques tends to hide many interacting arguments.

BALANCE EXPERIMENT METHODOLOGY

Feed presentation and the measurement of energy intake are amongst the most difficult aspects of balance experiments. If birds are given free access to the feed, a technique which still seems to have the widest acceptance, great care is required to avoid food loss, to recover lost food including that in the drinker, to avoid separation of food components, to monitor dry-matter changes and to take samples. In total these are difficult to control in a consistent way but specially designed systems have been described and used with apparent success (e.g. Terpstra and Janssen, 1975).

Such free-feeding methods are used in type (1) assays which form the greater part of the literature on ME determination for poultry. Farrell (1978) proposed that the advantages of a rapid, type (2), assay could be obtained by training birds to consume satisfactory intakes in 1 h after a 23 h fast. In this assay equal quantities of basal feed and test ingredient were combined, although pelleting of the feed was recommended to maintain intakes across a range of feeds. Several groups have reported difficulties in maintaining satisfactory intakes (Mutzar and Slinger, 1980; Jonsson and McNab, 1983; Parsons, Potter and Bliss, 1984; Kussaibati and Leclercq, 1985) but, notwithstanding this, several assays based on this method of feed presentation have been described with minor variations in starvation and feeding times (Chami, Vohra and Kratzer, 1980; Vohra, Chami and Oyawoye, 1982; Parsons, Potter and Bliss, 1984). Kussaibati and Leclercq (1985) propose an original assay which fits roughly into the same category but which has not, apparently, been tested elsewhere. Adult cockerels were starved for 24 h, fed *ad libitum* for 24 h and then starved for 24 h. Excreta were collected for the final 48 h of this 72 h assay.

Whilst assays of the type proposed by Farrell (1978) can obviously be successful it is clear that variations in food intake will occur with some feedstuffs. Extensions of the basic principle could include the use of different, but pre-determined, intakes (du Preez, Minnaar and Duckitt, 1984) or, if EEL is known, correction of the data to take account of different intakes. This latter route was followed by Jonsson and McNab (1983) but these modifications have not been adopted routinely.

The presentation of feed by tube in type (3) assays permits very accurate feeding and offers the opportunity for controlling some problems such as variations in dry matter content of the feed. Since dose size is reduced, problems of sampling become more important. The only real disadvantages of the technique are the limit on dose size and attitudes towards the acceptability of a procedure which is frequently called 'force-feeding'. Our experience is that, with practice, the technique is extremely rapid (15–30 s/bird with most feeds) and that there is little evidence of more stress beyond that involved in handling. Experience and skill must be attained, although this is easily done with practice by most operators. Wehner and Harrold (1982) raised the issue of stress and suggested the use of slurry feeding to reduce it. However they quoted feeding times of 8 to 12 min with dry feeds which would be very stressful, but which are unnecessary (Fraser and

Sibbald, 1983). Finely divided, hygroscopic or very bulky ingredients may present problems to inexperienced feeding teams but again these can be overcome with practice. In this laboratory, glucose monohydrate is routinely fed; this is a problem material and granulation is used to reduce the difficulties. Alternatives to intubation of dry feeds include use of slurries (Wehner and Harrold, 1982; Teeter *et al.*, 1984) or gelatin capsules (Bilgili, Arscott and Kellems, 1982).

Balance experiments with ducks and geese present many problems (Ostrowski-Meissner, 1984). His assay for ducks involves training birds to consume a normal intake in 1 h per day but then administering a test dose by tube, so as to avoid losses in the drinking water. It is suggested that training reduced the 'stress' of tube-feeding and allowed intakes to be increased from 30 to 70 g. Slurry feeding was used with ducks by Mohamed *et al.* (1984) who did not report any problems. Storey and Allen (1982) fed dry feed to geese by tube but did comment on some difficulties.

Excreta collection is another simple task which is difficult to do well in routine experiments. When done with trays under cages the problems include adherence to feathers, physical losses, contamination with food, feathers and scurf, fermentation and losses in collection and transfer. Sibbald (1986) lists useful precautions to be taken; frequent collection e.g. 12-hourly as in Dale *et al.* (1985), and continuous mechanical blowing to remove scurf are the sort of devices which might be judged beneficial.

The only alternative to collection trays is to attach a bag to the bird for collection and methods for doing this have been described. Sibbald (1983) proposed the use of adhesives to fasten the bags but preliminary experience in this laboratory was not encouraging so far as routine programmes were concerned. The use of a harness to hold the bag (Sibbald, 1986; Almeida and Baptista, 1984) has also been associated with problems in this laboratory and elsewhere (P.J. Gallimore, personal communication). In the only direct comparison of methods (Sibbald and Wolynetz, 1986) there was evidence of increased excreta output on trays as compared with bags. This led to lower estimates of TME_n, by about 0.5 MJ/kg across four feeds, when a conventional assay was used with zero intake plus one level of feeding. Other studies showed that the slope of the line relating excreta energy to input was independent of collection method so TME estimates in multilevel trials were not affected by collection method. It is not clear whether the use of trays leads to an overestimate or the use of bags to an underestimate of true excreta loss. Excreta in bags remains moist for up to 48 h and may be subject to fermentation loss. Addition of acid to the bags gave results similar to trays which suggests the second cause, but the acid caused further problems and accumulation of some non-excreta material on trays seems inevitable. Sibbald and Wolynetz (1986) also found a high level of data loss with bags which in these studies were fastened to the birds by adhesive.

For routine use it is difficult to see how trays can be avoided without considerable increase in cost although this may be justified for work on amino acid availability where feather and scurf contamination can lead to large, systematic errors.

USE OF DIGESTION MARKERS OR INDICATORS

Accurate feeding and excreta collection in ME assays create a lot of menial, repetitive and relatively unpleasant tasks but require impeccable standards to be maintained. The use of digestion markers or indicators may reduce these problems and also facilitate balance studies in less-than-ideal animal caging. The proposed requirement for food intake data is not met but this can be recorded with sufficient accuracy for correcting AME data without incurring the exacting demands of total collection. New problems and sources of variation are introduced with markers but the routine tasks are moved to the more congenial laboratory atmosphere and may be amenable to automation. Problems due to excreta contamination are not avoided by markers.

Marker methods were reviewed by Sibbald (1982) in relation to the present topic. Since that time the use of titanium dioxide (Peddie *et al.*, 1982) and magnesium ferrite (Neumark, Bielorai and Iosif, 1982) have been proposed for studies with poultry. Titanium dioxide was adopted as a routine marker in our laboratory because of safety restrictions on the use of chromium. All markers must be insoluble and are therefore difficult to analyse by chemical methods. Magnesium ferrite is determined by physical methods and polythene by gravimetric methods which do not require the markers to be brought into solution.

At the present time chromium sesquioxide (subject to very variable views on its safety to humans), acid insoluble ash (Vogtmann, Pfirter and Prabucki, 1975), 'fibre' fractions (Bolton 1954; Almquist and Halloran, 1971), titanium dioxide (Peddie *et al.*, 1982) and perhaps polyethylene (Roudybush, Anthony and Vohra, 1974) or magnesium ferrite (Neumark, Bielorai and Iosif, 1982) are candidates for use in routine, practical poultry studies depending on local circumstances. The use of radioactive markers, ^{51}Cr, ^{91}Y and ^{41}Ce (Sklan *et al.*, 1975) is attractive if suitable equipment for their detection is available. It is not possible to draw a general conclusion about markers. They are not a substitute for meticulous and carefully controlled work but do have a contribution to make. The simultaneous use of two or more markers would provide a continuous and very exacting check on procedures (Dr V. Petersen, personal communication) and such work would have a considerably enhanced value. In many ways the rapid assays remove the need for markers, especially in type (3) assays, because the problem of feeding and excreta collection are greatly simplified. With low dose levels in type (3) assays retention or loss of marker would be a very critical issue.

SAMPLE PREPARATION AND ANALYSIS

Grinding of feeds prior to evaluation is another unresolved technical issue. In type (3) assays grain fed whole will appear in the excreta and fineness of grinding may be a variable in relation to the ME value of some ingredients. In reports of slurry feeding considerable emphasis seems to be given to fine grinding.

Each of the measurements used to calculate an ME value is subject to potential bias or imprecision and this is not revealed in the ME data themselves except as poor repeatability. Again, the small intakes in type (3) assays make these more sensitive issues. It is surprising that a detailed evaluation of the potential errors in ME studies has not been reported so that particular attention could be devoted to the most important steps. It is generally agreed, for example, that more replication should be used to determine the combustion value of a feed than of excreta, but such issues are rarely reported. It is probably not too harsh a judgement to say that too little attention is paid to both the accuracy and precision of the various chemical analyses used in ME determinations. When ring tests are undertaken the results are sometimes very poor (see Carew, 1978 for chromium and Fisher, 1983b for feed analyses). Even a detailed analysis of a single measurement, such as combustion value, in one laboratory (Sibbald and Morse, 1983), underlines the need for constant vigilance.

Losses of energy from excreta during drying have been variable in different experimental studies (Sibbald, 1982) but careful oven drying (60°C for 18 or 24 h) may be equivalent to freeze-drying (Wallis and Balnave, 1983; Dale *et al.*, 1985) thus simplifying the requirements for ME assays. Probably the control of moisture content (Terpstra and Janssen, 1975) and the loss of moisture in grinding are sources of potential error at least as great as drying methods. Attempts to apply standardization in this field would present awesome problems but might be rewarding; in the meantime it is only possible to encourage high standards and journals might promote these by insisting on better evidence of internal checking.

ASSAY DESIGN, 'END EFFECTS', RATE OF PASSAGE OF FOOD

In type (1) assays test feeds are given for a preliminary period to establish 'equilibrium' conditions and any differences in gut-fill at the beginning and end of the assay ('end-effects') are assumed to cancel out. Any difference will be relatively small in comparison with the total balance if the usual 3–5 day collection period is used. Short periods of food restriction may be used to establish similar patterns of eating at the beginning and end of the assay. The amounts of ingesta carried over will be quite large and dependence on the cancelling of errors is not entirely satisfactory. Changes in the intake of rapidly growing birds or in the response to unpalatable feeds may cause systematic bias rather than random error. Studies of the variability of ME data, rather than on the residues themselves, suggest that equilibrium periods of 2 days and collection periods of 3–5 days are sufficient for variability to approach minimum values. As Sibbald and Price (1975) pointed out, required levels of error can be obtained by varying replication and/or length of experiment.

The design of a rapid assay is both more critical, because of small intakes, and more complex, because of several interacting factors. The basic assumption is that the digestive tract is free from residues or 'empty' at the beginning and end of the assay. Factors likely to be involved include the nature of the previous (holding) feed, starvation periods, nature of test feed, levels of input, length of collection period, water intake and random variation in caecal evacuation. Sibbald (1976) originally proposed 24 h starvation and excreta collection periods (24 h + 24 h assay) but now (1986) proposes (24 h + 48 h) for routine use. In this laboratory (48 h + 48 h) is used routinely (McNab and Fisher, 1984; McNab and Blair, 1987). The longer period is clearly more stressful and therefore further factors such as bird size and glucose feeding come into consideration.

Earlier work (Sibbald, 1982) showed that 12 h starvation prior to feeding was insufficient but that extension beyond 24 h had only small effects on determined TME values. Farrell's (1978) assay effectively involved 23 h starvation, but following a different feeding pattern. Direct investigation, however, shows a measurable difference between feed residues remaining after 24 h and 48 h starvation (*Table 4.1*). These observations, and the logic of equalizing the pre- and post-feeding starvation periods, encourages the use of the (48 h + 48 h) assay and the adjustment of other factors to deal with the increased stress. In general, clearance rates

Table 4.1   FEED RESIDUES IN THE GASTROINTESTINAL TRACT OF ADULT COCKERELS 24 AND 48 h AFTER THE REMOVAL OF A MIXED FEED PREVIOUSLY GIVEN *AD LIBITUM*[a]

Bilgili and Arscott (1982)[b]					
Time (h)	*Crop*	*Gizzard*	*Intestine*	*Caeca*	*Total*
24	0.04	4.10	1.46	0.34	5.94
48	0.03	2.15	1.25	0.31	3.74

McNab and Blair (1987)[c]		
Time (h)	*Total weight*	*Total energy*
24	1.59 (0.8–2.6)	20.3
48	0.17 (0.05–0.3)	2.1

[a] Similar data, but after tube feeding an aliquot of a single ingredient can be found in Mutzar and Slinger (1980)
[b] Values are total dry matter recovered from each segment
[c] Values are g lyophilized residue from the whole tract with range in parenthesis (n = 20) and total energy, kJ

are variable between feeds and rates of input (Sibbald, 1982) and therefore a constant holding diet of well digested components (low residue) should be used although this is not a very critical issue (Shires *et al.*, 1979). To some extent a correction is made for carry-over of energy from the previous feed in calculating TME in Sibbald's assay because a comparable error will occur in the fed and negative control birds. Reliance should not be placed on this however, and the proposal that dried excreta from the negative controls should be sieved to remove 'fibrous' residues (Bilgili and Arscott, 1982) is inappropriate for this reason.

The time required to ensure complete clearance of a test-feed, especially when single ingredients are fed in type (3) assays is a complicated and unresolved issue. The original suggestion of 24 h (Sibbald, 1976; Farrell, 1978) is known to be too short and all data collected on this basis should be ignored because of the uncertainty involved. Farrell (1980) now recommends 32 h and Sibbald (1986) 48 h for routine use. *Table 4.2* lists unpublished TME values comparing data from 48 h and

Table 4.2   COMPARISON OF TME_n VALUES COMPUTED AFTER 48 AND 72 H EXCRETA COLLECTION[a]

Ingredient	*No. samples*	TME_n (MJ/kg)	
		48 h	72 h (rel)
Full fat soya	4	14.44	14.42 (100)
Wheat	12	12.77	12.75 (100)
Fish meal	12	13.29	13.06 (98)
Blood meal	5	13.37	12.09 (90)
Meat and bone meal	7	10.77	10.46 (97)
Wheatfeed	12	8.59	8.56 (100)
Carrot	1	9.97	9.80 (98)
Cabbage	1	9.81	9.44 (96)
Pea hulls	1	1.79	1.63 (91)

(McNab, unpublished data)
[a] All evaluations by the method of McNab and Blair (1987)

72 h collections with input levels of 50 g. For some materials, e.g. blood meal, these data suggest that there is a persistent problem under these conditions. The results of Sibbald and Morse (1983), and Sibbald's work elsewhere, suggest that the use of lower intakes ameliorates the problem but at the cost of both reduced accuracy and increased influence of any residual end effects. Our experience is that this problem occurs mainly with high protein, and especially finely divided animal products. Wetting of the feed in the crop may be one factor but taste or 'palatability' may also be involved since, for example, wetting blood meal in the crop by feeding water appears to be distasteful to the bird. Gut stasis may be induced by the sudden introduction of some ingredients and attempts to evaluate coffee residues by tube-feeding had to be abandoned since no passage at all could be obtained (unpublished observations). At the present time it is only possible to counsel caution with unusual ingredients in tube-feeding experiments and to recommend a routine check for feed residues after excreta collection in doubtful cases. The routine extension of collection times to 72 h would provide an empirical solution but at considerable cost to the birds. Also the longer the collection period the higher the endogenous:exogenous energy ratio and greater the relative importance of error in the correction for EEL.

The assessment of these factors in Farrell's (1978) assay is also difficult. With high intakes, such as 70–80 g, 32 h collection may be too short for some feeds; for example, see Sibbald and Morse (1983) for alfalfa and oats. On the other hand the use of complete feeds and a feeding period of 1 h may reduce the problem in comparison with tube-feeding of single ingredients. It seems reasonable to suppose that because the crop does not become so tightly packed with dry food under these conditions, water intake might be more normal.

The importance of water intake in ME assays is yet another area where firm conclusions cannot be reached and which may be a significant source of variation. McNab and Blair (1987) observed that, despite the ready availability of water, tube-fed birds were rarely seen to be drinking and speculated that low and variable water intakes might explain erratic clearance rates. *Table 4.3* summarizes the findings which led to the routine administration of water (about 50 ml/bird) during the excreta collection period. This routine also provides an opportunity to palpate

Table 4.3   THE EFFECT OF WATER ADMINISTRATION ON ENERGY EXCRETION AND THE TME VALUE OF SOYABEAN MEAL

Water administration[a]		Soyabean meal fed (g)	Excreted energy (kJ)	TME (MJ/kg)
Before feeding	After feeding			
+	+	0	89.4 ± 20.4	–
–	–	25	257.3 ± 42.0	11.00
–	+	25	246.9 ± 33.5	11.40
+	+	25	236.4 ± 5.3	11.83
		SEM 17.9		

(McNab and Blair, 1987)
[a] Adult cockerels (6 per treatment) were starved for 48 h before being given soyabean meal by tube. Water was also given by tube, 50 ml per administration, after 8 and 32 h starvation (before feeding) and 32 h after feeding soyabean meal. The data are not corrected for N loss

the crop and to mix any food residues with water and has not, in our experience, led to any loss of food from the crop. It seems reasonable to argue that this practice will change the relationship between dose level and clearance rates (Sibbald and Morse, 1983) but this has not been demonstrated.

A direct investigation of the role of water:food ratios was reported by Van Kampen (1983) but with rather ambiguous results. In birds with free access to water a positive relationship was found between AME (y, %GE) and water:food ratio

$$y = 66.38 + 2.97x \ (P<0.01, \ r = 0.49) \tag{4.1}$$

This is an effect of considerable magnitude but it could not be demonstrated experimentally when water was administered by tube immediately after feeding (mixture of free- and tube-feeding, feeding time 15 min) in a rapid assay. However, excreta were collected for only 24 h which might have masked any treatment effects. Experiments of this sort also need to be done with a range of ingredients and not just with practical feeds.

Finally, the alternative assay design suggested by Kussaibati and Leclercq (1985) should be mentioned. This is a variation on Farrell's ideas but trained cockerels are not used and birds are allowed free access to feed for 24 h after 24 h starvation. Excreta are collected during feeding and for a further 24 h of starvation. High intakes were maintained, about 100–200 g/bird, but in spite of this the authors suggested an adjustment for endogenous energy loss during the final 24 h of the assay in the calculation of AME values. The argument for using this correction is important in assessing Farrell's assay for AME, and any other method involving feed withdrawal, since the situation is very similar. The argument in favour of making the correction (Kussaibati and Leclercq, 1985) is that when assays as described above, but involving 1, 3 and 6 days of *ad libitum* feeding, were compared, the uncorrected AME data varied with assay length whereas the corrected data did not.

Experimental methods and the EEC equation for ME

The EEC equation for controlling the AME_n content of poultry compound feeds is as follows (equation 4.2)

$$AME_n \ (kJ/g \ DM) = 0.3431\% \ FAT + 0.1551\% \ CPR + 0.1669\% \ STC$$
$$+ 0.1301\% \ SUG \tag{4.2}$$

where FAT, CPR, STC and SUG are the analysed fat, crude protein (N × 6.25), starch and sugar contents determined by appropriate methods. This equation was calculated from five data sets giving a total of 189 observations on AME_n and FAT, CPR, STC and SUG. Three of the data sets (those from laboratories 3, 4 and 5) have been published formally (Hartel *et al.*, 1977; Fisher, 1982; Leclercq, Provotel and Carré, 1984) and these sources can be checked for further details. All the experiments were done with adult birds, mostly cockerels, and four laboratories used techniques which were essentially similar and which fall into type (1). One data set (Fisher, 1982) is based on a type (3) assay.

There were no feeds in common between these experiments and therefore they can only be compared statistically by fitting constant terms for each laboratory in a given regression equation. In doing this in a discussion of experimental procedure it ·is important to remember that, in addition to different techniques, each laboratory worked with a different population of feeds and, to a small extent, used different analytical techniques. The pooled regression analysis that was used also ignored laboratory × analytical variable interactions, which, in a formal statistical sense, were present. To give some idea of the stability of the equation across laboratories the equations obtained by using each data set separately are shown in *Table 4.4*. It will be seen that the coefficients for FAT and STC were numerically reasonably stable whilst those for CPR and SUG varied more widely. Formal tests showed significant differences amongst the coefficients for SUG ($P<0.05$), STC and FAT ($P<0.001$). Forcing these data into a parallel analyses subsumes all these sources of variation into the laboratory constants (intercept values) and is obviously a very rough indication of the magnitude of procedural effects.

Table 4.4   EQUATIONS OBTAINED USING DATA FROM INDIVIDUAL LABORATORIES

Laboratory	Equation[a]	r^2	S
1	−0.125 +32.3 FAT +16.4 CPR +17.0 STC + 6.2 SUG	0.942	0.300
	32.0 FAT +16.3 CPR +16.8 STC + 5.8 SUG	0.942	0.293
2	−4.243 +40.9 FAT +17.0 CPR +21.5 STC +37.9 SUG	0.993	0.198
	36.3 FAT + 6.9 CPR +18.8 STC +25.0 SUG	0.991	0.208
3	0.019 +32.1 FAT +15.1 CPR +17.0 STC +10.9 SUG	0.974	0.289
	32.2 FAT +15.1 CPR +17.0 STC +11.0 SUG	0.978	0.295
4	0.898 +32.9 FAT +15.6 CPR +16.3 STC +11.5 SUG	0.979	0.286
	34.5 FAT +16.5 CPR +17.2 STC +15.9 SUG	0.978	0.295
5	−1.701 +41.1 FAT +18.0 CPR +17.4 STC +16.9 SUG	0.954	0.280
	38.7 FAT +15.7 CPR +15.4 STC +14.8 SUG	0.946	0.299

[a] Equations predict AME_n (kJ/g), composition variables are g/g

The equation (4.3) fitted to these data, prior to its simplification to remove the constant terms, is as follows.

$$AME_n = \begin{matrix} -0.57 \pm 0.26 \\ -0.69 \pm 0.26 \\ -0.47 \pm 0.23 \\ 0.26 \pm 0.26 \\ -0.63 \pm 0.26 \end{matrix} + 34.73\ FAT + 15.93\ CPR + 17.11\ STC + 13.4\ SUG$$

(lab. effects 1–5 ± SE)   (4.3)

Four out of the five intercept values were significantly different from zero (laboratories 1, 2, 3 and 5) and all of these were of similar negative magnitude. The intercept value for laboratory 4 was positive but not significant. The total range in intercept values ($0.26 + 0.69 = 0.95$ MJ/kg) indicates the variation across laboratories in measured AME_n value for a feed of given composition.

So far as technique is concerned these intercept values clearly distinguish laboratory 4 which used a type (3) assay from the others, which all used type (1) assays. The observation that type (3) gives the higher estimated AME_n values is consistent with the evidence produced by Härtle (1986). Beyond this observation it is probably not valid to draw any further conclusions about the importance of technique for ME determination in the derivation of the EEC equation.

The use of a single equation raises the question of fixed effects in ME determination of which those of species and age are the most important. The effect of species was not considered in the development of the equation. Two data sets were available which gave a comparison of adult cockerels and growing broilers but the analysis of these did not lead to recommendations for using separate equations. These analyses have not been reported and are briefly outlined here.

All of the feeds tested in laboratory 2 ($n = 18$) and 23 out of the 48 tested in laboratory 5 were evaluated with birds of two ages. The combined sets of data were used to estimate coefficients for FAT, CPR, STC and SUG in the EEC equation using a number of steps (*Table 4.5*) to compare the results obtained at the two ages.

Table 4.5 COMPARISON OF PREDICTION EQUATIONS FOR ADULT AND YOUNG BIRDS

A) *Each data set considered separately*

	AME_n (Mean ± SD)	FAT	CPR	STC	SUG	r^2	S
	(kg/g)		Coefficients (kJ/g)			(kg/g)	
Lab. 2:							
adults	13.17 ± 2.04	36.3	6.9	18.8	25.0	0.991	1.208
chicks	12.20 ± 1.85	30.3	2.7	18.7	18.7	0.979	0.292
Lab. 5:							
adults	13.92 ± 1.06	40.1	16.5	15.4	6.1	0.943	0.271
chicks	13.65 ± 1.01	31.1	14.2	16.1	15.5	0.825	0.454

B) *Nutrient × age interactions*

	Intercept	FAT	CPR	STC	SUG	r^2	S
Chicks:							
Lab. 2	−2.237 ⎫						
Lab. 5	−2.002 ⎭ −2.120	32.54	15.78	18.81	19.79		
						0.959	0.0347
Adults:							
Lab. 2	−1.817 ⎫						
Lab. 5	−1.582 ⎭ −1.700	40.07	18.51	17.76	6.84		
Significance of differences:							
Labs.	$P<0.05$	$P<0.01$	NS	NS	NS		
Ages	NS						

C) *Fat coefficient × age interaction only*

	Intercept	FAT	CPR	STC	SUG	r^2	S
Chicks:							
Lab. 2	−1.921 ⎫						
Lab. 5	−1.687 ⎭ −1.804	31.79					
			17.11	18.24	12.98	0.957	0.350
Adults:							
Lab. 2	−2.029 ⎫						
Lab. 5	−1.795 ⎭ −1.912	40.54					
Significance of differences:							
Labs.	$P<0.005$		$P<0.001$				
Age	NS						

In the first step of the analysis each of the four data sets was analysed separately (*Table 4.5A*). In both laboratories the mean observed AME values were lower in chicks than in adults but the difference was greater in laboratory 2 than in laboratory 5. Within both data sets there was a significant correlation between dietary fat level and the difference (adult AME_n – chick AME_n) (*Figure 4.2*). The difference in fat levels used explains some, but not all, of the tendency for the adult–chick difference to be greater in laboratory 2. The data in *Figure 4.2* suggest that, as expected, fat digestibility is a major component of the age effect on ME.

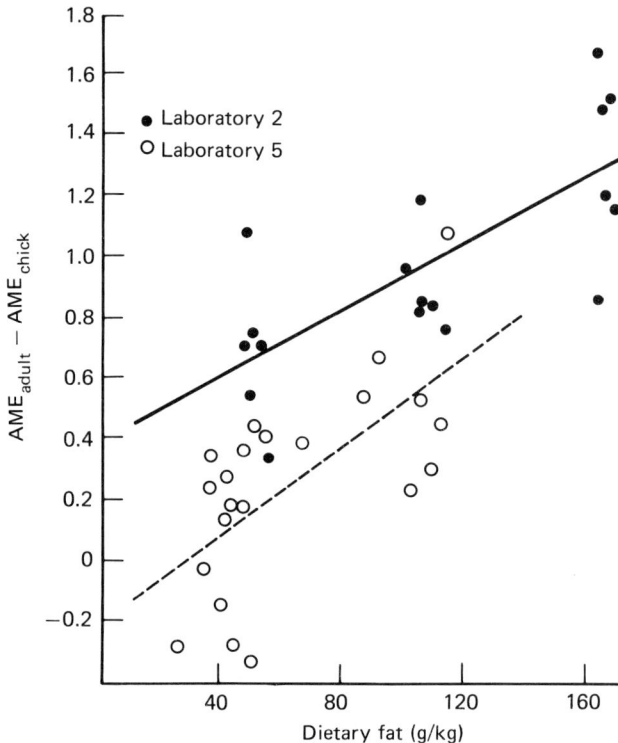

Figure 4.2   The effect of dietary fat on the difference in ME values measured in young and adult birds. Results from two laboratories (see text)

In the second stage of the analysis (*Table 4.5B*) intercept values were fitted for each laboratory × age data set and coefficients for each age. Comparison of the intercepts reveals overall significant effects of laboratory but not of age. Amongst the coefficients there are quite large numerical differences for FAT, CPR and SUG but only the age effect for FAT is significant. This result led to the final analysis (*Table 4.5C*) which was similar to the previous stage except that the only interaction included was for fat × age. It will be observed from the r^2 and S values that very little is sacrificed in terms of goodness-of-fit by using common coefficients for CPR, STC and SUG.

As in the analysis of the adult data the intercept terms in this final equation can be removed by combining them with each coefficient in turn. This is done separately for the chicks and adults and is only valid if FAT, CPR, STC and SUG together account for all of the AME_n. The coefficients (kJ/g) of equations obtained for chicks and adults by this procedure are as follows:

	FAT	CPR	STC	SUG
Chicks	30.0	15.3	16.4	11.2
Adults	38.7	15.2	16.3	11.1

This is a very satisfactory conclusion and description of the age effect but these equations were not adopted for use in the EEC because, in absolute terms, they do not fit in with expectation (the simultaneous requirement in all of the analyses done for the EEC, that the results should be estimated from data but also fit in with preconceived theoretical expectations is obviously difficult to satisfy). Thus, the coefficients for fat, 30 and 38.7 kJ/g, correspond to calculated digestibilities of 0.79 and 0.99 respectively (assuming digestible fat yields 37.875 kJ ME/g). The coefficient for fat in the adopted EEC equation is equivalent to a digestibility of about 0.90 which is probably more realistic. The low coefficients for SUG yielded by this analysis of chick and adult data have also been objected to during the discussions in the EEC. The higher value (13.04) in the adult-only equation is fortuitous and not significantly different from the results reported above.

It is not clear that the inclusion of additional data will, except fortuitously, clarify the issue of age effects any better. These results reinforce the sensible conclusion that the differences between adults and chicks are probably only concerned with fat digestibility and in this case theoretical assumed digestibilities could equally well be used. Of course, attempts to do this will quickly reveal the limitations of trying to reduce variations in fat digestibility to just two coefficients, which although better than one, still requires a gross simplification to be made of the real situation.

It is extremely difficult to carry out directly comparable balance experiments with birds of different ages to obtain an unbiased empirical estimate of any differences and such an approach is probably less likely to resolve the issue than the statistical one illustrated above. Data relevant to this discussion have been produced recently by Kussaibati, Guillaume and Leclercq (1982); Lessire and Leclercq (1982); Mollah et al. (1983); Sibbald and Wolynetz (1985); Härtle (1986). Although these all contain comparisons of young and adult birds consideration of the data only confirms that age effects do occur, that their incidence and magnitude is very variable and that when a range of techniques is used (e.g. Mollah et al., 1983) the assessment of age effects becomes peculiarly complex. Although these are all significant papers they do not lead to firm conclusions about the importance of age as a factor in energy declarations.

If it is accepted that age effects are a reflection only of differences in fat digestibility then the direct study of this topic might be more rewarding. A wide range of relevant literature was reviewed by Freeman (1984), Summers (1984), Wiseman (1984) and the papers by Kussaibati, Guillaume and Leclercq (1982) and Lessire and Leclercq (1982) contain valuable information. It is apparent that recognizing two age classes only is not sufficient; for poorly digested fats there is a continuation of change during the growing life of the broiler (Krogdahl, 1985). However, consideration of this area also reveals that age effects on fat digestibility, whilst generally found, are highly dependent on fat composition (Lessire and Leclercq, 1982) and with plant oils may be absent altogether. Level of dietary fat is another factor which may have a role as significant as age of bird in the prediction of ME.

It is concluded that the use of a single equation to predict ME levels for birds of all ages is unjustified, mainly because of variations in fat digestibility. However, to make the predictions more realistic not only age, but fat composition and level should be incorporated. This raises a lot of problems which have not yet been completely solved but is outside the present discussion. If a correction for age alone is made as an interim measure this should be done by defining suitable average digestibilities for fat in different classes of stock. Van Kampen (1983), in the work outlined above, also suggests that characteristic differences in water:food ratios, e.g. between adult laying fowls and cockerels will also affect ME values and would therefore indicate the use of different equations for this reason. This topic, however, requires further elucidation.

References

ALMEIDA, J.A. and BAPTISTA, E.S. (1984). *Poultry Science*, **63**, 2501

ALMQUIST, H.J. and HALLORAN, H.R. (1971). *Poultry Science*, **50**, 1233

BILGILI, S.F. and ARSCOTT, G.H. (1982). *Nutrition Reports International*, **25**, 613

BILGILI, S.F., ARSCOTT, G.H. and KELLEMS, R. (1982). *Nutrition Reports International*, **25**, 639

BOLTON, W. (1954). *World's Poultry Congress*, **10**, 94

CAREW, L.B. (1978). *Feedstuffs*, **50**(22), 31

CHAMI, D.B., VOHRA, P. and KRATZER, F.H. (1980). *Poultry Science*, **59**, 569

DALE, N.M., FULLER, H.L., PESTI, G.M. and PHILLIPS, R.D. (1985). *Poultry Science*, **64**, 362

DU PREEZ, J.J., MINNAAR, A. du P. and DUCKITT, J.S. (1984). *World's Poultry Science Journal*, **40**, 121

FARRELL, D.J. (1978). *British Poultry Science*, **19**, 303

FARRELL, D.J. (1980). In *Recent Advances in Animal Nutrition in Australia 1980*, p. 146. Ed. D.J. Farrell. Armidale; University of New England

FARRELL, D.J. (1981). *World's Poultry Science Journal*, **37**, 72

FISHER, C. (1982). Occasional Publication No. 2 from the Agricultural Research Council's Poultry Research Centre, 609 pp

FISHER, C. (1983a). In *Recent Advances in Animal Nutrition – 1982*, p. 113. Ed. W. Haresign and D. Lewis. Butterworths; London

FISHER, C. (1983b). Occasional Publication No. 3 from the Agricultural Research Council's Poultry Research Centre, 34 pp

FISHER, C. (1986). *The Feed Compounder*, January, 8

FRASER, D. and SIBBALD, I.R. (1983). *Poultry Science*, **62**, 2224

FREEMAN, C.P. (1984). In *Fats in Animal Nutrition*, p. 105. Ed. J. Wiseman. Butterworths; London

HARTEL, H. (1986). *British Poultry Science*, **27**, 11

HARTEL, H., SCHNEIDER, W., SEIBOLD, R. and LANTZSCH, H.J. (1977). *Archiv fur Geflugelkunde*, **41**, 152

JONSSON, G. and MCNAB, J.M. (1983). *British Poultry Science*, **49**, 349

KROGDAHL, A. (1985). *Journal of Nutrition*, **115**, 675

KUSSAIBATI, R., GUILLAUME, J. and LECLERCQ, B. (1982). *British Poultry Science*, **23**, 393

KUSSAIBATI, R. and LECLERCQ, B. (1985). *Archiv für Geflügelkunde*, **49**, 54

LECLERCQ, B., PREVOTEL, B. and CARRÉ, B. (1984). *British Poultry Science*, **25**, 561

LESSIRE, M. and LECLERCQ, B. (1982). *Animal Feed Science and Technology*, **7**, 365

MCNAB, J.M. and BLAIR, J.C. (1987), *British Poultry Science*, in press

MCNAB, J.M. and FISHER, C. (1981). *Proceedings of the 3rd European Symposium on Poultry Nutrition*, p. 45

MCNAB, J.M. and FISHER, C. (1984). *Proceedings of the XVII World's Poultry Science Congress*, p. 374

MOHAMED, K., LECLERCQ, B., ANWAR, A., EL-ALAILY, H. and SOLIMAN, H. (1984). *Animal Feed Science and Technology*, **11**, 199

MOLLAH, Y., BRYDEN, W.L., WALLIS, I.R., BALNAVE, D. and ANNISON, E.F. (1983). *British Poultry Science*, **24**, 81

MUZTAR, A.J. and SLINGER, S.J. (1980). *Nutrition Reports International*, **22**, 745

NEUMARK, H., BIELORAI, R. and IOSIF, B. (1982). *Journal of Nutrition*, **112**, 387

OSTROWSKI-MEISSNER, H.T. (1984). *Nutrition Reports International*, **29**, 1239

PARSONS, C.M., POTTER, L.M. and BLISS, B.A. (1984). *Poultry Science*, **63**, 1610

PEDDIE, J., DEWAR, W.A., GILBERT, A.B. and WADDINGTON, D. (1982). *Journal of Agricultural Science, Cambridge*, **99**, 233

PESTI, G.M. and EDWARDS, H.M. (1983). *Poultry Science*, **62**, 1275

ROUDYBUSH, T., ANTHONY, D.L. and VOHRA, P. (1974). *Poultry Science*, **53**, 1894

SHIRES, A., ROBBLEE, A.R., HARDIN, R.T. and CLANDININ, D.R. (1979). *Poultry Science*, **58**, 602

SIBBALD, I.R. (1976). *Poultry Science*, **55**, 303

SIBBALD, I.R. (1979). In *Recent Advances in Animal Nutrition — 1979*, p. 35. Ed. W. Haresign and D. Lewis. Butterworths; London

SIBBALD, I.R. (1982). *Canadian Journal of Animal Science*, **62**, 983

SIBBALD, I.R. (1983). Research Branch Contribution 83-20E/83-20F, Animal Research Centre, Agriculture Canada, 89 pp

SIBBALD, I.R. (1986). Research Branch Contribution 86-4E, Animal Research Centre, Agriculture Canada, 114 pp

SIBBALD, I.R. and MORSE, P.M. (1983). *Poultry Science*, **62**, 68

SIBBALD, I.R. and PRICE, K. (1975). *Poultry Science*, **54**, 558

SIBBALD, I.R. and WOLYNETZ, M.S. (1985). *Poultry Science*, **64**, 127

SIBBALD, I.R. and WOLYNETZ, M.S. (1986). *Poultry Science*, **65**, 78

SKLAN, D., DUBROV, D., EISNER, U. and HURWITZ, S. (1975). *Journal of Nutrition*, **105**, 1549

SQUIBB, R.L. (1971). *Journal of Nutrition*, **101**, 1211

STOREY, M.L. and ALLEN, N. K., (1982). *Poultry Science*, **61**, 101

SUMMERS, J.D. (1984). In *Fats in Animal Nutrition*, p. 265. Ed. J. Wiseman, Butterworths; London

TERPSTRA, K. and JANSSEN, W.M.M.A. (1975). Spelderholt Report 101.75, Spelderholt Institute for Poultry Research, Beekbergen, The Netherlands

TEETER, R.G., SMITH, M.O., MURRAY, E. and HALL, H. (1984). *Poultry Science*, **63**, 573

VAN KAMPEN, M. (1983). *British Poultry Science*, **24**, 169

VOGTMANN, H., PFIRTER, H.P. and PRABUCKI, A.L. (1975). *British Poultry Science*, **16**, 531

VOHRA, P., CHAMI, D.B. and OYAWOYE, E.O. (1982). *Poultry Science*, **61**, 766

WALLIS, I. and BALNAVE, D. (1983). *British Poultry Science*, **24**, 255

WEHNER, G.R. and HARROLD, R.L. (1982). *Poultry Science*, **61**, 595

WISEMAN, J. (1984). In *Fats in Animal Nutrition*, p. 277. Ed. J. Wiseman. Butterworths; London

WOLYNETZ, M.S. and SIBBALD, I.R. (1984). *Poultry Science*, **63**, 1386

5

THE IMPACT OF DECLARATION OF THE METABOLIZABLE ENERGY VALUE OF POULTRY FEEDS

B.C. COOKE
Dalgety Agriculture Ltd, Clifton, Bristol, UK

The subject of declaration of energy content of poultry feeds has been under discussion ever since, 25 years ago, Bolton (1962) proposed an equation. Within the EEC some members, such as Denmark, have operated a system of compulsory declaration of metabolizable energy (ME) values of feeds for a number of years. Other member states such as Holland have allowed compounders to make a declaration according to a state prescribed method. In both these countries control of the declared value is carried out by microscopic examination of the feed to determine its basic ingredient make-up together with examination of the mill formulation from which a calculation of energy content is made.

When the EEC introduced the Compound Feeds Directive (79/373/EEC) in 1979 a clause was inserted which banned any member state not currently operating an energy declaration from introducing one until such time as a universal EEC method of declaration was derived.

Within the UK, discussions were held with the NFU in the late 1970s with a view to trying to find methods of declaring energy values of all animal feed compounds. These discussions led to a series of studies on ruminant, pig and poultry compounds. The poultry work was carried out by Fisher and his team at the Poultry Research Centre, Edinburgh (PRC which is now the Institute of Grassland and Animal Production, Poultry Division). The results of this work were published by Fisher in 1982.

Shortly after the publication of the PRC work, the EEC brought together a number of experts under the auspices of the World Poultry Science Association in order to recommend an equation that would be suitable for use throughout the community for calculating the energy value of compound poultry feeds. The equation recommended by this group of experts was passed into EEC law by a Council Directive of 9 April 1986 (86/174/EEC).

The equation in this Directive is as follows:

$$ME \ (MJ/kg) = 0.3431 \ (\% \ fat) + 0.1551 \ (\% \ crude \ protein) + 0.1301 \ (\% \ total \ sugar \ (expressed \ as \ sucrose)) + 0.1669 \ (\% \ starch) \quad (5.1)$$

A feature of this equation is that the result expressed as ME (MJ/kg) is corrected to zero nitrogen retention. The EEC Directive lays down that member states must bring this method of calculating the energy value of poultry feeds into operation by

First published in *Recent Advances in Animal Nutrition – 1987*

30 June 1987 at the latest. Current indications are that the UK Government will bring in the regulations under the Agriculture Act somewhere around June 1987.

This EEC Directive lays down that member states must allow a minimum tolerance of ±0.4 MJ ME/kg on the figure declared. The compound feed trade associations of the UK (UKASTA) and of Europe (FEFAC) are both firmly of the opinion that if declaration of energy (by the EEC equation) in poultry feeds is to become a workable proposition a tolerance of at least ±1.0 MJ is necessary.

The importance of the parameters of the EEC equation

The four parameters in the equation have differing importance in the final ME value calculated for a feed.

(1) Oil which is to be measured by acid hydrolysis followed by petroleum ether extraction carries a constant of 0.3431 and thus a 0.5% discrepancy in an oil analysis is equivalent to 0.17 MJ ME/kg. Therefore in a compound with a 5% oil declaration a variation in this value of 10% would be equivalent to 0.5% oil or 0.17 MJ ME/kg.
(2) For crude protein the constant is 0.1551 and thus a 0.5% discrepancy, in an analytical result, would only be equivalent to 0.08 MJ ME/kg. However, as declared crude protein figures are higher than those for oil a 10% variation in analytical value would have a larger effect upon the ME calculation. For instance, at 20% crude protein a 10% variation would be equivalent to 2% protein or 0.31 MJ ME/kg.
(3) The sugar constant is 0.1301 and sugar levels in poultry feeds are fairly low. Thus, a 10% variation from a declared value of 5% sugar would only be equivalent to 0.07 MJ ME/kg.
(4) Starch has a constant of 0.1669 but as the level of starch is higher than any of the other parameters, variation in the measurement of starch will have a much greater effect on the calculated ME value. Feeds for laying hens would typically contain 35% starch and thus a 10% discrepancy in the value would be equivalent to 0.58 MJ ME/kg. Broiler feeds may contain in the region of 42.5% starch and thus the same 10% discrepancy would be equivalent to 0.71 MJ ME/kg.

From the foregoing, it is obvious that starch is the most critical parameter followed by crude protein, oil and then sugar.

Analytical problems

As has been stated above, the fat included in the equation is the total oil as measured by acid hydrolysis followed by petroleum ether extraction. A ring test has recently been carried out between 20 laboratories within ADAS and UKASTA, and the results are shown in *Table 5.1*. A similar ring test has been carried out with protein using the same compound sample, the results of which are shown in *Table 5.2*.

In relation to the analytical method for sugar, a ring test involving 11 laboratories has recently been conducted. The results are given in *Table 5.3*.

Ten laboratories have recently been involved in a ring test on starch analysis of a feed for laying hens and a turkey compound. The results are shown in *Table 5.4*.

Table 5.1   OIL DETERMINED BY ACID HYDROLYSIS IN ADAS/UKASTA COLLABORATIVE TRIALS 1986 (20 laboratories)

Mean oil (%)	6.4
Range (%)	5.6–6.9
CV^a (%)	5.1
SD^b	±0.32
95% Confidence limits	±0.64
0.64% oil equivalent to	±0.22 MJ ME/kg

[a] CV = coefficient of variation
[b] SD = standard deviation

Table 5.2   PROTEIN DETERMINED IN ADAS/UKASTA COLLABORATIVE TRIAL 1986 (19 laboratories)

Mean protein (%)	21.3
Range (%)	20.8–21.7
	(value of 18.9 excluded)
CV^a (%)	1.6
SD^b	±0.35
95% Confidence limits	±0.70
0.70% protein equivalent to	±0.11 MJ ME/kg

[a] CV = coefficient of variation
[b] SD = standard deviation

Table 5.3   SUGAR DETERMINED IN ADAS/UKASTA COLLABORATIVE TRIALS 1986 (11 laboratories)

	Laying hen compound	*Turkey compound*
Mean sugar (%)	3.9	4.6
Range (%)	2.3–4.6	4.1–5.3
CV^a (%)	21.2	8.0
SD^b	±0.70	±0.37
95% Confidence limits	±1.40	±0.74
1.40 or 0.74% sugar equivalent to	±0.18 MJ ME/kg	±0.10 MJ ME/kg

[a] CV = coefficient of variation
[b] SD = standard deviation

From the data given in *Tables 5.1* to *5.4* it can be seen that given a single batch of a compound feed, variation between laboratories could be as high as ±0.22 MJ ME/kg from the oil analysis, ±0.11 MJ ME/kg from protein analysis, ±0.18 MJ ME/kg from sugar analysis and ±0.30 MJ ME/kg from the starch analysis. At the worst total variation purely due to variation between laboratories would be ±0.81 MJ ME/kg.

Even more worrying are some recent experiences which have occurred in the field. On five different occasions samples of feeds for laying hens were taken from farms in different parts of the country. These were all submitted to a single laboratory for analysis and for calculation of ME according to the European

Table 5.4 STARCH DETERMINED IN ADAS/UKASTA COLLABORATIVE TRIAL 1986 (10 laboratories)

	Laying hen compound	*Turkey compound*
Mean starch (%)	31.84	30.53
Range (%)	29.9–33.3	28.8–32.6
CV[a] (%)	2.8	3.0
SD[b] (%)	±0.88	±0.92
95% Confidence limits	±1.76	±1.84
1.84% starch equivalent to	±0.29 MJ ME/kg	±0.31 MJ ME/kg

[a] CV = coefficient of variation
[b] SD = standard deviation

Table 5.5 RECENT FIELD EXPERIENCES WITH STARCH ANALYSIS OF FEEDS FOR LAYING HENS

Sample no.	*Original result*		*Revised result*		*Difference in ME between the results*
	Starch (%)	*Calculated ME (MJ/kg)*	*Starch (%)*	*Calculated ME (MJ/kg)*	*(MJ/kg)*
1	28.9	10.0	36.0	11.2	1.2
2	28.9	10.5	37.8	12.0	1.5
3	24.1	9.3	36.2	11.3	2.0
4	25.3	10.0 ± 0.7	35.2	11.6 ± 0.7	1.6
5	26.5	10.2 ± 0.7	34.9	11.6 ± 0.7	1.4

equation. The results obtained on first analysis for starch were so low that unrealistically low ME values were reported. This matter was taken up with the laboratory concerned and it was found that there was a small discrepancy in the method of starch analysis being used. Correction of the analytical technique and re-analysis led to a dramatic increase in the level of starch reported and of the ME consequently calculated. The original and revised results are given in *Table 5.5*. The discrepancy in calculated ME varied between 1.2 and 2 MJ/kg of the fresh feed. Whilst it is extremely lucky for the feed trade that the discrepancy in this particular laboratory was found and corrected prior to the introduction of a declaration of energy values of poultry feeds into UK legislation, the data serves to illustrate the dangers inherent in the use of any equation if the official methods of analysis are not strictly adhered to.

Formulation of products and declaration of energy

In order to formulate diets in the future, the feed compounder will not only have to consider the apparent metabolizable energy (AME) or true metabolizable energy (TME) value to which he wishes to formulate, but will also have to give consideration to the ME value to be declared on the bag or ticket. The simplest approach is

Table 5.6   FORMULATION — CALCULATION OF INGREDIENT ME FROM THE EUROPEAN EQUATION

	Oil % × 0.3431		+Protein % × 0.1551		+Sugar % × 0.1301		+Starch % × 0.1669		Calculated ME
	Oil (%)	ME (MJ/kg)	Protein (%)	ME (MJ/kg)	Sugar (%)	ME (MJ/kg)	Starch (%)	ME (MJ/kg)	(MJ/kg)
Wheat	1.9	0.65	11.5	1.78	5	0.65	59	9.85	12.93
Ext soya (Hi Pro)	2.3	0.79	48.0	7.44	10	1.30	1	0.17	9.70
Fat	98	33.62	—		—		—		33.62
Meat and bone (12 oil 48 Pro)	12.6	4.32	47.2	7.32	—		—		11.64
Fish (68 Pro)	8.4	2.88	67.8	10.52	—		—		13.40

Table 5.7   FORMULATION — EXAMPLE BASED ON CALCULATED ME OF INGREDIENTS

	Contributions				
Ingredient	Oil (%)	Protein (%)	Sugar (%)	Starch (%)	Calculated ME (MJ)
75% Wheat	1.43	8.63	3.75	44.25	9.70
15% Soya	0.34	7.20	1.50	0.15	1.45
2% Fat	1.96	—	—	—	0.67
5% Meat and bone	0.63	2.36	—	—	0.58
3% Fish	0.25	2.03	—	—	0.40
Total	4.61	20.22	5.25	44.40	12.80 MJ ME/kg

Alternative: 4.61% × 0.3431 + 20.22% × 0.1551 + 5.25% × 0.1301 + 44.40% × 0.1669 = 12.81 MJ ME/kg

to calculate the ME value according to the equation for each individual ingredient and to ascribe this value into the formulation system. Examples for some common ingredients are shown in *Table 5.6*. Once these values are entered in the computer matrix, formulation can proceed in the normal manner and the declared energy value can be formulated to, as illustrated in *Table 5.7*.

Alternatively, the compounder can take the oil, protein, sugar and starch values of the finished product and apply the constants from the equation to calculate the energy for declaration purposes. However, this method would mean that every time a diet was reformulated, the calculation would have to be carried out. Due to small changes in oil, protein, sugar or starch content within the formulated minimum/maximum small changes would occur in the calculated ME value. Thus it is better to formulate to a calculated ME in the first place.

If compounders do choose to ascribe the calculated ME value to each raw material as recommended above, it is essential that this value is only used to derive the declared energy figure. The relationship between one ingredient and another in terms of TME or AME is different from the relationship which will be obtained by the calculated ME derived from the European equation and thus inaccurate buying decisions could be made if this calculated ME value was used to ascribe a biological value for use in the finished feed. This change in the relationship is illustrated for six common ingredients in *Table 5.8*.

Table 5.8 BIOLOGICAL VALUE (AME) *VERSUS* CALCULATED ME IN RAW MATERIALS

Raw materials	AME		Calculated ME by equation	
	(MJ/kg)	(as % of wheat)	(MJ/kg)	(as % of wheat)
Wheat	13.0	100	12.93	100
Ext Soya	10.6	82	9.70	75
Fat 1	30.0	231	33.62	260
2	35.0	269	33.62	260
Meat and bone	10.8	83	11.64	90
Fish	14.2	109	13.40	104

Unfortunately, the EEC equation takes no account of the digestibility of the fat, protein, etc. For example, from the data in *Table 5.8* it can be seen that two fats, one of low digestibility to poultry and one of high digestibility, will both have the same relative value to wheat by the EEC equation, i.e. 2.6 times the wheat value, but in AME terms will have vastly different relationships with wheat, varying from 2.31–2.69 times the AME value. It can also be seen that the relationship between other ingredients and wheat would be inaccurate in both biological and economic terms if the energy value derived from the equation was used to describe the energy value of the ingredient.

Problems of practical application

Data on three finished products are presented in *Table 5.9* from which it can be seen that with compound feeds for older and mature poultry the calculated ME value using the EEC equation is lower than the AME value obtained from the sum of the ingredients. This discrepancy is due mainly to the fact that the equation is corrected to zero nitrogen retention whereas AME values are not normally corrected. Conversely the data in *Table 5.9* illustrate that for protein concentrates the calcu-

Table 5.9 BIOLOGICAL VALUE *VERSUS* CALCULATED ME IN FINISHED PRODUCTS

Finished products	Oil (%)	Protein (%)	Sugar (%)	Starch (%)	Calculated ME (MJ/kg)	AME (MJ/kg)
Broiler feed	8.1	22.9	5.2	37.5	13.27	13.55
Laying hen feed	5.4	18.0	5.2	34.8	11.13	11.50
Poultry grower protein concentrate	7.2	45.5	4.5	1.0	10.28	10.20

lated ME value is often higher than the AME value which is regarded as the best measure of predicting animal performance.

Species and age of bird

The EEC equation has been developed from data obtained on older or mature chickens, not young chicks, and for various reasons the value calculated is unlikely to be anywhere near the biological value of the feed for other ages or types of bird. For instance, in young chicks it is usual for fat digestibility to be lower than in older birds and logically fat should have a lower constant in the equation when the equation is being used to predict the energy value of a chick starter diet. However, no consideration is given to this fact within the EEC Directive and it is expected that a single method of calculating energy will be used for all poultry feeds within the UK. While it can already be stated that this will give misleading data for young chick diets, it is unknown how accurate the estimate will be for turkey, duck or goose diets. Nonetheless, it is expected that the law will require a declaration of energy according to the equation for all poultry feeds.

Enforcement

The EEC Directive allows member states to require or permit the declaration of energy value of compound feedstuffs. It is not yet clear whether the UK authorities will make the declaration mandatory or optional for the feed compounder. However, as the method of declaring energy of poultry compounds will be defined in law under the Regulations of the Agriculture Act, it is almost certain that it will be local trading standards officers who will be required to police this new declaration. Thus, when sampling and analysing feeds for oil, protein, etc, they will also be expected to obtain an analysis of starch and sugar content in order to calculate the ME by the equation. Any discrepancy outside of the legally allowed tolerance will then presumably be treated in the same way as a discrepancy in oil or protein.

Cost of implementation

It is difficult to define precisely the cost that the feed compounder will have to carry because of the introduction of energy declaration on poultry feeds. Obviously extra data will have to be held in computer matrices which will make formulation systems slightly less efficient. Attention will have to be given to ensuring that the declared values are kept up to date which will involve more management time. However, the major cost for the majority of feed compounders will be in building up a matrix of starch and sugar values on raw materials, and in regularly monitoring these parameters on both raw materials and finished products. The cost of a starch or sugar analysis is £10/sample. Thus, it might be reasonable to suppose that analytical costs will increase by about 5p/tonne on poultry feeds. Further to this, there will be a significant cost in constraining formulations to a particular declared value of energy. This cost will vary from compounder to compounder but could well amount to about 10p/tonne. Therefore, it is estimated that the introduction of the energy declaration will cost in the region of 20p/tonne of poultry feed.

Conclusions

The introduction of the EEC equation for the calculation and declaration of the ME content of poultry compound feeds must be regarded as a move in the right direction. Irrespective of the many reservations about the accuracy of the figure and the discrepancy between the figure and the AME or TME value, evidence suggests that the ME value derived by the equation gives a good estimate of the difference between two feeds. Declaration of this ME will therefore aid the farmer in comparing feeds either from the same or different compounders.

There are, however, a number of outstanding problems to be resolved. The first of these is explaining to the farmer customer the low value obtained by the equation due to the fact that it is corrected to zero nitrogen retention. The second is the problem of analytical techniques which have already been identified and further problems which undoubtedly will occur but have not yet been identified.

Table 5.10   VARIATION IN WHEAT TME MEASURED AT PRC

No of samples	7
Mean TME (MJ/kg)	13.11
Range (MJ/kg)	12.49–13.37
CV[a] (%)	2.4
SD[b]	±0.31
95% Confidence limits	±0.62

[a] CV = coefficient of variation
[b] SD = standard deviation

The final problem relates to the subject of tolerance, where the feed trade feels that ±1 MJ ME/kg is the lowest figure that can be meaningful. Data presented earlier illustrate that discrepancies between laboratories can account for over 0.6 MJ ME/kg on a single sample of feed and that if correct procedures are not followed in the laboratories, discrepancies of up to 2 MJ ME/kg can occur. In order to put the value of ±1 MJ ME/kg into context it is worthwhile considering the variation reported by the PRC (McNab, 1986) during 1986 on the TME value of wheat measured in chickens. The results of this work are given in *Table 5.10*. This shows that for seven samples of wheat the 95% confidence limits on TME value is ±0.62 MJ ME/kg. Thus, the tolerance required for an energy declaration by the EEC equation is probably little or no greater than the true variability in AME or TME between different batches of a feed made to a constant nutritional specification.

References

BOLTON, W. (1982). *Proceedings XIIth World Poultry Congress*, **2**, 38
FISHER, C. (1982). *Energy Values of Compound Poultry Feeds*. PRC Occasional Publication No. 2. Poultry Research Centre, Edinburgh
MCNAB, J. (1986). Personal communication

6

INFLUENCE OF FIBRE ON DIGESTIBILITY OF POULTRY FEEDS

W.M.M.A. JANSSEN
Spelderholt Centre for Poultry Research and Extension, Beekbergen, The Netherlands
and
B. CARRÉ
Station de Recherches Avicoles, INRA, Nouzilly, France

Characteristics of fibre in feedstuffs

Two sorts of fibre can be distinguished: water-insoluble fibre and water-soluble fibre.

Water-soluble fibre corresponds chemically with the non-starchy water-soluble polysaccharides. These polysaccharides can be extracted by cold or hot water and precipitated by ethanol or acetone. Water-soluble fibre includes β-(1→3) glucan from barley, arabinoxylan from rye, highly methylated pectin from fruits, galactomannan from leguminosae such as guar gum, polysaccharides from algae such as alginate (linear chains of mannuronic acid and guluronic acid) or carrageenan (linear chain of sulphated galactose).

Water-insoluble fibre consists of insoluble cell wall material and is composed of cellulose, hemicellulose, pectic substances, protein and lignin. The proportions of these components vary widely, depending on the origin of the plant material.

Plant cell walls can originate from 'primary' cells or from secondary cells; the 'primary' cells being little differentiated and the 'secondary' cells highly differentiated. This differentiation of cells leads to the deposition of additional cell wall components such as cellulose, hemicelluloses and lignin. The 'primary' cells are encountered, for example, in the starchy endosperm of cereal grains or in the parenchyma of roots and cotyledons. The 'primary' cell walls of the starchy cereal endosperm are characterized by a very low cellulose and lignin content and by the weakness of their ultrastructure.

The bulk of hemicelluloses of cereal 'primary' cell walls, which are mainly arabinoxylan, are readily soluble in water or dilute alkali (Mares and Stone, 1973). The 'primary' cell walls of dicotyledonous plants, such as cell walls located in cotyledon seeds of leguminosae, are characterized by the presence of pectic substances, which can represent up to 70 per cent of the cell wall material (Brillouet and Carré, 1983; Carré, Brillouet and Thibault, 1985). The pectic substances are made of a rhamnogalacturonan backbone bearing branched side chains mainly composed of arabinose and galactose (Aspinall, 1980). Pectic substances can sometimes be extracted by hot water, but, very often, the use of chelating agents such as ethylenediamine tetracetic acid (EDTA) or ammonium oxalate diluted in hot water, is needed to extract them (Aspinall *et al.*, 1967).

First published in *Recent Advances in Animal Nutrition – 1985*

The hemicellulose compounds of dicotyledonous 'primary' cell walls can be xyloglucan or type II arabinogalactan (Albersheim, 1976). The cellulose content of dicotyledonous 'primary' cell walls varies generally from 5 per cent (Carré, Brillouet and Thibault, 1985) to 30 per cent (O'Neill and Selvendran, 1980); their lignin content is normally very low (Brillouet and Carré, 1983).

The secondary plant cells are, for instance, the xylem vascular bundles, the pericarp cells of cereal grains, the palisade cells and the spool-shaped cells located in hulls of leguminosae seeds. The ultrastructure of secondary cell walls is generally more resistant than the ultrastructure of 'primary' cell walls. Cellulose, a major component of secondary cell walls, exhibits a high crystallinity in these types of cell walls. The major hemicellulose polymers of secondary walls are generally xylan-related polysaccharides such as arabinoglucuronoxylan of wheat bran (Brillouet _et al._, 1982) or glucuronoxylan of soya bean hull (Aspinall, Hunt and Morrison, 1966). Pectic substances are generally very minor components in secondary cell walls. Lignin, which is a characteristic component of secondary cell walls, rarely exceeds 20 per cent of the cell wall material. The importance of lignin arises from the possibility of connections between lignin and polysaccharides through ether or ester linkages (Neilson and Richards, 1982; Chesson, Gordon and Lomax, 1983; Scalbert _et al._, 1984). These linkages probably explain the resistance of cell wall to bacterial degradation (Morrison, 1972).

METHODS OF ANALYSIS OF FIBRE

Crude fibre

This is the commonest method of determination of fibre and is defined as the residue after successively boiling of the material in 0.26 N sulphuric acid and 0.23 N potassium hydroxide. The major component of crude fibre is cellulose, together with residual hemicelluloses (5–40 per cent of crude fibre), residual protein (2–10 per cent of crude fibre) and lignin (10–50 per cent of crude fibre; 50–100 per cent of the original lignin) (Rinaudo and Chambat, 1976).

Acid detergent fibre

Acid detergent fibre content (ADF) (Van Soest, 1963) is in general similar to crude fibre content (Baker, 1977; Nyman _et al._, 1984). Like crude fibre, ADF may be contaminated by non-cellulosic polysaccharides (2–50 per cent of ADF) and by residual protein (1–25 per cent of ADF) (Bailey and Ulyatt, 1970; Rinaudo and Chambat, 1976; Theander and Aman, 1980). Both ADF and crude fibre may also be contaminated by condensed tannins (Theander _et al._, 1977); probably the highest residual protein contents in ADF and crude fibre correspond to the presence of highly insoluble protein–tannin complexes. ADF and lignin contents in sorghums are positively correlated with tannin content (ITCF, 1982).

Cellulose

Cellulose can be determined more accurately by the measurement of glucose, specifically liberated by hydrolysis with concentrated sulphuric acid according to the method of Saeman _et al._ (1954).

Table 6.1 CRUDE PROTEIN, CRUDE FIBRE, NDF[a] AND ER[b] CONTENTS OF SOME IMPORTANT FEEDINGSTUFFS (% OF DRY MATTER)

	Maize	Wheat	Barley	Cassava root	Wheat bran	Lucerne	Sunflower meal	Rapeseed meal	Soya bean meal	Pea	Field bean	White lupin (Lutop)
Crude protein	10.4	13.7	11.0	3.1	15.9	16.9	35.5	37.7	54.0	25.6	29.1	45.0
Crude fibre	2.3	2.6	4.6	4.5	11.0	28.2	25.2	15.4	4.1	5.5	8.2	12.6
NDF	10.5	12.7	17.7	11.2	47.3	46.1	42.5	26.1	8.5	8.7	13.9	18.3
ER	10.3	11.8	16.7	12.7	44.7	45.0	40.2	32.0	17.1	13.1	17.4	30.9

[a]Neutral detergent fibre, obtained after destarching with a gelatinization step at 95°C (10 min) followed by α-amylase (from *Bacillus subtilis*; 3 × crystallized; Boehringer Mannheim) treatment (1 h; 50 °C in water). Residues are recovered and rinsed by centrifugation (40000 g) and not corrected for ash.
[b]Enzymatic residue measured as the residue of pronase and α-amylase treatments (Brillouet and Carré, 1983), recovered and rinsed by centrifugation (40000 g). Not corrected for ash.

Water-insoluble fibre

The total water-insoluble fibre can be estimated either by neutral detergent fibre (NDF) determination (Van Soest and Wine, 1967) or by isolation of the water-insoluble residue after enzymatic removal of protein and starch (Hellendoorn, Noordhoff and Slagman, 1975; Asp *et al.*, 1983; Brillouet and Carré, 1983). NDF and enzymatic methods generally lead to similar results, except with legumes, where the values for enzymatic residue (ER) are considerably higher than for NDF (*Table 6.1*). The cell wall pectic substances present in leguminosae are responsible for this difference as these substances are dissolved by EDTA present in the solution of Van Soest's method (Bailey and Ulyatt, 1970; Schweizer and Würsch, 1979; Carré, Brillouet and Thibault, 1985).

Prediction of feeding value of feeds and feedstuffs

In the past, several methods have been developed to predict the metabolizable energy (ME) value of feeds and feedstuffs for poultry.

METHODS BASED ON DIGESTIBLE NUTRIENTS

If complete information on chemical composition and on digestibility of the nutrients is available, ME can be estimated very accurately with formulae of the following basic form (Janssen, 1976):

$$ME = a \times \text{digestible CP} + b \times \text{digestible F} + c \times \text{digestible NFE} + d \times \text{digestible CF} \qquad (6.1)$$

where CP, F, NFE and CF represent respectively 1 g or 1 per cent of digestible crude protein, fat, nitrogen-free extract and crude fibre. A great variety of energy equivalents are found in the literature (Janssen, 1976). For many years in the USA and Japan, nutritive value was based on total digestible nutrients (TDN) which is defined as follows:

$$TDN = \text{digestible CP} + \text{digestible (NFE + CF)} + 2.25 \text{ digestible F} \qquad (6.2)$$

According to Carpenter (1962), Fraps and Carlyle found the following relation between 'effective digestible nutrients' (EDN) and ME:

$$ME \text{ (MJ)} = 17.6 \text{ EDN} \qquad (6.3)$$

The only difference between TDN and EDN is that EDN does not take crude fibre into account. Fraps and Carlyle considered the direct contribution of crude fibre to the nutritive value of a feed or feedstuffs as negligible. Nowadays, ME is used as a measure of feeding value for poultry all over the world.

METHODS BASED ON CHEMICAL ANALYSIS

Many attempts have been made to find (simple) relations between ME and one or more chemical or physical characteristics of a feedstuff, a group of feedstuffs or even all feedstuffs. In this connection, Carpenter (1961) pointed out '... In general it is true that the higher the fibre content the lower the metabolizable energy, and for a range of materials, such as wheat offals from a generally similar process, there is a high negative correlation between fibre content and metabolizable energy so that analysis for this alone in a sample can allow a more accurate assessment of its energy value'.

Table 6.2   THE RELATIONSHIP BETWEEN DIGESTIBILITY (%) OF THE ORGANIC MATTER (Y) AND CRUDE FIBRE (X) CONTENT (% IN DRY MATTER)

Cattle : $Y = 90.1 \ - 0.879\,X$	
Rabbit : $Y = 96.48 - 1.551\,X\ (r = -0.899)$	
Poultry : $Y = 86.06 - 1.995\,X\ (r = -0.746)$	

Mitchell (1942) found a good correlation between crude fibre and the digestibility of organic matter of feedstuffs (*Table 6.2*). This table shows that the negative influence of crude fibre on digestibility increases from cattle to rabbits and from rabbits to poultry. Mitchell (1942) calculated the following equation based on 60 experiments of Fraps:

$$Y = 86.90 - 3.624\,X \tag{6.4}$$
where Y = metabolizability (%) of the gross energy and
X = % crude fibre in dry matter

At Spelderholt Institute the following equations having an important relationship with fibre were found (Janssen *et al.*, 1979):

ME (MJ/kg DM) =

Wheat products	$16.539 - 0.769\,CF\%$	(6.5)
Barley and barley products	$12.88 - 0.378\,CF\% + 0.038\,starch\%$	(6.6)
Oats and oat products	$12.43 - 0.25\,CF\% + 0.489\,F\%$	(6.7)
Maize products	$17.74 - 0.144\,CP\% + 0.057\,F\%$	
	$- 0.668\,CF\%$	(6.8)
Rice products	$19.91 - 0.371\,CP\% - 0.534\,CF\%$	
	$+ 0.218\,F\%$	(6.9)
Soya beans (in meal feeds)	$11.59 - 0.247\,CF\% + 0.257\,F\%$	(6.10)
Soya beans (in pelleted feeds)	$11.03 - 0.233\,CF\% + 0.345\,F\%$	(6.11)
Tapioca root meal	$16.96 - 0.182\,ash\% - 0.431\,CF\%$	(6.12)
Groundnut products	$12.85 - 0.164\,ash\% - 0.199\,CF\%$	
	$+ 0.267\,F\%$	(6.13)
Sunflower seed products	$16.73 - 0.791\,ash\% - 0.245\,CF\%$	
	$+ 0.249\,F\%$	(6.14)
Cotton seed products	$9.01 - 0.133\,CF\% + 0.182\,F\%$	(6.15)

Carpenter (1961) derived the following equation from digestion data of 17 feedstuffs and compound feeds:

$$\text{ME(MJ/kg 90\% dry matter)} = 0.945 \times \text{gross energy} \left(1 - \frac{\text{CF\%}}{21}\right) \tag{6.16}$$

The residual standard deviation was 1.339 MJ/kg. Substituting the average gross energy of the wheat products (17.25 MJ/kg) used in the experiments by Carpenter and Clegg (1956) the following equation was derived:

$$\text{ME(MJ/kg dry matter)} = 18.113 - 0.862 \, \text{CF\%} \tag{6.17}$$

This shows a good agreement with Spelderholt's equation for wheat products. The difference in intercept can be explained by the fact that Carpenter and Clegg (1956) determined the 'classical' ME and Janssen *et al.* (1979) determined the ME value corrected for nitrogen retention.

Thomke (1960) found the following relationships for oats:

$$\text{ME(MJ/kg dry matter)} \quad = 16.694 - 0.481 \, \text{CF\%} \text{ (for laying hens)} \tag{6.18}$$
$$\text{or} = 16.234 - 0.498 \, \text{CF\%} \text{ (for young chicks)} \tag{6.19}$$

Hill, Carew and Renner (1957) studied the feeding value of five maize by-products of different quality. The variability in ME could be explained by the variability in fat and fibre content. An increase in 1 per cent fat (ether extract) and crude fibre resulted in an increase in ME of 0.368 MJ/kg and a decrease of 1.059 MJ/kg respectively. For cottonseed meals Watts and Davenport (1971) found:

$$\text{ME(MJ/kg)} = 3.552 + 0.097 \, \text{CP\%} + 1.254 \, \text{F\%} - 0.20 \text{CF\%} \tag{6.20}$$

Comparing this with the Spelderholt equation, the calculated ME values differ considerably. This is because of the very large value for fat in the equation of Watts and Davenport (1971).

The majority of these relationships clearly demonstrate the important influence of crude fibre on the feeding value of the products concerned, but also the large differences in regression-coefficients sometimes found for the same groups of products.

In order to improve the predictive value of fibres, several methods of fibre measurement have been tested (Carré, Prévotel and Leclercq, 1984): 48 mixed diets of different composition were analysed and their ME value determined. From this study it appears that the enzymatic residue (ER) content was the most efficient fibre indicator for predicting ME value. Crude fibre, acid detergent fibre (ADF) and neutral detergent fibre (NDF) exhibited similar efficiency. Lignin content had the poorest predictive value. The best equation was as follows:

$$\text{ME(MJ/kg air-dry matter)} = 0.914 \, \text{GE (MJ/kg)} - 0.061 \, \% \, \text{CP}$$
$$- 0.041 \, \% \, \text{ER}^{1.5} \tag{6.21}$$
$$(n = 48; \text{r.s.d.} = 0.20)$$

Analytical ranges of the 48 diets were on air-dry matter basis:

ME (MJ/kg)	: 9.43	− 14.66
Crude protein (%)	: 7.4	− 27.3
Fat (%)	: 1.6	− 11.0
Crude fibre (%)	: 1.5	− 8.4
ER (%)	: 7.6	− 22.2

The equation was in good agreement with a theoretical equation which was established on the principle that the water-insoluble fibres act as diluter of available nutrients.

THE RELATIONSHIP OF CRUDE FIBRE WITH THE DIGESTIBILITY OF THE NUTRIENTS IN SPELDERHOLT EXPERIMENTS

Due to the indirect nature of this relationship many regression coefficients are incomprehensible from a physiological point of view. In the Spelderholt experiments digestibility of the nutrients was also determined in order to gather more information on the basis of these relationships.

In these experiments mature Araucana × White Leghorn cocks were used. Every day a moderately restricted amount of food was given to each bird. The experimental feeds consisted of a basal diet and mixtures of the basal diet + the feedstuff under study. The method was based on quantitative faeces collection. A detailed description has been given by Terpstra and Janssen (1976).

The results give data on by-products of wheat, barley, rice, tapioca and cottonseed. In *Tables 6.3* to *6.7* the percentages of crude fibre, crude protein, fat, nitrogen free extract (NFE) and starch of the products concerned are shown. Also metabolizable energy (MJ/kg DM) and the digestibility of crude fibre, crude protein, fat and NFE are presented. The digestibility of starch was not determined. For the wheat, rice and cottonseed products, some experimental diets were given both in meal and in pellet form in order to study the effect of pelleting. These results are also included.

Wheat by-products

The products studied varied widely in composition as the ranges of crude fibre (1.0–13.9 per cent), of NFE (57.3–75.8 per cent) and of starch (9.7–66.1 per cent) demonstrate (*Table 6.3*).

The positive contribution of crude fibre to the feeding value is small since its digestibility is generally very low. The negative influence is substantial as there is a strong negative relationship with the digestibility of protein, fat and NFE. The last relationship is probably due to the high negative correlation between crude fibre and starch as with increasing fibre content, the percentage of starch in NFE diminishes (87 to 17 per cent for products 1 and 10 respectively).

There is a large pelleting effect on the digestibility of fat, a moderate effect on the digestibility of protein and only a minor effect on the digestibility of NFE. It is worth noting that the effects of pelleting are larger at the higher crude fibre contents.

Table 6.3 WHEAT BY-PRODUCTS

Product	Composition (% in dry matter)				
	Crude fibre	Crude protein	Fat	NFE	Starch
1	1.0	17.8	3.5	75.8	66.1
2	1.8	24.6	5.5	65.1	50.6
3	3.3	20.9	5.3	67.4	45.8
4	3.6	20.9	7.5	64.3	ND
5	3.9	20.2	7.9	63.7	37.8
6	8.6	20.1	6.2	60.0	25.0
7	9.4	18.9	4.5	60.5	ND
8	11.1	18.6	5.2	58.8	16.9
9	13.3	18.0	4.0	57.5	ND
10	13.9	16.5	5.0	57.3	9.7

Product	Pellet/ meal	ME content (MJ/kg DM)	Digestibility (%)			
1	P	16.07	8.4	87.1	89.0	93.1	ND
2	P	15.04	13.2	81.7	87.7	86.6
3	P	13.78	10.0	81.9	86.7	77.4
4	P	13.46	2.6	81.2	82.4	73.4
5	P	13.41	5.7	79.4	88.8	72.2
6	P	10.19	8.4	77.6	76.2	51.3
7	P	9.00	5.8	74.1	63.4	49.5
8	P	7.93	10.3	71.7	65.8	37.1
9	P	6.44	10.4	69.7	31.3	32.2
10	P	5.84	14.0	70.6	54.5	22.5
4	M	13.24	−14.2	80.8	80.2	73.4
7	M	8.59	4.0	68.2	38.2	50.5
9	M	5.41	10.6	57.6	7.8	28.9
	P̄[a]	9.70	6.3	75.0	59.0	51.7
	M̄[a]	9.08	0.3	68.8	42.1	50.9

Relationship with crude fibre		r	r.s.d.
Metabolizable energy (MJ/kg DM)	$= 16.43 - 0.76\,CF(\%)$	0.998	0.25
Protein digestibility (%)	$= 85.6 - 1.15\,CF(\%)$	0.97	1.58
Fat digestibility (%)	$= 97.2 - 3.53\,CF(\%)$	0.90	8.76
NFE digestibility (%)	$= 94.5 - 5.00\,CF(\%)$	0.99	2.82
Starch (%)	$= 60.0 - 3.86\,CF(\%)$	0.96	5.88

[a]Based on samples 4, 7 and 9.
ND, not determined

Barley by-products

The barley products showed considerable variation in composition (crude fibre, NFE and starch) (*Table 6.4*). The positive contribution of crude fibre to feeding value is negligible in these products. The high digestibility of crude fibre of product 1 is probably an artefact and due to the very low crude fibre level. At this level it is practically impossible to determine digestibility accurately. In this series of products no clear relationship was found between crude fibre and digestibility of fat and protein. There was a high negative correlation between crude fibre and the NFE digestibility, due to the negative correlation between the starch content in

Table 6.4 BARLEY BY-PRODUCTS (PELLETED)

| Product | Composition (% in dry matter) | | | | |
	Crude fibre	Crude protein	Fat	NFE	Starch
1	0.7	8.2	1.4	88.7	79.7
2	3.1	15.1	4.7	74.0	55.4
3	4.6	11.1	2.2	79.6	59.7
4	5.1	21.9	7.6	60.6	32.6
5	8.4	11.9	3.1	71.5	48.2
6	12.0	12.9	4.3	64.8	32.3
7	15.9	11.9	3.9	61.8	25.8
8	19.4	11.7	4.3	56.0	16.5

Product	ME content (MJ/kg DM)	Digestibility (%)			
1	15.69	24.6	67.4	73.1	92.6	ND
2	13.77	7.4	71.4	85.9	79.9	
3	13.39	6.0	69.5	60.1	82.0	
4	12.25	−6.0	73.9	85.9	64.7	
5	11.64	−2.8	71.3	76.0	73.6	
6	9.49	−4.6	69.3	81.0	58.6	
7	7.97	−0.5	69.2	76.2	49.5	
8	6.10	−1.1	65.7	71.4	36.5	

	Relationship with crude fibre			r	r.s.d.
Metabolizable energy (MJ/kg DM)	= 15.48 − 0.48 CF(%)			0.99	0.42
NFE digestibility (%)	= 90.4 − 2.69 CF(%)			0.95	6.07
Starch (%)	= 67.7 − 2.77 CF(%)			0.88	10.80

ND, not determined

NFE and crude fibre. The starch content decreased from 90 per cent in the product with the lowest crude fibre content to 29 per cent in the product with 19.4 per cent crude fibre. As a consequence there is a strong negative correlation between crude fibre and ME ($r = 0.99$).

Rice by-products

In this series the positive contribution of fibre seems to be larger than in the other series. The first three products had particularly high crude fibre digestibilities (*Table 6.5*). However, in absolute terms the fibre contribution is of only minor importance, as the levels in these three products were very low and subject to the inaccuracies already mentioned. The other products with considerably higher crude fibre content showed slightly positive to slightly negative digestibilities of fibre.

The relationship between crude fibre and feeding value is worse than that for wheat and barley. An explanation for this may be that the large variability in fat content is not related to fibre content. The negative correlation between crude fibre and ME seems to be due to the rather strong relationship between crude fibre and the digestibility of NFE and to the relationship between crude fibre and the starch content of NFE.

Table 6.5   RICE BY-PRODUCTS

Product			Composition (% in dry matter)				
			Crude fibre	Crude protein	Fat	NFE	Starch
1			0.5	8.1	0.7	90.2	88.6
2			0.9	11.4	6.4	77.7	71.0
3			1.1	8.4	2.3	87.0	82.5
4			2.9	13.5	12.7	65.2	53.9
5			6.8	14.9	14.0	56.4	37.0
6			9.1	16.0	20.3	45.8	22.2
7			11.2	13.3	14.6	48.7	27.1
8			14.1	19.9	1.7	44.8	15.9
9			15.0	16.4	1.3	49.2	24.3

Product	Pellet/ meal	ME content (MJ/kg DM)	Digestibility (%)				
1	P	16.84	77.6	83.8	88.0	98.5	ND
2	P	16.61	32.1	79.1	90.0	94.3	
3	P	16.72	23.9	79.6	81.3	97.8	
4	P	16.15	6.3	73.9	90.2	88.2	
5	P	13.64	−8.1	65.8	85.7	75.8	
6	P	13.60	1.7	62.0	87.2	62.8	
7	P	11.94	2.4	67.9	86.7	62.1	
8	P	5.36	−7.0	52.5	15.8	35.1	
9	P	6.74	7.9	62.7	25.9	54.0	
7	M	11.90	6.4	60.5	87.1	63.6	
9	M	6.42	4.2	61.2	21.3	52.3	
	$\bar{P}^a$	9.34	5.2	65.3	56.3	58.0	
	$\bar{M}^a$	9.16	5.3	60.8	54.2	58.0	

Relationship with crude fibre		
	r	r.s.d.
Metabolizable energy (MJ/kg DM) = 17.92 − 0.71 CF(%)	0.94	1.55
Protein digestibility (%) = 80.6 − 1.59 CF(%)	0.91	4.59
NFE digestibility (%) = 99.7 − 3.71 CF(%)	0.96	6.41
Starch (%) = 77.4 − 4.45 CF(%)	0.92	11.37
Starch (%) = 88.2 − 10.85 CF(%) + 0.43 CF (%)^2 0.98		6.48

^{a}Based on samples 7 and 9.
ND, not determined

Tapioca by-products

There is a very high negative correlation between crude fibre and ME, mainly due to the relationship between crude fibre and digestibility of NFE and the relationship between crude fibre and starch in NFE (*Table 6.6*).

This experiment once again clearly shows the problems of determining the digestibility of the nutrients present at a very low level.

Cottonseed by-products

Roughly one can distinguish two groups of products (*Table 6.7*): the first group with a low fibre and high protein content (products 1–5) and the second one containing

Table 6.6 TAPIOCA BY-PRODUCTS (PELLETED)

Product	Composition (% in dry matter)				
	Crude fibre	Crude protein	Fat	NFE	Starch
1	0.1	0.5	0.1	99.2	99.3
2	4.4	3.0	0.6	86.6	75.7
3	4.6	3.2	0.6	86.5	76.2
4	5.0	3.0	0.8	85.4	74.1
5	6.3	4.1	0.9	80.5	67.2

Product	ME content (MJ/kg DM)	Digestibility (%)			
1	17.23	−1205.5	−42.2	419.1	101.7	ND
2	14.23	5.2	50.6	27.5	94.4	
3	14.00	−0.7	42.3	17.1	93.4	
4	14.05	1.6	50.3	27.7	91.9	
5	12.97	−5.0	56.6	1.6	88.6	

Relationship with crude fibre				
			r	*r.s.d.*
Metabolizable energy (MJ/kg DM)	=	$17.28 - 0.68\,CF(\%)$	0.997	0.14
NFE digestibility (%)	=	$102.3 - 2.04\,CF(\%)$	0.99	0.88
Starch (%)	=	$99.7 - 5.19\,CF(\%)$	0.999	0.75

ND, not determined.

high levels of fibre and low levels of protein (products 6–9). The digestibility of crude protein was higher in the first group with a low level of crude fibre than in the other group. The digestibility of fat is mainly regulated by the fat content. At high levels the fat was well digested, at low levels the digestibility was quite variable. No clear pelleting effects could be demonstrated.

Discussion

The influence of fibre may be based on the following effects:

1 Positive effects —digestibility of fibre and thus direct contribution to the feeding value
2 Negative effects—increasing effects on the production of endogenous material
　　　　　　—diluting effect on other nutrients
　　　　　　—barrier to the penetration of digestive enzymes

POSITIVE EFFECTS

Digestibility of fibre

Most data in the literature tend to demonstrate that the cellulose complex from plant feedstuffs is not digested by poultry (Bolton, 1955a; Almquist and Halloran, 1971; Vogt and Stute, 1971). The results of the experiments presented in *Tables 6.3* to *6.7* confirm these data. Recently Duke *et al.* (1984) found that adult turkeys

Table 6.7 COTTONSEED BY-PRODUCTS

Product	Composition (% in dry matter)			
	Crude fibre	Crude protein	Fat	NFE
1	12.0	41.3	6.3	33.6
2	14.3	44.3	0.8	33.1
3	16.6	44.7	1.6	30.0
4	17.1	40.8	1.6	34.1
5	18.4	44.1	1.3	30.0
6	24.6	26.2	7.6	36.3
7	25.2	27.6	7.3	34.3
8	25.9	27.0	7.1	35.0
9	26.2	26.3	6.5	35.8

Product	Pellet/ meal	ME content (MJ/kg DM)	Digestibility (%)			
1	P	8.59	-6.2	62.0	83.9	41.0
2	P	7.31	-5.3	66.7	-2.5	34.8
3	P	6.91	-11.3	70.7	54.2	30.6
4	P	7.07	-7.1	70.8	18.3	37.2
5	P	6.77	-7.2	69.9	46.4	30.1
6	P	7.24	-11.9	59.6	87.0	35.0
7	P	6.84	-11.9	59.9	88.7	34.1
8	P	6.86	-4.1	59.4	87.7	26.6
9	P	7.11	-2.0	55.9	83.9	33.3
3	M	7.17	-8.4	70.6	47.6	33.8
8	M	6.77	-8.9	58.9	89.6	31.0
	$\bar{P}^a$	6.88	-7.7	65.0	71.0	28.6
	$\bar{M}^a$	6.91	-8.6	64.8	68.6	32.4

	Relationship with crude fibre			
			r	r.s.d.
Metabolizable energy (MJ/kg DM)	= 11.53 − 0.27 CF(%)		0.93	0.32
NFE digestibility (%)	= 56.7 − 1.40 CF(%)		0.78	3.33

[a]Based on samples 3 and 8

utilized about 10 per cent of a ^{14}C labelled purified cellulose. However, the cellulose preparation used was able to form a gel (Duke *et al.*, 1984); therefore this cellulose preparation was quite different from naturally occurring cellulose which normally exhibits a high crystallinity and water insolubility.

Hemicellulosic compounds measured as 'whole pentosan' are not digested by young birds (Bolton, 1955a; Antonion, Marquardt and Cansfield, 1981). However, it has often been reported that adult birds are able to digest a substantial amount of 'whole pentosans'; but in the procedure for determining pentosans no fractionation was made to distinguish water-soluble and water-insoluble polysaccharides (Bolton, 1955a; Vogt and Stute, 1974). Carré (1983) demonstrated that the water-insoluble xylosyl polymers of white lupin cotyledon were not digested by adult cocks. So the digestible fraction of the pentosans would correspond with the water-soluble pentosans. This is supported by the observation that the highest digestibility of pentosans was found in the cereal grain with the largest proportion of endosperm (wheat) (Bolton, 1955b) in which, according to Mares and Stone (1973), the water-soluble pentosans are concentrated. Carré (1983) found that the digestibility of the water-insoluble pectic substances for adult birds was very low.

NEGATIVE EFFECTS

Added fibre

Numerous experiments have been conducted to study the influence of added fibre on the digestibility of protein and lipids. Thus, they investigated the effects due to the physicochemical properties of fibre such as water-binding capacity, cation-exchange capacity and bile acid adsorption capacity. Mostly it has been found that the addition of water-insoluble fibre has little or no effect on digestibility of protein and lipids (Kibe, Tasaki and Saito, 1964; Sibbald, 1980; Akiba and Matsumoto, 1980; Parsons, Potter and Brown, 1982; Carré, 1983).

The decrease of apparent protein digestibility sometimes observed with added fibre, might be due to an increase of endogenous losses, thus decreasing apparent digestibility (Farrell, 1981; Raharjo and Farrell, 1984).

Fibre occurring naturally in feedstuffs

Water-insoluble fibre (WIF), being undigestible, may act as a diluter of the nutrients. Carré, Prévotel and Leclerq (1984) determined several estimates of fibre (NDF, ADF, crude fibre, etc.). For 13 cereal-based diets, an average of 22 per cent of the cell wall fraction (CW) determined by the enzymatic residue method was crude fibre.

When crude fibre was substituted in the Spelderholt equation for wheat by-products by 0.22 CW, it gave the equation:

$$ME\ (MJ/kg\ DM) = 16.539 - 0.169\ CW \tag{6.22}$$

This equation indicates a slight negative ME value for cell wall material ($-0.36\,MJ/kg$), although it is statistically not valid to extrapolate to 100 per cent CW.

Substituting crude fibre by 0.22 CW in the equation of Carpenter and Clegg (1956) the following equation, for material standardized to 90 per cent dry matter, is derived

$$ME\ (MJ/kg\ 90\%\ DM) = 0.945 \times GE \times \left(1 - \frac{CF(\%)}{21}\right) \tag{6.23}$$

This also gives an estimated value for cell wall content which is slightly negative but close to zero. It can be concluded that the cell wall fraction does not have any positive contribution to the energy value. An objection to this approach may be that the crude fibre content in the cell wall is based on mixed feeds used by Carré, Prévotel and Leclerq (1984). For further evidence, determination of crude fibre as well as of cell wall needs to be done in ME studies.

The plant cell walls may act as a barrier to the attack of intracellular compounds by enzymes of the gastrointestinal tract. The fibre content in wheat by-products was indicative of its origin; from the inside to the outside of the grain, fibre content increases. The higher the fibre content, the stronger the barrier effect. This latest point can explain the negative relationship between fibre and digestibility of crude protein and fat.

These effects have also been described by Saunders, Walker and Kohler (1969) using microscopic examination of plant tissues in diets and excreta.

Fibre also seems to be an indicator of the available nutrient content. For example, a highly significant negative relationship existed between crude fibre and starch level in the nitrogen free extractives of the grain by-products studied. The starch content increased from the outside part to the inside part of the grain.

Sometimes, the effect of crude fibre seems to be very complicated as described by Scheele (1983). He found an increase in digestibility of added fat when increasing the fibre content originating from sunflower seed meal or alfalfa meal in the diet, but a decrease in digestibility of the fat when the increased fibre content of the basal ration was based on wheat bran. Thus the fibre of wheat bran seems to have a negative effect on the digestibility of added fat while the fibre of sunflower seed meal and alfalfa meal seem to have a positive effect.

Conclusions

The positive contribution of fibre to the feeding value of feedstuffs for poultry seems to be of negligible importance as the cell wall fraction is undigestible.

It is strongly suggested that in several kinds of feedstuffs, fibre may be a good predictor of the feeding value. This is due to a diluting effect of the cell wall fraction and to the strong negative correlation between crude fibre content and the digestibility of crude protein and fat.

In the grain by-products studied, there also appeared to be a strong negative correlation between crude fibre and the starch content in the nitrogen free extract, probably causing the decreasing digestibility of this fraction with increasing crude fibre content.

References

AKIBA, Y. and MATSUMOTO, T. (1980). *Journal of Nutrition,* **110**, 1112

ALBERSHEIM, P. (1976). In *Plant Biochemistry*, p. 225. Ed. J. Bonner and J.E. Varner. Academic Press; New York

ALMQUIST, H.J. and HALLORAN, H.R. (1971). *Poultry Science,* **50**, 1233

ANTONION, T.C., MARQUARDT, R.R. and CANSFIELD, P.E. (1981). *Journal of Agricultural and Food Chemistry,* **29**, 1240

ASP, N.G., JOHANSSON, C.G., HALLMER, H. and SILJESTRÖM, M. (1983). *Journal of Agricultural and Food Chemistry,* **31**, 476

ASPINALL, G.O. (1980). In *The Biochemistry of Plants,* **12**, 473. Ed. J. Preiss. Academic Press; New York, London, Toronto, Sydney, San Francisco

ASPINALL, G.O., HUNT, K. and MORRISON, I.M. (1966). *Journal of the Chemical Society,* Part C, 1945

ASPINALL, G.O., BEGBIE, R., HAMILTON, A. and WHYTE, J.N.C. (1967). *Journal of the Chemical Society,* Part C, 1065

BAILEY, R.W. and ULYATT, M.J. (1970). *New Zealand Journal of Agricultural Research,* **13**, 591

BAKER, D. (1977). *Cereal Chemistry,* **54**, 360

BOLTON, W. (1955a). *Journal of Agricultural Science,* **46**, 420

BOLTON, W. (1955b). *Journal of Agricultural Science,* **46**, 119

BRILLOUET, J.M. and CARRÉ, B. (1983). *Phytochemistry, 22*, 841

BRILLOUET, J.M., JOSELEAU, J.P., UTILLE, J.P. and LELIÈVRE, D. (1982). *Journal of Agricultural and Food Chemistry, 30*, 488

CARPENTER, K.J. (1961). In *Nutrition of Pigs and Poultry*, p. 29. Ed. J.T. Morgan and D. Lewis. Butterworths; London

CARPENTER, K.J. and CLEGG, K.M. (1956). *Journal of the Science of Food and Agriculture, 7*, 45

CARRÉ, B. and LECLERCQ, B. (1985) *British Journal of Nutrition, 54,* 669

CARRÉ, B., BRILLOUET, J.M. and THIBAULT, J.F. (1985). *Journal of Agricultural and Food Chemistry, 33*, 285

CARRÉ, B., PRÉVOTEL, B. and LECLERCQ, B. (1984). *British Poultry Science, 25*, 561

CHESSON, A., GORDON, A.H. and LOMAX, J.A. (1983). *Journal of the Science of Food and Agriculture, 34*, 1330

DUKE, G.E., ECCLESTON, E., KIRKWOOD, S., LOUIS, C.F. and BEDBURY, H.P. (1984). *Journal of Nutrition, 114*, 95

FARRELL, D.J. (1981). *World's Poultry Science Journal, 37*, 72

HELLENDOORN, E.W., NOORDHOFF, M.G. and SLAGMAN, J. (1975). *Journal of the Science of Food and Agriculture, 26*, 1461

HILL, F.W., CAREW JR., L.B. and RENNER, R. (1957). *Feedstuffs, 29* (37), 84

ITCF (1982). *Le Sorgho grain. Composition Chimique et Valeur Alimentaire. Résultats récolte 1980.* Institut Technique des Céréales et des Fourrages; Paris

JANSSEN, W.M.M.A. (1976). In *Proceedings International Symposium on Computer Use in Feed Formulation.* National Renderers Association; Brussels

JANSSEN, W.M.M.A., TERPSTRA, K., BEEKING, F.F.E. and BISALSKY, A.J.N. (1979). *Feeding values for poultry*, 2nd edn. Mededeling 303. Spelderholt Institute for Poultry Research, Beekbergen; The Netherlands

KIBE, K., TASAKI, I. and SAITO, M. (1964). *Japanese Journal of Zootechnical Science, 35*, 159

MARES, D.J. and STONE, B.A. (1973). *Australian Journal of Biological Sciences, 26*, 793

MITCHELL, H.H. (1942). *Journal of Animal Science, 1*, 159

MORRISON, I.M. (1972). *Journal of the Science of Food and Agriculture, 23*, 455

NEILSON, M.J. and RICHARDS, G.N. (1982). *Carbohydrate Research, 104*, 121

NYMAN, M., SILJESTRÖM, M., PEDERSEN, B., BACHKNUDSEN, K.E., ASP, N.G., JOHANSSON, C.G. and EGGUM, B.O. (1984). *Cereal Chemistry, 61*, 14

O'NEILL, M.A. and SELVENDRAN, R.R. (1980). *Carbohydrate Research, 79*, 115

PARSONS, C.M., POTTER, L.M. and BROWN, R.D. (1982). *Poultry Science, 61*, 939

RAHARJO, Y. and FARRELL, D.J. (1984). *Animal Feed Science and Technology, 12*, 29

RINAUDO, M. and CHAMBAT, G. (1976). *Revue française des corps gras, 23*, 605

SAEMAN, J.F., MOORE, W.E., MITCHELL, R.L. and MILLETT, M.A. (1954). *Tappi, 37*, 336

SAUNDERS, R.M., WALKER JR., H.G. and KOHLER, G.O. (1969). *Poultry Science, 48*, 1497

SCALBERT, A., BRILLOUET, J.M., ROLANDO, C. and MONTIES, B. (1984). In *International Workshop on Plant Polysaccharides, Structure and Function*, Nantes, p. 164. Institut National de la Recherche Agronomique, Centre National de la Recherche Scientifique; Paris

SCHEELE, C.W. (1983). In *Proceedings of the 4th European Symposium on Poultry Nutrition, Tours.* Ed. M. Larbier. INRA; Nouzilly, France

SCHWEIZER, T.F. and WÜRSCH, P. (1979). *Journal of the Science of Food and Agriculture, 30*, 613

SIBBALD, I.R. (1980). *Poultry Science,* **59**, 836

SOEST, P.J. VAN (1963). *Journal of the Association of Official Agricultural Chemists,* **46**, 829

SOEST, P.J. VAN and WINE, R.H. (1967). *Journal of the Association of Official Agricultural Chemists,* **50**, 50

TERPSTRA, K. and JANSSEN, W.M.M.A. (1976). *Report 101.75,* Spelderholt Institute for Poultry Research, Beekbergen; The Netherlands

THEANDER, O. and AMAN, P. (1980). *Journal of the Science of Food and Agriculture,* **31**, 31

THEANDER, O., AMAN, P., MIKSCHE, G.E. and YASUDA, S. (1977). *Journal of Agricultural and Food Chemistry,* **25**, 270

THOMKE, S. (1960). *Archiv für Geflügelkunde,* **24**, 557

VOGT, H. and STUTE, K. (1971). *Archiv für Geflügelkunde,* **35**, 29

VOGT, H. and STUTE, K. (1974). *Archiv für Geflügelkunde,* **38**, 117

WATTS, A.B. and DAVENPORT, R.F. (1971). *Poultry Science,* **50**, 1643

7

RECENT ADVANCES IN DIETARY ANION–CATION BALANCE IN POULTRY

P. MONGIN

Station de Recherches Avicoles, INRA–CRVZ, Nouzilly, 37380 Monnaie, France

Introduction

The fundamental approach to the problem of anion–cation balance was recently reviewed by Mongin (1980a). The main conclusion drawn was that in order to keep its acid-base homeostasis as close as possible to normal, the animal has to regulate the input and/or the output of acidity, i.e. to control the balance of acid within the body. The net acidity intake is measured by the difference between fixed anions and cations, $(An-Cat)_{in}$. Fixed anions and cations are defined as those which cannot be metabolized or broken down further during digestive or metabolic processes. Likewise the net acidity output will be measured by the balance of ions excreted in the urine, $(An-Cat)_{out}$. However, a third term must be also considered because some acidity is produced within the body, mainly by the metabolism of the dietary proteins. This component is called the endogenous acid production, (H^+_{endo}).

When the sum of the net acidity intake plus the endogenous acid production is equal to the net acidity excreted in the urine the animal is in a steady state, and this can be represented by:

$$(An-Cat)_{in} + H^+_{endo} - (An-Cat)_{out} = 0 \qquad (7.1)$$

Under such conditions the blood pH is 7.4, the plasma bicarbonate is 25 meq/ℓ and the base-excess is zero. However, if net acidity intake plus endogenous acid production is different from the net acidity excreted the so-called alkali reserve or base-excess (BE) of the blood will be modified accordingly to enable the animal to achieve a steady state. This new steady state can be represented by:

$$(An-Cat)_{in} + H^+_{endo} - (An-Cat)_{out} + BE = 0 \qquad (7.2)$$

The extent of the modification of the base excess can be represented by a rearrangement of this relationship:

$$(Cat-An)_{in} - (Cat-An)_{out} - H^+_{endo} = BE \qquad (7.3)$$

First published in *Recent Advances in Animal Nutrition – 1981*

An example will serve to make equation (7.3) more understandable. If it is assumed that for some reason the excretion of net acidity increases without any modification of the diet, the term $(Cat–An)_{out}$ will then decrease with the result that the left-hand side of equation (7.3) will increase, and hence the base excess will also increase. The consequence of this is that the animal will become alkalotic. On the other hand, if it is assumed that part of dietary calcium carbonate is replaced by calcium chloride the term $(Cat–An)_{in}$ will decrease because only the fixed anions are considered. Consequently BE will decrease and the animal will become acidotic. Thus it is clear that the acid-base status of the blood gives a good picture of the balance between anions and cations and can therefore be used as an experimental criterion.

Relationship between acid-base balance and mineral balance in practice

Under the normal conditions of livestock nutrition it is difficult to measure both mineral excretion and endogenous acid production. However, the composition of the mineral intake and the acid-base status of the animal are much easier to monitor and control. The experimental approach which has to be adopted to measure acid-base balance is therefore to change the mineral composition of the diet and then measure the acid-base balance of the animal in order to determine when the base-excess is equal to zero.

Within the context of livestock ration formulation only the $(Cat–An)_{in}$ is of interest. However, this is quite complex and is represented by the following equation:

$$(Cat–An)_{in} = meq(Na^+ + K^+ + Ca^{2+} + Mg^{2+}) - meq (Cl^- + SO_4^{2-} + H_2PO_4^- + H\,PO_4^{2-}) \quad (7.4)$$

In practice, it is often difficult to control intakes of all of the eight elements listed. The two anions of phosphate, for example, are included in the diet in both organic (phosphoprotein and phospholipid) and inorganic (mineral phosphate) form and usually it is not known whether the organic phosphates are in the mono-acid, tri-acid or phytate form. Furthermore, inorganic phosphate is largely incorporated as dicalcium phosphate, not to balance cations and anions in the diet but to meet the requirement for available phosphorus.

On the other hand, the intake of magnesium is determined by ingredient composition and is largely in excess. Likewise sulphate is included simply as the anion for essential trace elements or alternatively to prevent the breakdown of methionine. Finally calcium is to a large extent included as the carbonate salt and is primarily involved in skeletal development rather than acid-base homeostatis.

Since it appears that only sodium, potassium and chloride are important in determining acid-base balance in practical circumstances it is perhaps more appropriate to rewrite equation (7.4) as shown in the following equation:

$$(Cat–An)_{in} = meq(Na + K - Cl) + meq (Ca + Mg - SO_4 - H_2PO_4 - H\,PO_4) \quad (7.5)$$

The final term of this equation can then be included under the term 'endogenous production', allowing equation (7.3) to be modified such that $(Cat-An)_{in}$ is replaced by $(Na + K - Cl)_{in}$:

$$(Na + K - Cl) = (Cat-An)_{out} + H^+_{endo} + BE \qquad (7.6)$$

The practical importance of mineral balance in poultry

It is, however, necessary to determine whether the fundamental assumptions made in deriving equation (7.6) actually hold good in the practical situation.

ACID-BASE BALANCE AND ITS RELATIONSHIP TO THE $(Na + K - Cl)$ BALANCE

The acid-base status of the animal should reflect the ionic balance of the diet, and this is confirmed by the results of two experiments. In the first, poultry were fed one of nine synthetic diets in which the sum $(Na + K - Cl)$ concentration ranged from -20 to $+40$ meq/100 g. Plasma bicarbonate concentration was shown to be linearly related to $(Na + K - Cl)$ concentration (Mongin and Sauveur, 1973) (*Figure 7.1*).

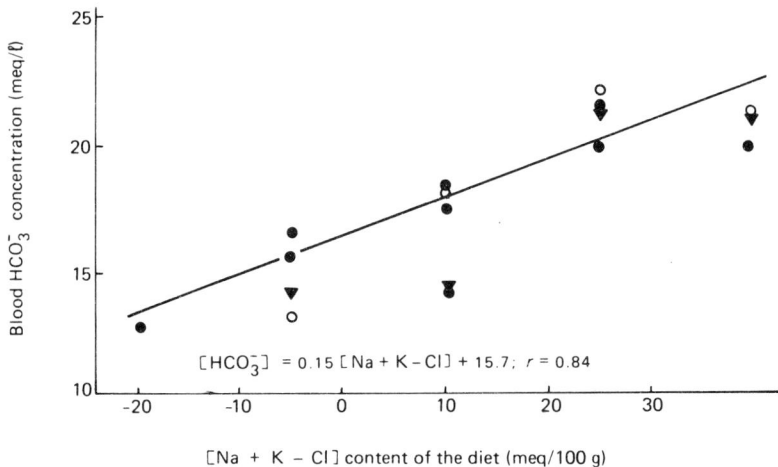

$$[HCO_3^-] = 0.15 [Na + K - Cl] + 15.7; \; r = 0.84$$

$[Na + K - Cl]$ content of the diet (meq/100 g)

Figure 7.1 Relationship between the $[Na + K - Cl]$ content of the diet and plasma bicarbonate concentrations in growing chicks ($\bigcirc$, Na/K = 0.4; $\bullet$, Na/K = 1.0; $\blacktriangledown$, Na/K = 2.5) (From Mongin and Sauveur, 1973)

In the second experiment chicks were fed eight natural diets, all having the same Na and K contents but varying Cl concentrations. It was found that blood pH had a marked dependence on dietary chloride (Hurwitz *et al.*, 1973). Although the curve was sigmoidal rather than linear this form was thought to result from the fact that some of the diets were deficient in chloride as judged by the growth responses. Both experiments, however, confirm that the acid-base balance of the animal reflects the ionic balance of the diet.

MINERAL BALANCE AND BODY GROWTH

Systematic experiments have been conducted to determine the exact relationship between the sum of cations minus anions in the diet and body growth, using either synthetic diets (Nesheim *et al.*, 1964; Melliere and Forbes, 1966; Mongin and Sauveur, 1973, 1977) or natural diets (Hurwitz *et al.*, 1973; Mongin and Sauveur, 1977). The results of most experiments confirm that when (Na + K − Cl) content of the diet is higher or lower than 25 meq/100 g growth is depressed, as demonstrated by the data in *Figure 7.2*. This is thought to result from the fact that when the acid-base

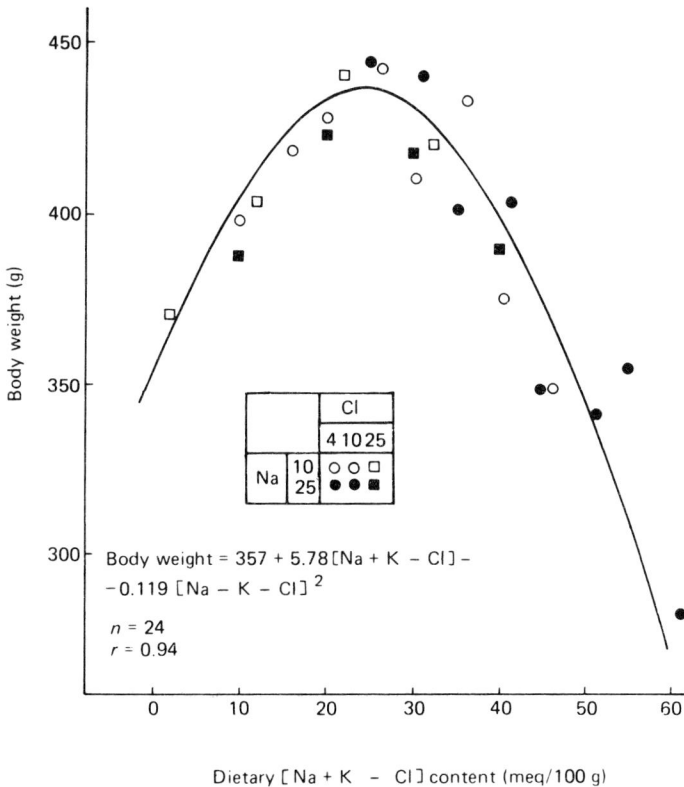

Figure 7.2   The effect of the [Na + K − Cl] content of the diet of body weight at 4 weeks of age in chicks (From Mongin and Sauveur, 1977)

balance deviates towards either an alkalotic or an acidotic situation most of the metabolic pathways cannot work under optimal conditions and are more involved in homeostatic regulation than in the growth process.

ENDOGENOUS ACID PRODUCTION AND PROTEIN SOURCES

Endogenous acid production (H^+_{endo}) is primarily dependent on the nature of the protein sources for three reasons. Firstly, the composition of the nitrogen products varies; proteins of purified soya give 3.9 meq H^+/g nitrogen, while those of beefsteak give only 2.9 meq/g. Secondly, the organic phosphorus present in proteins and lipids increases the H^+ production; egg yolk phospholipids and soya phosphatides both give 13.8 meq H^+/g N. Thirdly, the mineral composition of natural protein sources is highly variable.

From a practical standpoint the feed compounder cannot control the endogenous acid produced during the metabolism of protein, but he can, none the less, estimate the mineral composition of protein feedstuffs and hence adjust the (Na + K − Cl) content of the diet to make the relative base excess equal to zero.

The main protein source in poultry nutrition is often soya, but for several reasons other protein sources are used (e.g. rapeseed or meat meal). Substitution of soya by another protein source leads to a modification of the anion–cation balance, and usually feed manufacturers are not aware of this. In *Table 7.1* different situations are presented to illustrate how the

Table 7.1   VARIATION IN THE (Na + K − Cl) CONTENT OF THE DIET WHEN PART OF THE SOYA IS SUBSTITUTED BY ANOTHER PROTEIN SOURCE[a] (From Mongin, 1980b)

Protein sources used	*Percentage of soya/percentage of other protein source in the overall diet*				
	20/0	17/2	14/4	11/6	8/8
	(Na + K − Cl) content of the diet (meq/100 g)				
50% crude protein soya/ 75% crude protein fish meal	17.42	16.35	15.27	14.20	13.13
50% crude protein soya/ 50% crude protein meat meal	17.42	16.42	15.41	14.41	13.41
50% crude protein soya/ 42.5% crude protein sunflower or peanut meal	17.42	16.82	16.26	15.68	15.10

[a]Assuming that the overall protein content of the diets remains constant (i.e. soya + other protein = 10 per cent crude protein), the cereal plus protein sources represent 88.5 per cent of the diet and other mineral sources are not included.

(Na + K − Cl) content of the diet for a laying bird varies when part of the soya is replaced by another protein source. For example, when half of the 50 per cent crude protein soya is substituted by a 75 per cent crude protein fish meal the (Na + K − Cl) content is decreased by approximately 3 meq/100 g diet. A similar decrease in the (Na + K − Cl) content occurs with a weight-to-weight substitution of 50 per cent crude protein soya and 50 per cent crude protein meat meal. While the effect of replacing part of the soya by either sunflower or peanut meals is a little less marked, the trends are still the same.

The next point to be considered is the large variability in the mineral composition of protein sources other than soya (Mongin and Sauveur, 1976). To some extent the so-called quality of such nitrogen sources can be improved by adjustment of their mineral balance. For example, Miller (1969) showed that it was possible to improve the quality of menhaden fish meal fed to chicks by the addition of various mineral mixtures including chloride, carbonate and sulphate. When the sum (Na + K − Cl − SO_4)

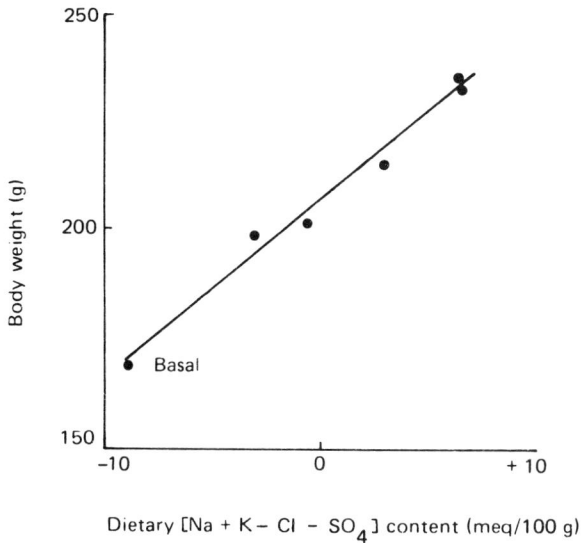

Figure 7.3 The effect of mineral content of the diet on the body weight of chicks at 2 weeks of age. The different diets were produced by adding different mineral mixtures to a basal diet to improve the so-called 'quality' of that protein source (From Miller, 1969)

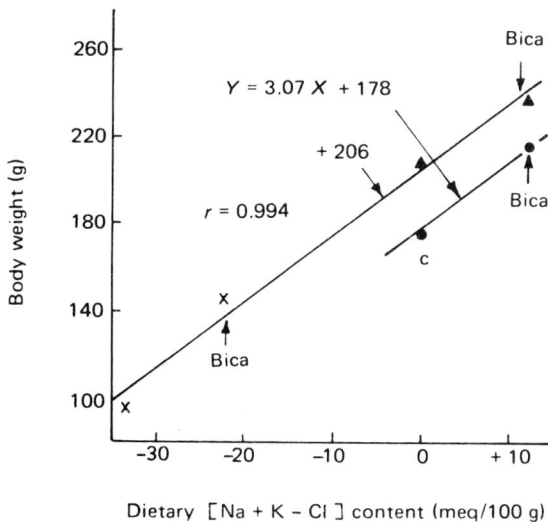

Figure 7.4 The relationship between the addition of glutamic acid, in either the free form (▲) or as the hydrochloride (X), and bicarbonate (Bica) on the [Na + K − Cl] content of the diet, and their effects on the body weight of chicks at 2 weeks of age. The sole source of protein in the basal diet was menhaden fish meal, and c represents the body weight of control chicks fed the diet in which the fish meal was supplemented with neither glutamic acid nor bicarbonate (From Miller and Kifer, 1969)

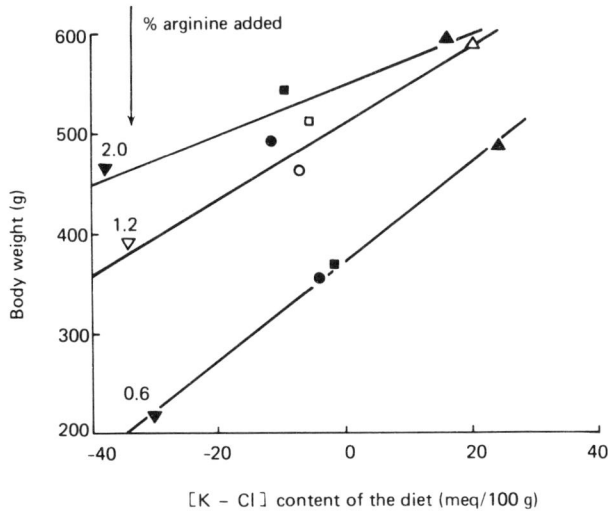

Figure 7.5 The effect of adding arginine, lysine and potassium acetate to a casein basal diet on the body weight of chicks at 4 weeks of age, and their relationship to dietary mineral balance. (● and ○, casein basal diet; ▼ and ▽, addition of 2 per cent lysine hydrochloride; ▲ and △, addition of 2.7 per cent potassium acetate; ■ and □, addition of 2 per cent lysine hydrochloride and 2.7 per cent potassium acetate) (From O'Dell and Savage, 1966)

contents of these various diets are calculated they can be shown to be linearly related to the body weight of chicks at 2 weeks of age (*Figure 7.3*).

In another experiment (Miller and Kifer, 1969), the sole addition of 1 per cent bicarbonate to menhaden fish meal improved growth by 23 per cent (*Figure 7.4*). The growth response to addition of glutamic acid was either negative when included as the hydrochloride or positive when included in the free form. As shown in *Figure 7.4* the imbalance between anions and cations explains 93 per cent of the variability in growth response, while glutamic acid accounts for only 5.8 per cent of that variability.

From these data it is clear that amino acid requirements obtained from experiments using synthetic diets or by the addition of pure amino acids as the hydrochloride have to be reconsidered. In most cases excess of chloride can prevent the beneficial effect of a single amino acid being fully expressed. A good illustration is given by the data of O'Dell and Savage (1966) in *Figure 7.5*. It can be seen that body growth in chicks is modified by variations of the anion–cation balance as well as by the addition of amino acids to a basal diet of casein. To achieve a given body growth, it is evident from these data that the arginine and lysine requirements depend on the mineral balance. The addition of 2.7 per cent potassium acetate to a diet with 0.6 per cent added arginine gives the same improvement in body growth as the addition of 2 per cent arginine plus 2 per cent lysine.

DIVALENT CATION INTAKE AND MINERAL BALANCE

Although earlier in this chapter faecal excretion was considered in the fundamental approach to mineral balance in poultry, it was ignored when developing the more practical approach to poultry feeding because it is usually quite constant. In addition, the causes of mineral partition between urine and faeces were largely ignored, particularly for divalent cations. With monovalent ions this approach is justified but it is not entirely correct for divalent ions, as shown in the following example. Let it be assumed that, on a milliequivalent basis, part of NaCl or KCl in a diet is replaced by $CaCl_2$. All three of these are neutral salts and should not therefore induce either alkalosis or acidosis. However, $CaCl_2$ does make birds acidotic, and the reason for this is that part of the calcium is not absorbed to the same extent as sodium, while chloride absorption remains constant. Indeed in the distal part of the digestive tract chloride is exchanged for bicarbonate instead of being absorbed with the calcium, with the result that calcium is excreted in the faeces as $CaCO_3$. The net result is that blood bicarbonate ions are exchanged for chloride ions and this makes the bird acidotic. The consequences of such ion exchange on chick growth are shown in *Figure 7.6*. Hydrochloric acid was mixed into a diet in increasing amounts to give

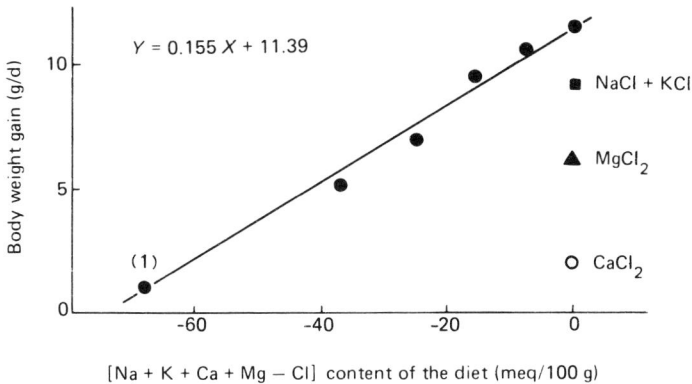

Figure 7.6 Effect of mono- and di-valent cations on body-weight gain in chicks. The mineral content of the diet was adjusted by adding varying amounts of hydrochloric acid to the diet (●) up to a maximum of 2.41 per cent added chloride (1). The effect on body-weight gain of adding the 2.41 per cent chloride ions as a mixture of NaCl + KCl (■), $MgCl_2$ (▲) and $CaCl_2$ (○) are shown (From Melliere and Forbes, 1966)

up to 2.1 per cent of added Cl and body-weight gain was shown to correlate well with the sum of dietary cations minus anions. When the same amount of chloride (2.41 per cent) was added as a mixture of NaCl and KCl, the daily body growth was similar to that obtained with no added chloride. However, addition of the 2.41 per cent Cl as calcium chloride had almost the same effect on growth as when added in the form of HCl, while the effect of adding the chloride as $MgCl_2$ was intermediate. It can be calculated that only 12 per cent of the Ca and 52 per cent of the Mg were absorbed, while 95 per cent of (Na + K) was absorbed.

Such results justify *a posteriori* why, in developing equation (7.6), calcium and magnesium were not included in the group of dietary cations. However, the effect of these ions has to be considered and the relationship must be adjusted when necessary.

INTAKE OF IONS AND WETNESS OF DROPPINGS

In birds, water intake depends on salt intake (Mongin, 1980a). Although in practice there is a well-known relationship between excess sodium intake and wetness of the litter, less attention has been paid to the other dietary ions. Potassium, as well as sodium, is another main cause of wet droppings, whereas chloride intake is unimportant (Vogt, 1971). Thus any increase in the dietary (Na + K) intake will enhance water consumption and thus water content of the excreta.

This is exemplified by the results of James and Wheeler (1949). By substituting part of the maize by soya bean meal to give diets containing 15, 20 and 25 per cent protein, they were able to increase protein intake. They observed an increase in both water intake and the amount of droppings

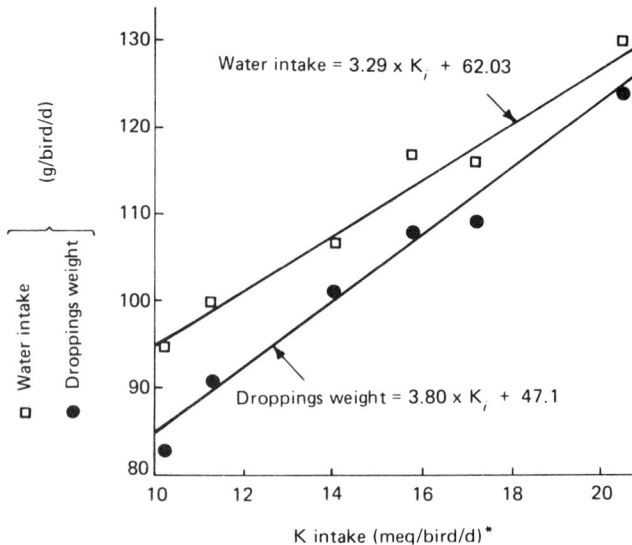

Figure 7.7   Water intake and weight of wet droppings as a function of potassium intake. Immature birds were fed 15, 20 or 25 per cent crude protein diets, the 20 and 25 per cent crude protein diets being prepared from the 15 per cent diet by substituting part of the maize with soya bean meal. The increase in both water intake and the amount of wet droppings can be explained just as satisfactorily by the increased K intake as it can by the increased protein intake suggested by the original authors. (*Assuming that the 15 per cent protein diet was achieved with 80 per cent maize + 16 per cent soya which gives 0.7 per cent K) (After James and Wheeler, 1949)

produced by growing chicks with increasing protein intake, and postulated a relationship between nitrogen and water metabolism. However, since the potassium content of soya bean meal was 6.5 times higher than that of maize, potassium intake almost doubled (*Figure 7.7*) and this may explain the observed effects on water intake and weight of dropping produced. Among the possible causes of wet droppings, dietary K load is largely ignored, although in practice it may be very important.

Limitations

As shown in this chapter the control of the sum of (Na + K − Cl) in a poultry diet can be a useful aid to improve the growth of poultry. However, this approach does have its limitations. The first is a situation where the rate of inclusion in the diet of one of the minerals in question is less than the bird's minimum requirement. In this situation the mineral deficiency is the more important phenomenon and any adjustment of (Na + K − Cl) intake will have little effect unless or until this deficiency is rectified. Adjustment of the dietary (Na + K − Cl) content will also have little effect when one of the minerals is present in such excess that it becomes toxic. In relation to this question of mineral toxicity, Sauveur and Mongin (1974) have found that chicks can tolerate K better than Na at a given value for (Na + K). Recently Talbot (1978) found that the toxic effect of excess sodium, as measured by mortality, can be overcome by the addition of potassium or chloride (*Figure 7.8*). He concluded that the ratio of K/(Na −Cl) must be higher than unity in order to avoid mortality in chicks. Since

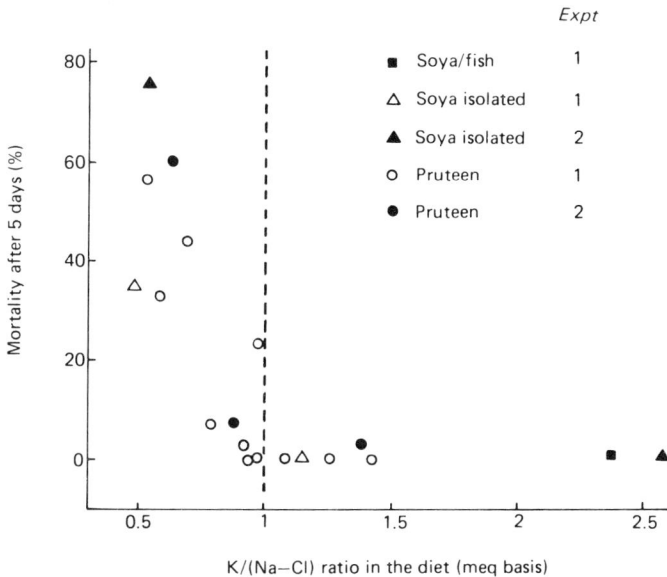

Figure 7.8   The effect on chick mortality of adding different amounts of potassium and chloride to high sodium diets. The high sodium diets were obtained by using high inclusion rates of soya isolate (20 per cent) and Pruteen (25 per cent), rather than the more normal inclusion rates of approximately 5 per cent (After Talbot, 1978)

earlier sections have advised that it is important to keep (Na + K − Cl) as close as possible to 25 meq/100 g of diet and that it is also necessary for the K/(Na − Cl) ratio to be >1, these two conditions can only be met when the K content of the diet is higher than 12.5 meq/100 g and the (Na − Cl) content lower than 12.5 meq/100 g. This conclusion is borne out by appropriate calculation from the data of Talbot (1978) and presented in *Figure 7.8*.

It is clear, therefore, that the $(Na + K - Cl)$ content of the diet can have marked effects on the growth rate of poultry. Care must, however, be taken to ensure that there is neither a deficiency nor an excess of any of these minerals since either of these situations will override the beneficial growth-promoting effects of keeping the $(Na + K - Cl)$ balance as close as possible to 25 meq/100 g diet.

References

HURWITZ, S., COHEN, I., BAR, A. and BORNSTEIN, S. (1973). *Poult. Sci.*, **52,** 903–906

JAMES, E.C., WHEELER, Jr., R.S. (1949). *Poult. Sci.*, **28,** 465

MELLIERE, A.L. and FORBES, R.M. (1966). *J. Nutr.*, **90,** 310

MILLER, D. (1969). *Poult. Sci.*, **48,** 1535

MILLER, D. and KIFER, R.R. (1969). *Poult. Sci.*, **48,** 1327

MONGIN, P. (1980a). In *Proc. 3rd Ann. Int. Minerals Conf.*, Orlando, Florida

MONGIN, P. (1980b). In *Proc. Florida Nutr. Conf.*, Orlando, 213

MONGIN, P. and SAUVEUR, B. (1973). In *Journées Avicoles et Cunicoles*, Paris

MONGIN, P. and SAUVEUR, B. (1976). *Ind. Alim. Anim.*, **9,** 49

MONGIN, P. and SAUVEUR, B. (1977). In *Growth and Poultry Meat Production*. Eds. K.N. Boorman and B.J. Wilson. *Proc. Br. Poult. Sci.*, **4,** 235

NESHEIM, M.C., LEACH, R.M., ZEIGLER Jr., T.R. and SERAFIN A. (1964). *J. Nutr.*, **84,** 361

O'DELL, B.L. and SAVAGE, J.E. (1966). *J. Nutr.*, **90,** 364–370

SAUVEUR, B. and MONGIN, P. (1974). *XVth World Poultry Congress*, New Orleans, 180

TALBOT, C.J. (1978). *Proc. Nutr. Soc.*, **37,** 53A

VOGT, H. (1971). *Arch. Geflügelk.*, **4,** 151

8

A BIOECONOMIC MODEL OF TURKEY PRODUCTION

R.E. SALMON
Research Station, Research Branch, Agriculture Canada, Swift Current, Saskatchewan

K.K. KLEIN
Research Station, Economics Branch, Agriculture Canada, Lethbridge, Alberta

Research in poultry nutrition has consisted mainly of the generation and interpretation of data on the responses of avian species to different levels of selected variables. Such response data provide guidelines for defining the dietary input specifications or nutritional requirements employed in formulating practical poultry diets. However, nutritional response data do not, by themselves, provide sufficient information to determine the most profitable diets and feeding programmes. It is quite possible that, depending on the relative costs of feedstuffs and the prices received for the final products (meat or eggs), the most profitable feeding programme will not necessarily yield the most rapid rate of growth or the greatest efficiency of feed conversion.

A major aim in poultry nutrition research should be to increase the profitability of poultry production. Research results are seldom adopted at the farm level unless they increase expected net returns. Many researchers have tended to ignore the economic factors, concentrating instead on maximizing output. Others have advised producers to minimize feed costs, without fully recognizing that different combinations of inputs can affect both the quantity and the quality of output. However, to the working nutritionist, the expected level of net returns should be the overriding concern in making decisions on diets and feeding programmes. Net returns are the difference between gross returns and cost of production. Gross returns are determined by slaughter weights, grades, prices and the level of mortality experienced, while the cost of production includes the costs of feedstuffs, poults or chicks, labour and various overhead costs.

The task of simultaneously evaluating the many factors that influence the profitability of poultry production is formidable. However, the immense calculating capacity of electronic computers has made feasible the recent development of mathematical models of animal production systems that are capable of evaluating economic data together with biological responses. Such models can aid in making rational decisions on nutrition as well as other aspects of management. This chapter describes the development of one such model and its use in the decision-making process, and suggests its further application.

First published in *Recent Advances in Animal Nutrition – 1978*

Response data

Nutritional response studies involve feeding diets containing a series of levels of the component under examination, added to a basal diet that should satisfy the known requirements of all other nutrients for the level of performance expected. Response may be measured in terms of live-weight gains, efficiency of feed conversion, carcass quality characteristics or various biochemical measures such as plasma amino acid or tissue enzyme levels. Response data are often asymptotic in form (*Figure 8.1*), although the shape of the curve may be modified by nutritional interactions or toxicity at high levels. Because of the innate variability of biological data, statistical techniques must be employed to estimate the probable point of maximum response on the curve. Their nature is outside the scope of this discussion.

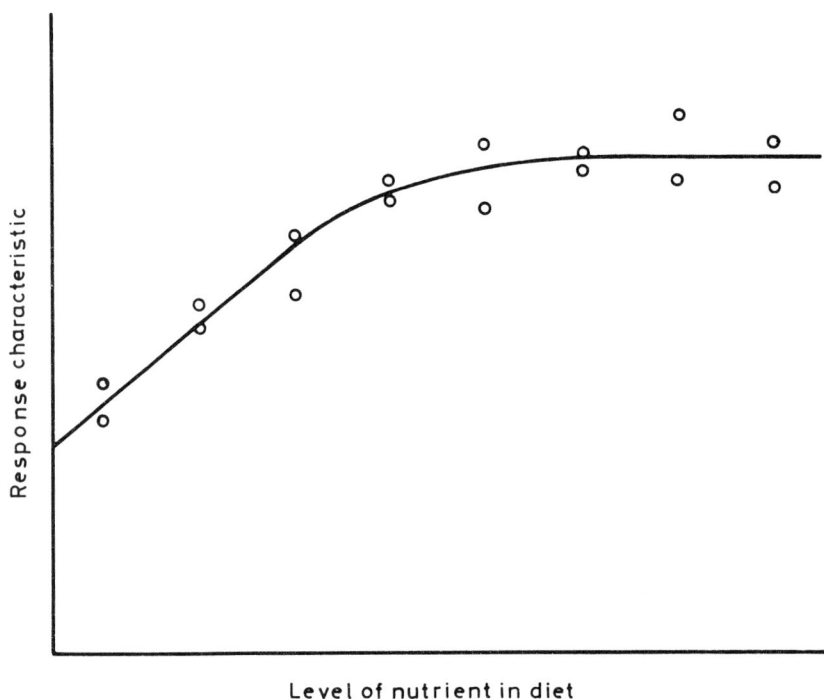

Figure 8.1 Hypothetical asymptotic response curve

The response to two factors may be determined simultaneously using a factorial experimental design. This useful technique can reveal interactions or interdependence between factors, which may be represented in the form of a three-dimensional response surface (*Figure 8.2*).

The response curves depicted in *Figures 8.1* and *8.2* may be defined by mathematical equations, which could then be used to predict the probable future response to the same experimental variables.

More complex multifactor relationships may be studied, but cannot readily be portrayed graphically. They may be expressed instead as regression equations, and thought of, if not easily visualized, as multidimensional response surfaces.

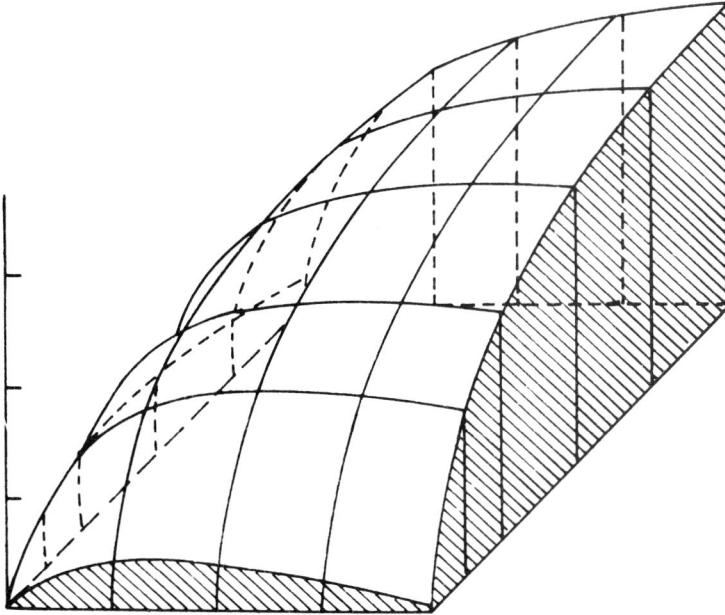

Figure 8.2 A three-dimensional resonse surface (from Ezekial and Fox, 1959)

Many practical feeding experiments are difficult to interpret in terms of simple nutritional relationships. The response to variable levels of a single feed-stuff may be influenced, not only by the amino acid levels in the diet, but also by differences in amino acid availability or amino acid interactions that are not always recognized. The method of substitution of feedstuffs may introduce differences in available energy. Carcass quality may be altered. Anti-nutritional or toxic factors, if present, may affect performance or increase mortality at high levels of some feedstuffs. Nevertheless, the complex responses observed in such trials, even if imperfectly understood, may be useful to nutritionists as guides in formulating practical poultry diets.

Attempts have now been made to incorporate the economics of poultry meat production into the analysis of nutritional data by treating costs as additional variables, which may be superimposed upon the already complex responses of growing birds to dietary variables.

The problem

The immediate objective of this study was to evaluate canola meal (low erucic acid, low glucosinolate rapeseed meal) in diets for growing turkeys. Canola meal is a domestically produced protein source that offers important economic bene-fits to Canada if it can replace a substantial portion of the country's imported soya bean meal requirements.

The purpose of this chapter, however, is to demonstrate a method of analysis with more general application. Although discussed in terms of canola meal in diets fed to turkeys, the principles may be applied to any suitable body of data.

An analysis of the competitiveness of canola meal relative to soya bean meal required two steps:

(1)   a feeding trial to obtain production response data from alternative feeding programmes, and
(2)   development of a model to evaluate the production response data in the presence of economic variables. Only by developing an analytical procedure that included economic as well as biological data could a proper evaluation of canola meal take place.

Each step will be discussed.

RAPESEED/CANOLA MEAL

Rapeseed meal (RSM) has been available in Canada since the early 1950s, as a byproduct of the oilseed crushing industry. Early attempts to use RSM as a protein source in animal feeds encountered problems, due both to improper processing and handling of the meal and to problems inherent in the meal itself.

Improved methods of processing on the part of the crushers helped to preserve the nutritive value of the meal, but its use was still restricted by the goitrogenic glucosinolates that it contained. Furthermore, RSM was relatively low in energy value in comparison with soya bean meal.

Oilseed breeding research before about 1972 was directed towards the near elimination of the long-chain fatty acid, erucic acid, from rapeseed oil. Erucic acid was suspected of various deleterious effects, which could be demonstrated by feeding at high levels to laboratory animals. More recently, however, having succeeded in achieving that objective, breeders have turned their attention to the feeding quality of RSM. Recent rape varieties contain much reduced levels of glucosinolates and are greatly improved as animal feedstuffs. Moody *et al.* (1978) reported that diets containing 250 g/kg low glucosinolate RSM derived from the variety Tower, had no adverse effect on the performance of growing turkeys to 112 days of age, in contrast with diets containing Target, a high glucosinolate type, which depressed weight gain and feed intake. In a study conducted at Swift Current, RSM of Tower and Candle varieties, fed to growing turkeys to eight weeks of age at up to 300 g/kg, had little effect on rate of gain. Weight gain was, however, depressed by RSM of the high glucosinolate type, Midas (*Figure 8.3*). Feed efficiency decreased with increasing levels of RSM, but diets containing Midas were converted to live weight less efficiently than those containing either Tower or Candle meal (Salmon, 1979). The difference may be explained by the greater true metabolizable energy (TME) of low in comparison with high glucosinolate RMS. Our own assays of Candle, Tower and Midas meals produced TME values of 10.7, 10.3 and 8.4 MJ/kg DM, respectively (unpublished data).

In more recent work, Candle rapeseed meal caused no growth depression at up to 450 g/kg of the diet of growing turkeys, to six weeks of age (Salmon, unpublished data).

Recent improvements in nutritional characteristics of rapeseed have been such, that a new name has been introduced to identify the newer varieties; hence, the term, 'canola' which refers only to rapeseed (and its products) that is low in both erucic acid and in glucosinolate.

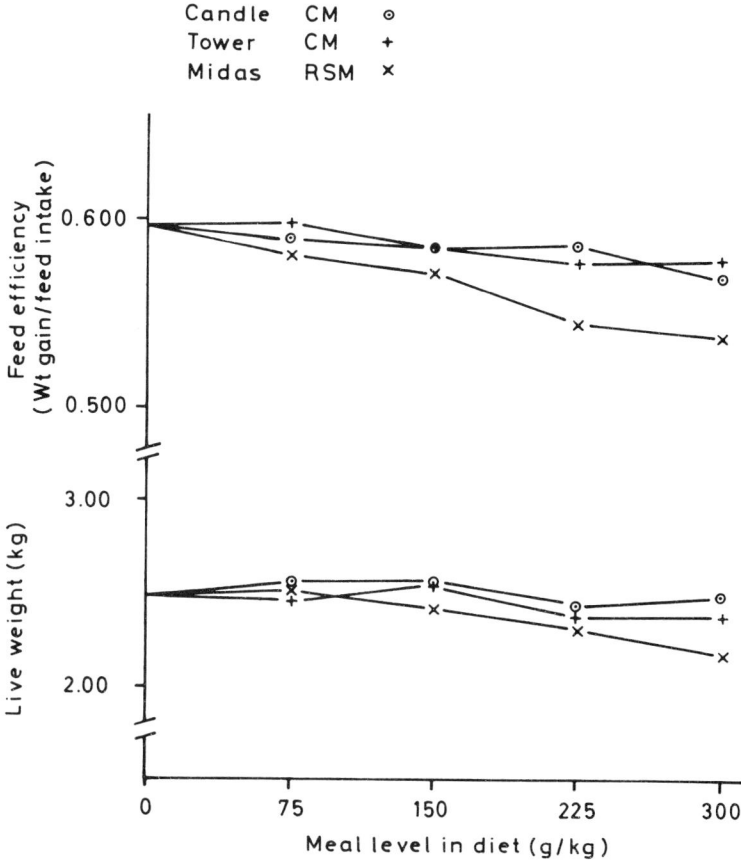

Figure 8.3 Live weight and feed efficiency of poults fed rapeseed or canola meal diets from 6 to 56 days of age

The feeding trial

As full details of the experiment have already been published elsewhere (Salmon, Klein and Larmond, 1979) the plan and results of the trial will be described only briefly here.

Candle canola meal (CM) was incorporated into diets for about 1000 small white turkeys, from day-old to market weight, at a series of levels up to 225 g/kg. The ten experimental treatments consisted of a control treatment (wheat–soya bean meal) plus three levels of CM and three nutrient density treatments, in a factorial arrangement. The nutrient density treatments were included to assess the feasibility of incorporating CM without adjusting for its low metabolizable energy content by means of added dietary fat. All diets (within each feeding period) were formulated with the same nutrient to ME ratios. Subsequent determinations of the true metabolizable energy (TME) values of the ingredients actually used, revealed that the energy value of CM had been underestimated relative to the other ingredients: the diets of the higher CM treatments therefore deviated from the ratios of energy to protein, as specified and embodied in the control diets.

The crude protein contents of the diets in each series were adjusted downwards at 3, 6, 9 and 11 weeks of age. All birds were weighed weekly and one-third of the birds in each pen were slaughtered at 12, 13 and 14 weeks and graded for finish according to Canadian grading regulations.

The live-weight gains observed were similar on all treatments to nine weeks (slightly reduced at the highest CM level), but from 9 to 14 weeks, live-weight gains were progressively reduced with each additional increment of CM in the diet, resulting in final weights that were lighter as the level of CM increased, *viz.*, 5.76, 5.70, 5.60 and 5.46 kg (*Figure 8.4*). Final 'weights were not influenced by the nutrient density treatments. The reduced gains associated with CM in the late growing period could have been related to differences in amino acid availability or to the inadvertently wider than specified energy to protein ratios of those diets. No such effect was observed by Moody *et al.* (1978) in their experiment.

Feed intake and conversion efficiency were related to the dietary TME values, although efficiency of feed conversion was somewhat improved at the low level of CM (75 g/kg) over that of the control groups.

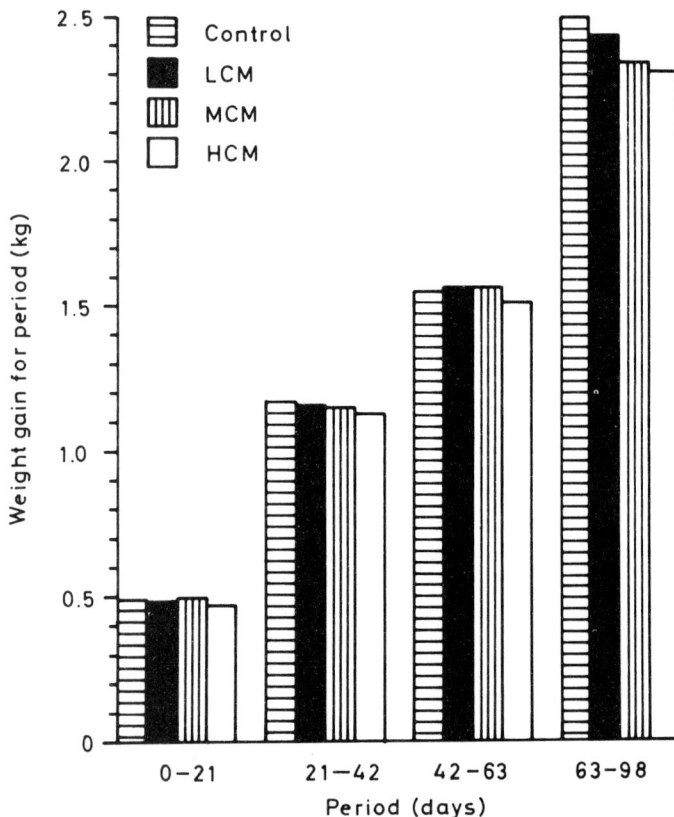

Figure 8.4 Live-weight gains, by periods, of small white turkeys fed low (LCM), medium (MCM) or high (HCM) levels of Candle canola meal. (From Salmon, Klein and Larmond, 1979)

The percentage of Grade A carcasses for finish was lower at 12 weeks with the high compared with the lower levels of CM, and also lower with the low rather than with the medium or high nutrient density treatments. However, carcass finish improved rapidly and by 14 weeks averaged over 99 per cent Grade A for all treatments.

The model

Much is read of models in research today. A model, in the present context, may be broadly defined as a mathematical representation of a set of responses. Models may be simple, as for example, equations representing the responses depicted in *Figures 8.1* and *8.2*, or very complex. Their common function is the prediction of future events on the basis of past experience.

Least-cost feed formulation, as employed by many feed manufacturers, is a form of applied modelling. It attempts to reduce the cost of animal production by meeting a set of dietary specifications as economically as possible. However, this application of the procedure does not assess the relative profitability of alternative diets or feeding systems.

Models of various types have been used in bioeconomic research, the most common being regression models, system simulation models and mathematical programming models. Each type has particular strengths: each is subject to peculiar shortcomings. The most important considerations in choosing a model of a production process are:

(1)   to select a type that considers enough detail on both the input and output sides to permit a realistic analysis, and
(2)   to select a type that does not seriously compromise the spirit in which the data were collected, through the various assumptions required to operate the model.

The linear programming (LP) procedure is often adopted because:

(1)   most agricultural experiments have a small number of discrete treatment levels, thus conforming to the structure of LP;
(2)   LP solution algorithms are available on most computer systems; they permit rapid solutions which are fairly understandable to scientists and extension personnel;
(3)   It permits an economical use of resources: the structure of a LP model is fixed and standard, thus reducing modelling time, and it can be used with available matrix generator and report writer routines.

The linear programming structure probably does fewer injustices to the interpretation of research results than do other model structures, because of the research procedures employed. If one is worried about linear combinations between two treatment levels, the relevant activities can be integerized and solved with a mixed integer programming (MIP) algorithm. If one is prepared to accept interpolations, then linear combinations may be least objectionable, because there are often too few observations on the exploratory variables to gain much confidence in any estimated curvilinear functions.

In accordance with the above principles, a LP model of a beef cattle production system was developed to maximize net returns from alternative feeding programmes (Sonntag and Hironaka, 1974). Their model considered the prices of feedstuffs and finished animals, overhead costs and the effects of the feeding programmes on animal performance and carcass quality. In poultry research, a LP model was devised to assess the influence of dietary energy levels and selling weights on the profitability of broiler chickens (Allison, Ely and Amato, 1978).

A LP model was also adopted as appropriate for the present analysis (Klein, Salmon and Larmond, 1979). The model maximizes net returns per lot to a turkey-producing operation by simultaneously considering economic and biological data. Economic data include feed ingredient prices, labour use and price, poult, energy and other overhead and operating costs, as well as prices received for each grade and weight class of turkeys. Biological data include feed intake and dietary specifications, live weights at time of sale, carcass grades and mortality. This specification differs from the standard 'least-cost' formulation system, in that different quantities and qualities of output are explicitly considered. The model essentially duplicates, in numerical form, the results that have been obtained experimentally, determining the most profitable of the options available (or combinations thereof) for any set of costs desired.

Mathematically, the model selects one or more feeding programmes that maximize net revenue (NR) where:

NR = (price per kg of each weight-grade class of turkey × weight produced of each grade) − (price of each feedstuff × quantity used in each feeding period) − (price of each type of labour × quantity used) − (cash and operating costs)

subject to:

(1)   adequate nutrients being available to satisfy minimum, exact or maximum specifications of each diet in each feeding period;
(2)   upper and lower limits on some ingredients in each feeding period;
(3)   the total quantity of feed (feed intake) being fixed for each diet (forcing the model to formulate diets with similar proportions of ingredients to those used in the feeding trial);
(4)   equal numbers of males and females at start (numbers differ at the end because of differential mortality);
(5)   an upper limit on productive capacity (number of birds).

The ten feeding programmes embodied in the feeding trial comprise the productive activities in the model. Each feeding programme has specific nutrient, ingredient and quantity (feed intake) constraints for each of five time periods, viz. 0−3, 3−6, 6−9, 9−11 weeks and either 11−12, 11−13, or 11−14 weeks. Thus, the birds can be marketed at 12, 13 or 14 weeks of age or linear combinations thereof, permitting the live weight at marketing to be specified in advance. The quantity (live weight) and quality (carcass grade) of output varies by feeding programme and by age at marketing (*Table 8.1*). An additional 15 per cent undergrades for dressing and conformation defects is assumed by the model; this is based on experience of the industry.

Table 8.1 FEED, INTAKE PER BIRD, LIVE WEIGHT AND CARCASS
GRADE DATA, CONTROL FEEDING PROGRAMME ONLY. (For complete
data see Klein, Salmon and Larmond, 1979).

Data	Period (weeks)	Male	Female
Feed intake (kg)	0−3	0.70	0.64
	3−6	1.97	1.80
	6−9	3.54	2.94
	9−11	3.32	2.79
	11−12	1.42	1.12
	12−13	1.85	1.42
	13−14	2.25	1.52
Live weight (kg)	12	5.50	4.07
	13	6.13	4.48
	14	6.78	4.76
Grade A (%)[a]	12	75.5	96.4
	13	91.2	99.8
	14	99.8	97.7

[a] Grades for finish only.

The dietary requirements of each of the feeding programmes in each time
period must be satisfied by purchase of feed ingredients, of which 13 are available in the model. The dietary specifications in the model are basically similar
to those employed by Salmon and O'Neil (1975) except that the specifications
for metabolizable energy were adjusted to reflect the values actually determined
on the ingredients used.

Feed intakes are adjusted by the model to take into account the mortality
rates experienced. Mortality was unrelated to the dietary treatments.

Table 8.2 ECONOMIC SUMMARY OF LP OPTIMAL SOLUTION

No. sold	
Males	90.554
Females	92.836
Total	183.389
Av. wt (kg)	4.785
% Grade A birds sold	69.693
Feed efficiency (kg feed/kg sold)	2.188
% mortality	8.305
% condemnation	1.000
Av. selling price ($/bird sold)	5.077
Cash costs ($/bird sold)	2.340
Cash costs ($/bird purchased)	2.150
Labour cost ($/bird sold)	0.346
Labour cost ($/bird purchased)	0.318
Feed costs ($/bird sold)	2.026
Feed costs ($/bird purchased)	1.858
Feed costs ($/kg bird sold)	0.423
Net return ($/bird sold)	0.360
Net return ($/bird purchased)	0.330
Net return ($/kg bird sold)	0.075
Age bird sold (weeks)	12.144

Table 8.3 DIET FORMULAE IN LP OPTIMAL SOLUTION

Ingredient	Cost $/tonne	Ration (g/kg feed)				
		Stage 1	Stage 2	Stage 3	Stage 4	Stage 5
Wheat	100.	493.43	562.06	629.43	735.46	773.68
Soya bean meal	250.	274.74	227.49	188.61	87.07	50.09
Canola meal	150.	75.03	75.00	75.06	74.98	74.82
Meat meal	250.	48.88	48.88	48.98	48.98	48.98
Fish meal	660.	48.88	29.38			
Dical. phosphate	267.	19.47	16.09	14.80	12.18	8.43
Calcium carbonate	28.	7.35	8.42	8.54	8.14	9.31
Salt	86.	2.45	2.45	2.45	2.45	2.45
Fat	500.	24.47	24.65	24.80	24.80	25.87
PMX72A	1329.	4.24	4.41	4.56		
PMX72B	619.				1.97	1.99
DL-methionine	3410.	1.06	1.18	1.38	1.23	1.18
Lysine HCl	3925.			1.40	2.73	3.21
Cost of ingr. ($/tonne)		200.87	182.89	166.65	151.04	146.86
Feed cost ($/tonne)		230.37	212.89	196.65	181.04	176.86

A report writer programme was appended to the model (based on a matrix generator/report writer computer package obtained from Haverly Systems Inc., Denville, New Jersey, USA). It extracts the useful information from the LP output and displays the information in easy-to-read tables. This greatly simplifies the procedure of extending results to potential users.

At present, five tables are prepared and printed. Those of greatest interest include an economic summary of the optimal solution, detailing the major cost categories and net returns (*Table 8.2*), and the composition of the optimal diet for each feeding period (*Table 8.3*). Others itemize in detail the ingredients and total weight of feed required per bird in each period, give a calculated analysis of each diet in the optimal solution, and provide information on penalty costs of alternative (non-optimal) feeding programmes.

THE ANALYSIS

A retrospective analysis of the competitiveness of CM in turkey broiler diets demonstrates the use of the model. Four price scenarios representative of prevailing price levels of the major feed ingredients in 1978, 1976, 1973 and 1966 were used. These scenarios exhibited marked differences in price ratios among the ingredients (*Table 8.4*). Diets containing 75 g/kg CM were optimal under all price scenarios, although the energy series, ages and weights at selling varied slightly.

Demand functions for CM in turkey broiler diets were derived by parametrically varying the price within each price scenario (*Figure 8.5*). The ratio of CM to soya bean meal (CM/SBM) prices, where CM entered the optimal basis, ranged from 0.82 to 1.06, much higher than the price ratios of 0.439–0.702 experienced during a 69-month period from 1969 to 1975. Over the recorded range of CM/SBM price ratios (0.439–0.702), optimal diets contained:

(1)   for the 1978 scenario: 44–193 g/kg CM, low nutrient density;
(2)   for the 1976 scenario: 75 g/kg CM, medium nutrient density;

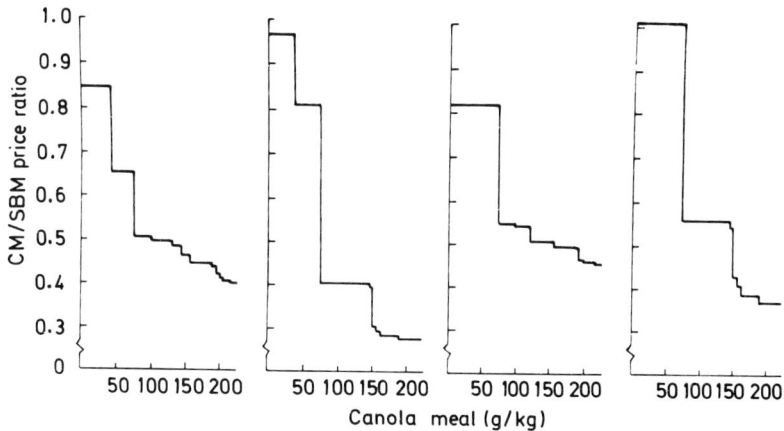

Figure 8.5 Optimal proportion of canola meal under various price ratios for canola meal and soya bean meal. Price scenarios, left to right; 1978, 1976, 1973, 1966 (from Klein, Salmon and Larmond, 1979)

(3) for the 1973 scenario: 75–225 g/kg CM, low nutrient density;
(4) for the 1966 scenario: 75–100 g/kg CM, medium nutrient density for price ratios above 0.55 and high nutrient density for price ratios below 0.55.

Although the optimal level of CM varied with market conditions, under all reasonable price ratios considered, CM could profitably replace at least some portion of the other high-protein feeds in the production system represented by the model.

Table 8.4 PRICE SCENARIOS FOR MAJOR FEED INGREDIENTS[a] AND TURKEYS ($/kg)

Item	Year			
	1978	*1976*	*1973*	*1966*
Ingredients				
Wheat	0.100	0.120	0.060	0.060
Soya bean meal	0.275	0.170	0.320	0.110
Canola meal	0.175	0.105	0.240	0.073
Meat meal	0.285	0.180	0.340	0.112
Herring fish meal	0.660	0.285	0.500	0.200
Fat	0.500	0.300	0.300	0.140
Turkey (live weight)				
Grade A	1.102	1.012	0.836	0.550
Grade B	1.013	0.920	0.770	0.480

[a] All other ingredient prices were held at 1978 levels

The optimal nutrient density levels selected by the model are of interest. In years of relatively high-priced fat (1978) or SBM (1973), the model selected the low-density option as most profitable. A higher-priced cereal component, relative to SBM (1976), favoured an increase in nutrient density, as did lower-cost fat (1966). Many additional model runs have favoured the low-density option, in spite of the increased feed intake and slightly longer time required to reach the specified slaughter weight. The implication for feed manufacturers is that it may not always be necessary to compensate fully for the lower ME content of CM by means of increased fat in the diet. Similarly, even if CM is not a factor under consideration, diets of lower nutrient density may often be more profitable than those of higher nutrient density.

The output of the model also considers the influence of varying ingredient levels and costs on net returns to the producer. To illustrate, with wheat, SBM and CM priced at $100, $250 and $150/t, respectively, the optimal solution included CM at 75 g/kg, and yields net returns above costs of $330/1000 birds, some $54 greater than the control (no CM) diet. By forcing the solution to accept CM at a series of levels ranging from 0 to 200 g/kg, and varying the cost of CM, we obtain a family of strongly curved responses (*Figure 8.6*).

$$y = -3.921 + 0.01787C + 2.498L - 0.009952CL - 0.004562L^2$$

where C = cost ($/t); r = 1.000; L = CM (g/kg)

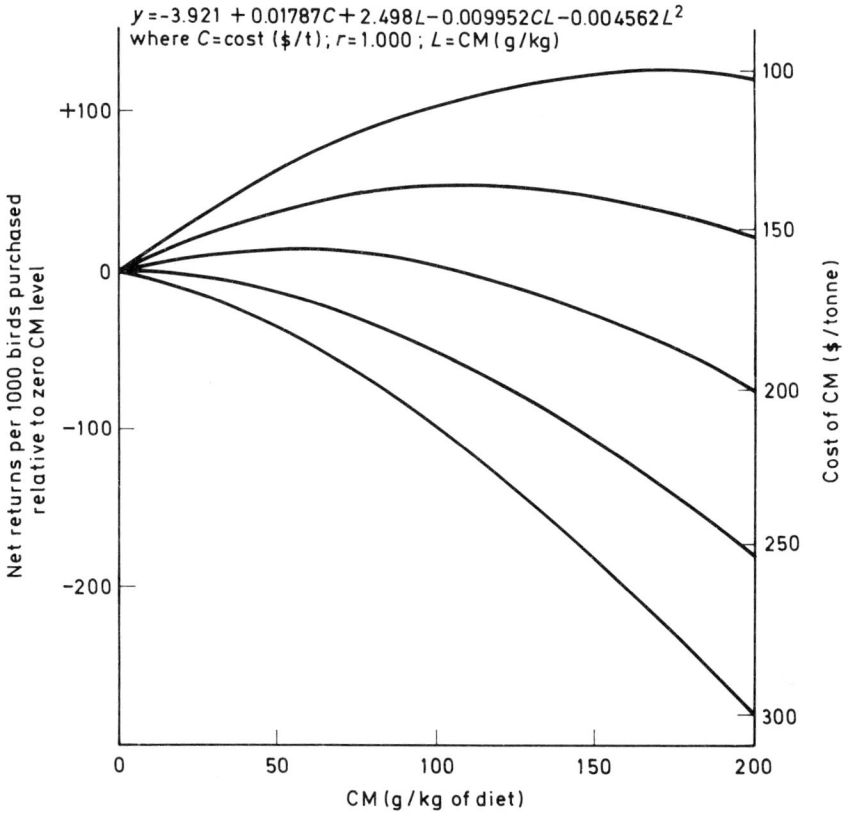

Figure 8.6 Effect of cost and level of includsion of canola meal on relative net returns of turkey broilers

The pronounced curvature of the responses may be attributed largely to two features of the data from which the model was derived: (1) the improved feed efficiency, relative to the controls, that was observed at the low level of dietary CM, and (2) the decreased growth rate and therefore increased feed intake needed to reach slaughter weight at higher CM levels. Further trials could lead to revision of those aspects of the data and modify the conclusions derived from the model. More detailed analysis of the value of CM in turkey diets would therefore be premature, pending refinement and validation of the model.

Conclusions and applications

At the present stage, the development of a method of analysis is more important than the specific conclusions derived therefrom. The questions raised above, regarding the validity of aspects of the model, demonstrate both the weakness and strength of deterministic models that are based on biological data rather than theoretical considerations. The input data may be subject to the vagaries of biological phenomena so familiar to researchers. Things happen in experiments that may not always be readily explained or duplicated. Models must therefore be validated by comparison with other data. If appropriate data are not available, further experiments are needed. As the validity of the model is confirmed, or its structure modified in the light of new information, its conclusions may be regarded with increasing confidence. On the other hand, complex responses involving performance and product quality variables are not readily predicted from known nutritional relationships, and require the intervention of living animals for their expression. Bioeconomic models then provide a vehicle for examining the economics of production systems with consideration of those variables.

The use of a matrix generator (in this case, also based upon the Haverly Systems Inc. operating procedure) greatly simplifies the data manipulations involved in building the model. It permits data to be entered in tabular form, directly from summarized experimental results. With only minor modifications, the matrix generator can accommodate a wide range of poultry nutrition experiments.

The items that must be specified in the matrix generator include:

(1) number of feeding periods;
(2) number and nutrient composition of ingredients in each feeding period;
(3) number of nutrient constraints;
(4) number of feeding programmes;
(5) number of marketing ages;
(6) nutrient constraints (including feed intake) by feeding period for each feeding programme;
(7) mortality rates by feeding programme;
(8) ratio of males to females;
(9) labour and cash use by feeding programme;
(10) production capacity;
(11) prices of ingredients and products (by grade and weight range).

A potentially valuable use of a model of this type is in establishing priorities for research. Nutritionists can examine the economic potential of hypothetical feeding programmes in areas where they lack specific knowledge. For example, it may be advantageous to vary the proportion of CM in poultry diets by feeding period. By freeing the constraints enforcing the same concentration of CM in each feeding period, one could determine the utility or futility of committing research resources to a costly feeding trial.

The model has important application in the extension area. Too often, poultry management and feed nutrition recommendations have been made in the absence of economic criteria and, when economic criteria have been used, they have often been of a least-cost rather than a maximum-profit nature.

Future plans for this modelling approach include, in addition to the validation and further exploration of the present model:

(1)   construction of a similar model for broiler chickens (now being developed);
(2)   development of a more general turkey broiler model from existing data on nutritional variables (protein, fat, energy) as they affect performance and carcass quality;
(3)   a model incorporating data on feeding programmes and protein levels in diets for heavy-type turkeys.

The invitation from the editors of this book requested a chapter illustrating the use of response data in making economic judgments about nutrition. It is hoped that ideas discussed here serve to stimulate increased awareness of the importance of considering all costs, as well as the quantity and quality of output, when making nutritional judgments and recommendations.

References

ALLISON, J.R., ELY, L.O. and AMATO, S.V. (1978). Broiler profit maximizing models. *Poultry Sci.,* 57, 845–853

EZEKIAL, M. and FOX, K.A. (1959). *Methods of Correlation and Regression Analysis,* 3rd Edn. New York; John Wiley and Sons

KLEIN, K.K., SALMON, R.E. and LARMOND, M.E. (1979). A linear programming model for determining the optimal level of low glucosinolate rape seed meal in diets of growing turkeys. *Can. J. Agric. Econ.,* 27, 61–73

MOODY, D.L., SLINGER, S.J., LEESON, S. and SUMMERS, J.D. (1978). Utilization of dietary Tower rapeseed products by growing turkeys. *Can. J. Anim. Sci.,* 58, 585–592

SALMON, R.E. (1979). Rapeseed meal in turkey starter diets. *Poultry Sci.,* 58, 410–415

SALMON, R.E. and O'NEIL, J.B. (1975). Evaluation of turkey broiler feeding programmes by biological assay and chemical analysis. *Can. J. Anim. Sci.,* 55, 451–459

SALMON, R.E., KLEIN, K.K. and LARMOND, E. (1979). Low glucosinolate rapeseed meal in turkey broiler diets of varying nutrient density. *Poultry Sci.,* 58, 1514–1523

SONNTAG, B.H. and HIRONAKA, R. (1974). A computer model for ration formulation and selection of a feeding program for cattle. *Can. Farm Econ.,* 9, 9–15

9

THE NUTRITIONAL REQUIREMENTS OF TURKEYS TO MEET CURRENT MARKET DEMANDS

C. NIXEY
British United Turkeys Ltd, Tarvin, Chester, UK

Market opportunities

Many of the feed mills in Britain produce little if any turkey food. They may be ignoring a market opportunity that already exists and certainly ignoring a potentially very significant market in the future. Of the various feed sectors, turkey feed had by far the largest percentage increase in 1986 over the previous year (*Table 9.1*). It is an area which could be exploited by general farmers but without encouragement from outside forces such as feed companies, they will miss out on the opportunity. Very few hotels or restaurants have turkey meat regularly on the menu, yet it is now an inexpensive meat, with a broad popular appeal. The reason it is not on menus is that restaurants cannot be assured of a regular supply of turkeys delivered to their door. It is not an area that the larger operations are interested in because of transport costs. However, a very nice income can be obtained by building up a local trade in large turkeys. The interest from the viewpoint of the feed industry is that it is a growth area offering all the year round business.

The weakness of empirical experimentation

A review of the published literature reveals that not much work has been published on turkey nutrition and what has been is open to criticism. The situation is further complicated by the tremendous genetic progress achieved in the last 20 years (*Table 9.2*). The little work that has been carried out on females in the past must be viewed in the light of the fact that they now grow as fast as large type males of 1969.

The initial comments on nutritional requirements relate to turkeys in general and are followed by comments on more specific requirements.

The turkey is a high protein, low fat containing animal. It is also a rapidly growing animal with the males multiplying their hatching weight by 300 times in 140 days. It is not surprising that, in a variety of situations all over the world, a growth response to increasing protein intake, or more specifically the first limiting amino acid, can normally be seen. Unfortunately very little work on the growing turkey's requirements for specific amino acids has been carried out. Most of that has been open to criticism for faulty design or interpretation. Most of the papers are on the lysine requirements with a lesser number on the sulphur amino acids and very few on the others.

First published in *Recent Advances in Animal Nutrition – 1988*

Table 9.1 UK COMPOUND FEED PRODUCTION (MILLION TONNES)

Type of food	1985	1986	% change 1985/86
Calf	0.41	0.40	− 2.5
Dairy	3.33	3.63	+ 9.0
Other cattle	0.66	0.74	+ 12.1
Pig starters	0.18	0.20	+ 5.9
Pig breeding	0.67	0.67	+ 0.2
Other pig	1.22	1.27	+ 4.3
Broiler chicken	1.27	1.36	+ 6.8
Layer	1.09	1.09	—
Turkey	0.40	0.51	+ 27.5

Source: MAFF

Table 9.2 GENETIC PROGRESS OF TURKEYS

Year	Brand name	18-week liveweight (kg) Males	Females
1966	Triple 6	9.45	6.75
1969	Triple 6	9.77	6.95
1972	BUT 6	10.27	7.27
1974	BUT 6	10.68	7.56
1977	BUT 6	10.96	7.80
1981	BIG 6	12.67	8.78
1982	BIG 6	12.80	8.84
1984	BIG 6	13.40	9.13
1986	BIG 6	13.96	9.88

(Based on the published performance goals of British United Turkeys Ltd)

LYSINE EXPERIMENTS

There are very few papers on lysine requirements of the female turkey. *Table 9.3* shows work on male turkeys with the author's conclusions expressed as g amino acid/ MJ ME. A decreasing requirement with age is seen but there is considerable range in the indicated lysine requirements. However, if the highest indicated requirements are plotted against age, the correlation is very good, decreasing by 5.7 mg lysine/MJ/day.

SULPHUR AMINO ACID EXPERIMENTS

It will be seen in *Table 9.4* that even fewer papers were found on total sulphur amino acids (TSAA). There is a further complication in that there is a requirement for methionine in addition to its contribution to the TSAA. However, it was not found feasible to separate out the methionine data. A contribution can be made to the total sulphur amino acids by cystine. Behrends and Waibel (1980) estimated that between 55 and 58% of the total sulphur amino acids, depending on age, could be cystine. There has been very little work on other amino acids but these are referred to later in this chapter.

Table 9.3 LYSINE REQUIREMENTS OF MALE TURKEYS

Approximate age range (days)	No. of data sets	Mean indicated lysine requirement (g/MJ ME)	Range of indicated requirement (g/MJ ME)	
			Low	High
0–28	6	1.236	1.135	1.328
28–56	3	1.254	1.219	1.278
56–84	10	0.896	0.729	1.110
84–112	3	0.836	0.646	0.978
112–140	9	0.557	0.478	0.717
140–168	2	0.512	0.470	0.569

Table 9.4 TOTAL SULPHUR AMINO ACID (TSAA) REQUIREMENTS OF TURKEYS

Approximate age range (days)	No. of data sets	Mean indicated TSAA requirement (g/MJ ME)	Range of indicated requirement (g/MJ ME)	
			Low	High
0–28	5	0.783	0.664	0.876
0–56	2	0.865	0.814	0.915
28–56	2	0.740	0.696	0.785
56–84	4	0.568	0.492	0.647
84–112	0	—	—	—
112–140	2	0.312	0.282	0.341
140–168	0	—	—	—

The Reading model

The Reading model produced by Fisher, Morris and Jennings (1973) to analyse and describe the response curve of a laying flock of hens to essential amino acids is well known. It has since been proposed for use with growing birds by Clark *et al.* (1982) and Fisher and Emmans (1982). If the broken line method is used to fit data, the indicated requirement for maximum bodyweight gain occurs at a lower level than that indicated by the Reading model. In the Reading model, the amino acid requirement for maximum bodyweight gain is calculated as $AA = a.\Delta W + b\bar{W}$ where ΔW is the potential bodyweight gain per day and $\bar{W}$ is the mean bodyweight. The constant a is defined as the amino acid intake (mg/day) per unit (g/day) bodyweight gain. The constant b is defined as the amino acid intake (mg/day) per unit (g) of bodyweight maintained. These constants are derived from fitting a curve as described by Fisher, Morris and Jennings (1973).

Very few of the papers reviewed gave sufficient data for use in the Reading model for lysine response and even fewer for TSAA. Those that did gave the a and b values as shown in *Table 9.5*. These confirm that more lysine than TSAA is required to produce 1 g of bodyweight gain. On the other hand, more TSAA than lysine is required to maintain 1 g of bodyweight for one day. The relatively greater significance of the maintenance requirement as the birds get older is reflected in the increased TSAA requirement.

Table 9.5 READING MODEL *a* AND *b* VALUES[1] DERIVED FROM SUITABLE DATA IN PUBLISHED PAPERS

Age (days)	a		b (× 10³)	
	Lysine	*TSAA*	*Lysine*	*TSAA*
0–28	22.32	13.75	22.8	64.1
28–56	19.91	ND[2]	17.2	ND
56–84	21.16	11.281	19.3	98.0
84–112	ND	ND	ND	ND
112–140	22.96	10.941	10.5	118.6

[1] *a* is the amino acid intake (mg/day) per unit (g/day) bodyweight gain.
  b is the amino acid intake (mg/day) per unit (g) of bodyweight maintained
[2] ND, no data

Input and output procedures

The advantage of the Reading model is that the constants *a* and *b* can be used to predict the requirements of birds with different genetic potentials for bodyweight gain. An example of the input and output predictions that can be obtained is shown in *Tables 9.6, 9.7* and *9.8*.

The model predicts the bodyweight gain to be expected from amino acid intake or—vice versa—amino acid required to produce the bodyweight gain. This type of information has a very practical value provided that the bodyweight gain and food intake are measured to an age. If one of these amino acids is the first limiting amino acid, it will pinpoint which. It will indicate the likely growth response to be expected

Table 9.6 READING MODEL: AMINO ACID INPUT AND BODYWEIGHT GAIN OUTPUT PREDICTIONS FOR BIG 6 MALES, 0–4 WEEKS OF AGE

Bodyweight gain (g/bird day)	Lysine required (g/bird day)	TSAA required (g/bird day)	TSAA as % of lysine
14	0.328	0.246	75.0
16	0.372	0.263	70.7
18	0.418	0.290	69.4
20	0.462	0.318	68.8
22	0.508	0.346	68.1
24	0.550	0.373	67.8
26	0.596	0.400	67.1
28	0.640	0.428	66.9
30	0.685	0.456	66.6
32	0.730	0.483	66.2
34	0.775	0.512	66.1
36	0.820	0.540	65.9
38	0.870	0.570	65.5
40	0.920	0.602	65.4
42	0.973	0.634	65.2
44	1.023	0.665	65.0
46	1.080	0.703	65.1
48	1.320	0.860	65.1
48.2	1.390	0.900	64.7

Based on *a* and *b* values shown in *Table 9.5*, a 55 g day old poult and a potential daily weight gain for the period of 48.2 g

Table 9.7 READING MODEL: AMINO ACID INPUT AND BODYWEIGHT GAIN OUTPUT PREDICTIONS FOR BIG 6 MALES, 8–12 WEEKS OF AGE

Bodyweight gain (g/bird day)	Lysine required (g/bird day)	TSAA required (g/bird day)	TSAA as % of lysine
80	1.812	1.506	83.1
84	1.892	1.552	82.0
88	1.980	1.596	80.6
92	2.066	1.642	79.5
96	2.150	1.688	78.5
100	2.236	1.735	77.6
104	2.322	1.784	76.8
108	2.410	1.834	76.1
112	2.498	1.886	75.5
116	2.592	1.942	74.9
120	2.690	1.992	74.1
124	2.790	2.054	73.6
128	2.890	2.105	72.8
132	2.985	2.154	72.2
136	3.083	2.206	71.6
140	3.200	2.290	71.5
144	3.620	2.585	71.4
146	3.950	2.770	70.1
146.4	4.210	2.950	70.1

Based on *a* and *b* values shown in *Table 9.5*, 8 week liveweight of 4.1 kg and a potential daily weight gain for the period of 146.4 g

Table 9.8 READING MODEL: AMINO ACID INPUT AND BODYWEIGHT GAIN OUTPUT PREDICTIONS FOR BIG 6 MALES, 16–20 WEEKS OF AGE

Bodyweight gain (g/bird day)	Lysine required (g/bird day)	TSAA required (g/bird day)	TSAA as % of lysine
74	1.845	2.465	133.6
78	1.938	2.508	129.4
82	2.030	2.554	125.8
84	2.075	2.576	124.1
88	2.168	2.630	121.3
92	2.260	2.667	118.1
96	2.353	2.713	115.3
100	2.445	2.760	112.9
104	2.542	2.808	110.5
108	2.640	2.858	108.3
112	2.743	2.912	106.2
116	2.850	2.966	104.1
120	2.960	3.027	102.3
124	3.065	3.094	100.9
128	3.163	3.172	100.3
132	3.284	3.270	99.6
136	3.590	3.536	98.5
139	4.130	3.980	96.4
139.3	4.370	4.170	95.4

Based on *a* and *b* values shown in *Table 9.5*, 16 week liveweight of 12 kg and a potential daily weight gain for the period of 139.3 g from increasing the intake of an amino acid. This can be achieved by altering the formulation or increasing the food intake somehow.

LYSINE AND TOTAL SULPHUR AMINO ACID RELATIONSHIPS

The relationship between TSAA and lysine is of importance. At maximum growth rate it is 64.7% for 0–4 weeks. However, from 8–12 weeks, the TSAA requirement relative to lysine has increased to 70.1%. This is even more pronounced from 16 to 20 weeks (*Table 9.6*). If these values are correct, it means that there is concern about TSAA levels because no one appears to be using similar TSAA levels to lysine levels at these ages.

A further reason for looking at TSAA levels, is the effect seen if the turkeys have a setback and do not grow to their potential for a period. This has the effect of increasing the relative influence of maintenance as opposed to growth. This increases the TSAA relative to the lysine requirement. It was also seen at the other ages. In situations of high stocking density where the effect becomes progressively more stressful, it would seem logical to use a higher TSAA level relative to lysine than in situations where good growth rate is expected. To make the best use of our knowledge, more information is required during the growing cycle, doing check weighings and measuring feed consumption. A lot of information is collected on breeding birds and it is time to do the same with growing birds.

Amino acid profiles

It is important to become conscious of the amino acid profiles of turkey diets using lysine value as the marker at 100 as the intake of one amino acid cannot be viewed in isolation. Interrelationships exist between amino acids.

Two groups of workers, one in Israel (Hurwitz *et al.*, 1983) the other in Edinburgh (Fisher and Emmans, 1983) have tried to overcome the problems of empirical experimentation by producing calculated models for the growing turkey. Their predictions (*Table 9.9*) serve as a useful introduction to the subject of amino acid profiles. The obvious amino acid on which to base the profile would seem to be lysine, being often first limiting and lacking the complications of total sulphur amino acid utilization.

The significance of these profiles is that they may indicate the likely first limiting amino acids with the use of particular ingredients. The profile of common ingredients is shown in *Table 9.10*. It will be seen that in high protein diets used early in life which

Table 9.9 A COMPARISON OF THE AMINO ACID PROFILES PREDICTED BY THE EDINBURGH (EDIN) AND ISRAELI (ISR) MODELS FOR LARGE MALE TURKEYS

Age (weeks)	Methionine Isr	Edin	Total SAA Isr	Edin	Tryptophan Isr	Edin	Threonine Isr	Edin
4	36.0	36.0	70.8	64.0	14.7	17.3	78.9	70.3
8	36.5	36.5	79.3	64.5	15.1	17.2	82.5	70.0
12	37.2	37.4	90.3	65.4	16.0	17.1	87.0	69.5
16	39.1	38.6	101.0	66.7	16.9	17.0	91.3	68.8
20	39.3	40.4	94.2	68.6	16.2	16.7	89.0	67.9

Table 9.10 THE AMINO ACID PROFILES OF INGREDIENTS COMMONLY USED IN
TURKEY DIETS

Ingredient	Lysine	Methionine	TSAA	Tryptophan	Threonine
Maize	100	83.3	145.8	37.5	162.5
Wheat	100	48.3	119.4	38.7	103.2
Soya	100	22.2	45.7	22.2	61.7
Fish (white)	100	37.1	53.6	14.8	56.7
Meatmeal	100	25.0	47.0	12.0	58.0
Sunflower	100	50.0	100.0	45.0	105.0
Indicated profiles for turkey diets					
4 weeks					
Israeli	100	36.0	70.8	14.7	78.9
Edinburgh	100	36.0	64.0	17.3	70.3
20 weeks					
Israeli	100	39.3	94.2	16.2	89.0
Edinburgh	100	40.4	68.6	16.7	67.9

NRC (1984)

contain large quantities of soya, fishmeal and meatmeal, there is a very real possibility
of the threonine level being well below even the Edinburgh predictions. Tryptophan is
unlikely ever to be the first limiting amino acid.

TYPES OF DELETERIOUS AMINO ACID PROFILES

There are three types of deleterious amino acid profiles. The most common causes a
decrease in food intake and growth rate which can be prevented by a supplement of
the limiting amino acid.

The second problem which is amino acid toxicity may be common in a mild form.
In this problem an excess of an amino acid reduces food intake and hence growth
rate. The effects of toxicity wear off in that food intake gradually comes back to
normal. Each time the food is changed, the amino acid profile will change and risk the
effects of a different amino acid being in excess. It is an argument for not changing
diets frequently. The third problem which is amino acid antagonisms is more
complicated. Sometimes the adverse effects of an excess of an amino acid can be
rectified by the addition of another amino acid. Of most practical significance in the
UK is an excess of lysine which can be alleviated by increasing the arginine level. This
relationship has been shown to exist in other animals and was illustrated in young
turkeys by D'Mello and Emmans (1975). Their results are shown in *Table 9.11*. It will
be seen that as the dietary lysine level increases so the optimum arginine level
increases. There are insufficient data points to assess the precise relationship required
between arginine and lysine but the ratio would appear to be around 1.1 arginine to
1.0 lysine. There is a poverty of other work on the subject so it is an area which needs
investigation because practical starter diets are usually below this ratio.

Commercial recommendations

The commercial nutritionist must take decisions about his formulations and *Table*

Table 9.11 WEIGHT GAIN (g/BIRD DAY) OF TURKEY POULTS FED DIETS OF DIFFERENT LYSINE AND ARGININE CONTENTS FROM 7 TO 21 DAYS

Dietary arginine content (%)	Dietary lysine content (%)		
	1.05	1.30	1.55
1.00	15.0	18.3	20.8
1.25	14.1	23.3	24.2
1.50	15.1	24.1	26.6
1.75	13.7	21.5	29.4

D'Mello and Emmans (1975)

Table 9.12 SUGGESTED DIETS FOR BIG 6 MALES WHERE THE CONDITIONS AND PELLET QUALITY ARE GOOD

Age fed (weeks)	0–4	4–8	8–12	12–16	16–20	20–24
Nutrient (g/MJ ME)						
Lysine	1.57	1.34	1.10	0.89	0.75	0.65
Methionine	0.57	0.53	0.47	0.44	0.43	0.42
Methionine + cystine	1.02	0.94	0.82	0.76	0.75	0.71
Tryptophan	0.27	0.23	0.19	0.15	0.13	0.11
Threonine	1.00	0.86	0.75	0.58	0.48	0.42
Arginine	1.69	1.46	1.21	1.02	0.88	0.80
Calcium	1.10	1.05	0.95	0.85	0.80	0.70
Available phosphorus	0.62	0.60	0.55	0.50	0.45	0.40
Sodium	0.13	0.13	0.13	0.13	0.13	0.13
NACL	0.30	0.30	0.30	0.30	0.30	0.30
Essential fatty acids	1.27	1.09	—	—	—	—
ME (MJ/kg)	12.0	12.1	12.2	12.4	12.6	12.8

9.12 shows the basis of the author's formulations for use with Big 6. These recommendations however are not written in stone. They must be adjusted according to the situation. This is best assessed by comparing the liveweight for age against the breed's target liveweights.

Food intake factors

The most common formulation fault is the use of too wide ME to amino acid ratio. Even the best formulation cannot give optimum results if it is not consumed. Liveweights are often reduced because of adverse feed intake factors.

Turkey growers are most sensitive to pellet quality. Recent work in the USA (Hassibi, personal communication) illustrates the influence the form of the feed can have on results (*Table 9.13*).

High stocking densities will reduce liveweights. The main effect is probably via reduced food intake. This may occur because the temperature in the microclimate surrounding individual birds is high or because of the competition to get to feeders and water points.

FACTORS AFFECTING FOOD INTAKE OF YOUNG TURKEYS

Recent work (Nixey, 1987) shows the influence of various factors on four week

Table 9.13 THE INFLUENCE OF THE FORM OF THE FEED ON GROWTH PERFORMANCE

	Liveweight at 20 weeks (kg)	*Food conversion ratio* (kg feed/kg liveweight gain)
Mash	12.58	3.31
Poor pellet	12.77	3.22
Good pellet	13.08	3.18

liveweights and also the influence of four week liveweights on subsequent growth rates. In this experiment, all birds received the same formulation. Three types of influence were investigated (a) crumb versus mash feeding, (b) day-old versus seven-day debeaking, and (c) supplementary food in floor trays for seven days versus plastic hanging feeders alone. The crumb-fed birds weighed 23% heavier than the mash-fed birds at 28 days. A 16% difference was present even at seven days. Day-old debeaking had an adverse effect of 6% with mash-fed birds and no effect on crumb-fed birds. Most surprisingly supplementary food in floor trays gave an 8% improvement in liveweight. There was a 30% difference in 28-day liveweight between the best and worst combinations which illustrates the influence of food intake factors.

Early bodyweight and subsequent growth rate

The logical question to ask was 'what effect does 28-day liveweight have on subsequent growth rate?' To investigate this, at 28 days each treatment was split into either a high or low protein feed programme. The results are shown in *Table 9.14*.

It will be seen that there is no evidence of compensatory growth even on the high protein programme. The explanation for the difference compared with the classic work by Auckland, Morris and Jennings (1969) probably lies in the fact that the Big 6 males grow more than twice as fast as the birds used in their work and are also later maturing.

Table 9.14 INFLUENCE OF 4 WEEK WEIGHT ON SUBSEQUENT GROWTH RATE OF BUT BIG 6 MALE TURKEYS

		Liveweights (kg)				
	4 weeks	*8 weeks 1 day*	*12 weeks 2 days*	*16 weeks 1 day*	*20 weeks*	*24 weeks 1 day*
(a) High protein programme						
Protein % 0–28 days	30	27	23	18	18	18
Crumbs	1.02	4.16	8.76	12.73	16.54	20.00
Mash	0.78	3.67	8.21	11.85	15.63	18.79
Diff.	0.24	0.49	0.55	0.88	0.91	1.21
(b) Low protein programme						
Protein % 0–28 days	30	23	18	18	14	14
Crumbs	1.02	3.96	8.12	10.95	14.34	16.74
Mash	0.78	3.42	7.34	10.06	12.91	14.79
Diff.	0.24	0.54	0.78	0.89	1.43	1.77

Nixey (1987)

Leg problems

If however a strain of turkey or a particular farm has a history of leg problems, an improvement in the leg problems and hence subsequent growth will be seen if the growth rate is slowed down prior to 12 weeks, particularly from six to 12 weeks. The results however will be inferior to those achieved if the cause of the leg problems can be removed and the birds grown fast throughout.

One of the keys to the problem would appear to be to prevent diarrhoea particularly between six and 12 weeks. The skeleton of the turkey, particularly the male is growing very fast at that time. The adverse effect may be malabsorption of nutrients or it may be that the weakness of the birds, illustrated by the shaky legs and reluctance to walk far, is causing abnormal bone growth. Martland (1984) created leg problems just by damping down the litter so that diarrhoea may also be having an influence by causing wet litter. The suggested causes of diarrhoea are several. It may be of infective origin (Saif, 1987). Nutritional aspects can also be the cause. Among the nutritional aspects under suspicion are (1) marked changes in the ingredients used between formulations, (2) the use of new season cereals, (3) the use of manioc, (4) poor quality fats, (5) an adverse balance between K, Na and Cl, (6) excess protein. Doubtless there are other suggestions for causes of diarrhoea.

Nutrition and meat yields

Increasingly the turkey is being used as a source of meat and not sold as a whole bird. In this situation, the yield of breastmeat is very significant as it is usually worth at least twice the value of dark meat. The nutritional programme will have a marked influence on the breastmeat yield. In general the better the growth rate achieved the better will be the breastmeat yield. However the yield of breastmeat is very sensitive to the rate of growth in the latter period. This is illustrated in *Table 9.15*. The small breastmeat yield of the males grown poorly between 16 and 20 weeks was not unexpected because at this age the breast muscles develop rapidly under a normal growth regime. A common problem occurs when the males are moved to a finisher type diet too early. The definition of too early will of course depend upon the nutrient content of the finisher diet. However a 15% protein finisher type diet is bad for a large type turkey before 20 weeks of age.

Traditional farm-fresh turkeys

GROWTH RATE

Of most interest is probably the Christmas turkey market. The most common complaint that the feed industry receives from this type of customer is the turkeys have grown too big. Customers order a turkey of a particular weight, most commonly around 14 lb (6–6.5 kg) plucked weight. They do not take kindly to being offered a bird 25% bigger than they ordered which may be the case if the farmer only has one flock being grown for Christmas. Often he has done his planning wrong, or was unable to obtain poults on the date required. Also, almost invariably, turkeys grown on the general farm for Christmas will exceed their breed target liveweights by a considerable margin. They will have received a lot of attention, been given ample food

Table 9.15 GROWTH PATTERN AND BREASTMEAT YIELD IN LARGE TYPE MALES

Age (weeks)		*Growth pattern*[a]	
0–16	*Good*	*Good*	*Poor*
16–20	*Good*	*Poor*	*Poor*
Liveweight (kg)	15.9	14.7	13.0
Breastmeat as percentage of liveweight	27.1	24.3	23.9

Nixey (1983)
[a]Good and poor growth rates were achieved by different protein intakes

and floor space and been grown at a favourable time of the year with cool temperatures. Their feed representative can help them with planning, encouragement or even help to do sample weighing at various ages and adjust the feed programme accordingly. Sample weighing is essential to forewarn of problems. Too often the first time the farmer realizes his birds are very heavy is on the day he comes to kill them.

SUBCUTANEOUS FAT COVER

The main nutritional problem we face with this market is to achieve a good finish on the birds. Finish is a term used to describe the subcutaneous fat cover which should be particularly evident on the breast and thighs of the bird giving a creamy white appearance. A poorly finished bird will appear bluish and red and look unattractive. The older the bird gets, the greater the fat cover that would be laid down naturally. The problem is occurring because the modern industrial strains of turkey are growing extremely fast and so to achieve the required weights they must often be killed around 15 weeks of age instead of 20 weeks as in the past. It is more difficult to induce the younger bird to lay down subcutaneous fat by nutritional manipulation.

Moran (1979) working with chickens found that finish could be improved by widening the energy to protein ratio but liveweights and fleshing were worsened. Salmon (1974) found that carcass finish was more strongly influenced by the level of added dietary fat than by varying the energy to protein ratio. Subsequently Salmon (1986) found that progressively increased nutrient density with age was effective in improving carcass finish.

Rose and Michie (1985) found it difficult to alter the amount of subcutaneous fat by nutritional means even though abdominal fat amounts were altered. It was hypothesized that any changes in total fatness of the bird would be mostly reflected in the fat depots with a better blood supply, which does not include those in the feather tract of the breast. Their work, however, did find the suggestion that high energy starter feed could subsequently have a long-term influence to make it easier to lay down subcutaneous fat.

In the author's experience the most predictable change that can be accomplished is to alter the type of fat laid down. It is directly influenced by the fatty acid profile of the diet (Neudoerffer and Lea, 1967; Salmon, 1967). Unsaturated fatty acids as found in vegetable oils produce an oily yellow fat whereas animal fats or fat metabolized from carbohydrates will result in a solid white pork-type fat. The latter gives a much more attractive finish and will also be visually more apparent. As a result even though the amount of subcutaneous fat has not been changed, the visual appearance, which is the point of the exercise, has been influenced beneficially.

References

AUCKLAND, J.N., MORRIS, T.R. and JENNINGS, R.C. (1969). Compensatory growth after under nutrition in market turkeys. *British Poultry Science*, **10**, 293–302

BEHRENDS, B.R. and WAIBEL, P.E. (1980). Methionine and cystine requirements of growing turkeys. *Poultry Science*, **59**, 849–859

CLARK, F.A., GOUS, R.M. and MORRIS, T.R. (1982). Response of broiler chicken to well balanced protein mixtures. *British Poultry Science*, **23**, 433–446

D'MELLO, J.P.F. and EMMANS, G.C. (1975). Amino acid requirements of the young turkey: lysine and arginine. *British Poultry Science*, **16**, 297–306

FISHER, C. and EMMANS, G.C. (1983). Calculated amino acid requirements for growing turkeys. *Turkeys*, **31** (No. 1), 39–43

FISHER, C., MORRIS, T.R. and JENNINGS, R.C. (1973). A model for the description and prediction of the response of laying hens to amino acid intake. *British Poultry Science*, **14**, 469–484

HURWITZ, S., FRISCH, Y., BAR, A., EISNER, U., BENGAL. I. and PINES, M. (1983). The amino acid requirements of growing turkeys. 1. Model construction and parameter estimation. *Poultry Science*, **62**, 2208–2217

MARTLAND, M.F. (1984). Wet litter as a cause of plantar pododermatitis, leading to foot ulceration and lameness in fattening turkey. *Avian Pathology*, **13**, 241–252

MORAN, E.T.J. (1979). Carcass quality changes with the broiler chicken after dietary protein restriction during the growing phase and finishing period compensatory growth. *Poultry Science*, **58**, 1257–1270

MORRIS, T.R. (1988). The place of the turkey in the animal industry of the future. In *Proceeedings of 21st Poultry Science Symposium on Recent Advances in Turkey Science*. Ed. Nixey, C. and Grey, T.C. Butterworths, London (in press)

NEUDOERFFER, T.S. and LEA, C.H. (1967). Effects of dietary polyunsaturated fatty acids on the composition of the individual lipids of turkey breast and leg muscle. *British Journal of Nutrition*, **21**, 691–714

NIXEY, C. (1983). Feeding turkeys on the farm. In *Proceedings of 7th Colborns Feed Industry Conference*, Colborn Dawes Ltd, Eastbourne

NIXEY, C. (1988). Nutritional responses of growing turkeys. In *Proceedings of 21st Poultry Science Symposium on Recent Advances in Turkey Science*. Ed. Nixey, C. and Grey, T.C. Butterworths, London (in press)

NRC (1984). National Research Council (US) Sub-Committee on Poultry Nutrition. *Nutrient Requirements of Poultry*, 8th Revised Edition. National Academy Press, Washington

ROSE, S.P. and MICHIE, W. (1985). Improving the fatness of turkeys by diet. Research and Development note No. 29 September 1985. The Scottish Agricultural Colleges

SAIF, Y.M. (1988). Enteric conditions of turkeys. In *Proceedings of 21st Poultry Science Symposium on Recent Advances of Turkey Science*. Ed. Nixey, C. and Grey, T.C. Butterworths, London (in press)

SALMON, R.E. (1974). Effect of dietary fat concentration and energy to protein ratios on the performance, yield of carcass components and composition of skin and meat of turkeys as related to age. *British Poultry Science*, **15**, 543–560

SALMON, R.E. (1976). The effect of age and sex on the rate of change of fatty acid composition of turkeys following a change of dietary fat source. *Poultry Science*, **55**, 201–208

SALMON, R.E. (1986). Effect of nutrient density and energy to protein ratio on performance and carcass quality of small white turkeys. *British Poultry Science*, **27**, 629–638

10

INFLUENCE OF DIET AND GENOTYPE ON CARCASS QUALITY IN POULTRY, AND THEIR CONSEQUENCES FOR SELECTION

F.R. LEENSTRA
Spelderholt Centre for Poultry Research and Extension,
Ministry of Agriculture and Fisheries,
The Netherlands

Introduction

One of the major concerns in the broiler industry is carcass fatness. The desired degree of fattening is a complicated parameter. On the one hand there are undesirable, even wasteful, fat depots like abdominal and crop fat. On the other hand some fat depots are desirable, even necessary: a certain quantity of subcutaneous fat gives the slaughtered broiler a good appearance and a certain amount of fat in the meat gives it palatability. There are no clear indications about the optimal fat content in the various parts of the broiler carcass from a consumer's point of view. The general idea is that abdominal fat and crop fat, and to some extent subcutaneous fat, can give the broiler an obese image. The fat content of legs seems to be optimal to moderately high, while breast meat contains enough or too little fat.

The direct concerns of broiler processors about fatness are more defined. High amounts of abdominal fat can reduce processing yields since it is often removed from the carcass together with the intestines during mechanical dressing.

Although the broiler industry is interested in fat deposition, there still is no system by which delivering overfat broilers is penalized. Unlike cattle, sheep and pigs, which are graded and paid for according to carcass quality, broilers are only paid for per amount of live weight delivered at the slaughter-plant. This is probably due to the fact that no practical methods are available by which broiler carcasses can be judged quickly and accurately on fat content. Until such a method is developed and used, ways to alter fat deposition in the broiler carcass will only be applied if they do not raise the costs of production. The main factors determining production costs of broilers are growth rate and feed costs per unit of growth. Thus a method which alters fat deposition has to have a beneficial effect on growth rate, feed conversion or feed costs or combinations of these factors.

Fat deposition can be influenced somewhat by environmental factors and to a larger extent by nutritional and genetic factors (Lin, Friars and Moran, 1980). Among environmental factors influencing fat deposition are housing systems: broilers reared in cages have a higher fat content than broilers reared on litter (Deaton *et al.*, 1974); temperature: moderately high temperatures give fatter broilers compared to low temperatures (Kubena *et al.*, 1974); and lighting systems: continuous light causes more fat deposition than intermittent light (Van Es, 1981).

First published in *Recent Advances in Animal Nutrition – 1984*

The influence of nutrition and genetics on fat deposition is large compared to these environmental factors. Genetic factors are mainly involved in the quantity of fat deposited, while nutritional factors influence both the quantity and the quality of fat (Bartov and Bornstein, 1976; McLeod, 1982). Only quantitative aspects of fat deposition will be discussed here.

Changes in fat deposition that can be achieved, in practice, are dependent on the genetic variation present, on correlations of fat deposition parameters with other traits, on the influence of nutrition and on the influence of genotype.

These four topics and their implications for selection programmes in broiler stock will be discussed.

Variation in body composition

The amount of variation present in body composition is an important predictor of the changes in body composition that are possible. A large amount of variation between individuals reared in the same environment can indicate genetic variation, while large variation between groups reared under different circumstances indicates the possibility of changing the parameter involved by changing the environment. Fat content is the most variable body composition trait in different species (Ricard, 1975; Lohman, 1973; Simpson and Goodwin, 1979). Lohman (1973) indicated that the amount of fat expressed as a percentage of total body weight can vary between 1 and 60% across species, sexes, ages and treatments. Even with diets that are likely to be used in practice for broilers, nearly twofold differences in total fat percentages and more than twofold differences in abdominal fat content could be attained (Scheele, Van Schagen and Ten Have, 1981). Between breeds or strains of chickens reared on the same diet differences of 50% and more occurred in total and abdominal fat contents (Edwards and Denman, 1975; Leclercq, Blum and Boyer, 1980).

Figure 10.1 illustrates, within strain and treatment, the variation in body composition parameters expressed as coefficients of variation. Total body fat, and especially the amount of abdominal fat, are highly variable characters compared to percentages of water, protein or ash and body weight of the chicken (Leenstra, 1983). The variation expressed as coefficients of variation found for body weight, abdominal fat and total fat content from a number of sources is summarized in *Table 10.1*. The coefficient of variation of the absolute amount of abdominal fat of broilers reared in the same environment varied between 24 and 47%, the coefficient of variation of the percentage of abdominal fat between 21 and 47%, that of total fat percentage between 7 and 18% and that of body weight between 6 and 12%. In all cases, the percentage of abdominal fat showed far more variation compared to the percentage of total fat, while the latter in general was more variable than body weight.

There is not much information available about variation in fat content of valuable broiler parts like breast and leg meat. Neupert and Hartfiel (1978) found less variation in fat content of legs than in total body fat content. Ricard (1983) found, for subcutaneous fat of the leg, a coefficient of variation of 26–28%, while intermuscular fat of the leg had a coefficient of variation of only 13–16%.

The fact that total percentage of body fat varies less than percentage of abdominal fat, supports the idea that percentage of fat in broiler meat varies less than total percentage of fat. The large differences in variation in fat content

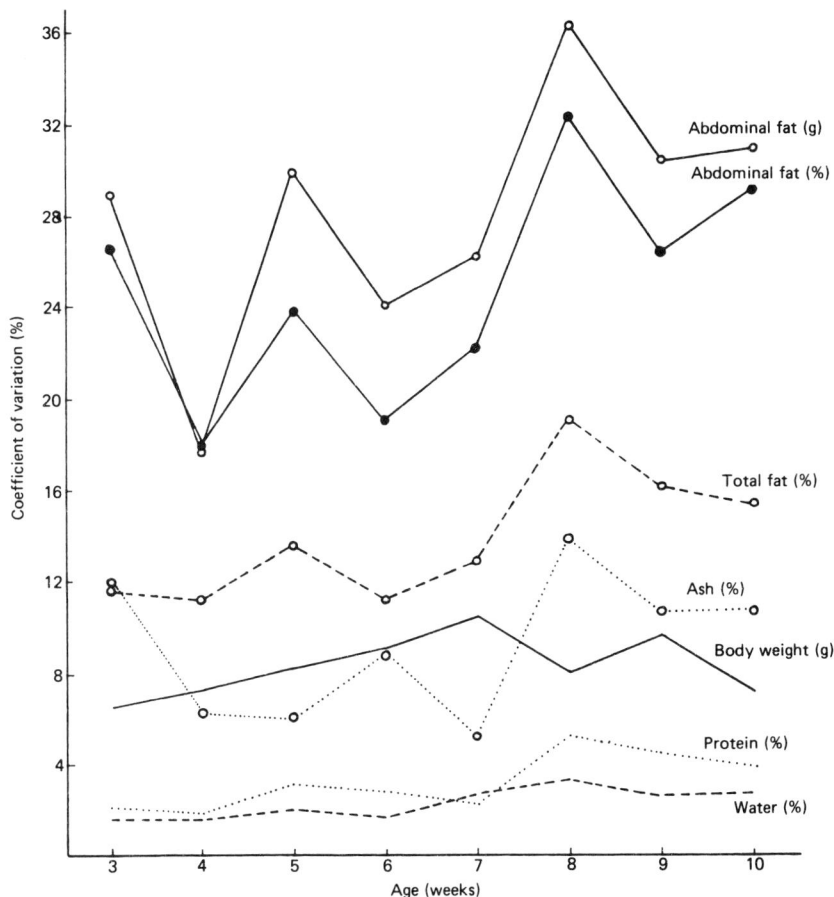

Figure 10.1 Variability, expressed as coefficients of variation, of body weight and body composition parameters of male broilers (From Leenstra, 1982)

Table 10.1 VARIABILITY, EXPRESSED AS COEFFICIENTS OF VARIATION (%), IN BODY WEIGHT AND FAT DEPOSITION PARAMETERS

Source	Sex	Body weight	Abdominal fat weight	% Abdominal fat	% Total fat
Becker et al. (1979)	♂♂	9.9	33.8	29.8	18.0
	♀♀	8.6	30.4	27.4	17.0
Becker et al. (1981)	♂♂	6.5	27.9	25.3	13.0
	♀♀	6.0	29.8	27.3	15.0
Griffiths, Leeson and Summers (1978)	♂♂	9.4		26.0	7.3
Leenstra (1983)	♂♂	8.3	28.0	24.7	13.9
	♀♀	7.8	24.0	21.6	12.5
Ricard, Leclercq and Marché (1982)	♂♂	11.4	30.0	24.8	
	♀♀	7.6	24.4	20.6	
	♂♂	9.1	35.7	33.2	
	♀♀	8.8	28.0	23.3	
	♂♂	11.6	27.5	25.5	
	♂♂	10.2	47.2	46.9	

between broiler parts signify that considerable changes in percentage of abdominal fat and probably subcutaneous fat do not necessarily have a strong effect on the amount of intermuscular fat.

Relationship between fat deposition and other parameters

The relationship between body fat content and other parameters can be divided into two groups:

(1) Relationships that consider body fat content and parameters that are important in broiler production for economic reasons like body weight, feed conversion, distribution of fat in the broiler:
(2) Relationships between body fat content and parameters that could be used to estimate broiler fat content in a not too laborious and/or non-destructive way. These relationships could facilitate research towards improving body composition and selection against fatness.

FAT CONTENT AND BROILER TRAITS

Body weight

It is suspected that broilers are fat because of intensive genetic selection for body weight at a fixed age within an *ad libitum* feeding regimen (Lin, 1981). Such a selection system favours the animals with large appetites, which are capable of overeating to such an extent that feed intake exceeds the birds' genetic capacity for lean tissue growth (Summers and Leeson, 1979). This is supported by the fact that broilers could only be force fed 15% above their *ad libitum* intake, while chickens of a layer type strain could be force fed 43% above their *ad libitum* intake (Summers and Leeson, 1979).

Table 10.2 CORRELATION COEFFICIENTS BETWEEN LIVE BODY WEIGHT AND FAT DEPOSITION PARAMETERS

Sex	Abdominal fat (g)	Abdominal fat (%)	Total fat (%)
♂♂	0.49[b]	0.29[b]	0.10[b]
♀♀	0.53[b]	0.36[b]	0.28[b]
♀♀	0.55[e]	0.18[e]	0.20[e]
♂♂	0.29[i]	0.40[a]	0.43[a]
♀♀	0.41[i]	0.18[a]	0.20[a]
♂♂	0.50[g]	0.33[h]	0.35[h]
♀♀	0.42[g]	0.24[h]	0.33[h]
♂♂	0.31[d]		0.20[f]
♀♀	0.28[d]		0.33[f]
♂♂	0.44[d]		0.01[f]
♂♂	0.29[d]		0.48[f]
♂♂	0.49[c]		0.20[c]
♂♂	0.60[c]		−0.04[c]

[a]Becker *et al.* (1979), [b]Becker *et al.* (1981), [c]Burgener, Cherry and Siegel (1981), [d]Cherry, Siegel and Beane (1978), [e]Friars *et al.* (1983), [f]Griffin, Whitehead and Broadbent (1982), [g]Griffiths, Leeson and Summers (1978), [h]Leenstra (1983), [i]Ricard and Rouvier (1969)

Within groups of birds, body weight and amount of fat deposited are positively correlated. However the correlation coefficient between body weight and percentage of body fat is not high and seldom exceeds 0.4 (*Table 10.2*). Van Middelkoop Kuit and Zegwaard (1977) and Ten Have and Scheele (1981) found that strains differing in body weight could have the same fat content and strains differing in fat content the same body weight. This confirms the conclusion from *Table 10.2* that the correlation between body weight and fat deposition is rather weak.

Feed conversion

Brody (1935) stated that efficient strains would gain more lean and less fat than less efficient strains. The deposition of fat tissue requires far more energy than the deposition of lean tissue, due to the fact that lean tissue contains about 70% water, while fat tissue is nearly all energy rich dry matter. A broiler converting its feed to lean tissue will therefore gain more per unit of feed than a broiler converting feed to fat tissue and thus have a better feed conversion (Thomas, Glazener and Blow, 1958).

Table 10.3 CORRELATION COEFFICIENTS BETWEEN FEED CONVERSION AND % ABDOMINAL FAT OR % TOTAL FAT

	Sex	*r*	*Source*
% abdominal fat	♂♂	0.12	Griffiths, Leeson and Summers
	♂♂	0.12	(1978)
	♂♂	0.45	
	♂♂	0.28	
	♂♂ + ♀♀	0.13	Leenstra (1982)
	♀♀	0.22	Friars *et al.* (1983)
% total fat	♀♀	0.24	
	♂♂ + ♀♀	−0.10	Pym and Solvyns (1979)
	♂♂ + ♀♀	−0.07	
	♂♂ + ♀♀	0.65	Washburn, Guill and Edwards
	♂♂ + ♀♀	0.36	(1975)

This of course is only true if the leaner animal has the same appetite, heat loss and activity as the fatter bird (Webster, 1978). The correlation coefficients between body fat content and feed conversion are in most cases reported to be positive (*Table 10.3*; Leclercq, Blum and Boyer, 1980; Whitehead and Griffin, 1982). One may expect, therefore, a better feed conversion if broilers are less fat, but, as with body weight, the relation is weak.

Body composition

Table 10.4 summarizes correlations found between body composition parameters. In general, percentage abdominal fat and percentage total fat are highly correlated (*r* between 0.41 and 0.98). Part of this high correlation is, however, caused by the fact that abdominal fat is an important part of total fat. Between groups the ratio of abdominal to total fat can vary considerably (Scheele, Van Schagen and Ten Have, 1981; Griffiths, Leeson and Summers, 1978).

As parts of body composition are expressed in percentages of body weight, a low percentage of fat does imply a higher percentage of one of the other components.

Table 10.4 CORRELATION COEFFICIENTS BETWEEN BODY COMPOSITION PARAMETERS

Abdominal fat % and total fat %	Total fat % and protein %	Total fat % and water %
0.66[d]	−0.89[d]	−0.79[d]
0.78[f]	−0.44[i]	−0.97[i]
0.90[f]	−0.50[i]	−0.94[i]
0.76[e]	−0.46[k]	−0.96[k]
0.88[e]	−0.57[k]	−0.96[k]
0.56[e]	−0.75[h]	−0.96[h]
0.41[e]		−0.98[j]
0.79[g]	−0.53[g]	
0.79[g]	−0.55[g]	
0.51[a]		−0.71[a]
0.71[a]		−0.89[a]
0.79[c]		
0.83[b]		

[a]Becker *et al.* (1981), [b]Bougon, Jacquet and Lecuyer (1983), [c]Elwinger (1980), [d]Friars *et al.* (1983), [e]Griffiths, Leeson and Summers (1978), [f]Leclercq, Blum and Boyer (1980), [g]Leenstra (1983), [h]Neupert and Hartfiel (1978), [i]Pym and Solvyns (1979), [j]Velu, Baker and Scott (1972), [k]Washburn, Guill and Edwards (1975)

The correlation coefficient between percentage of fat and protein is negative (*r* between −0.44 and −0.89), and the correlation between fat and water is even stronger (*r* between −0.71 and −0.97). A lower fat content does imply therefore more protein, but especially more water deposition.

INDIRECT MEASUREMENTS OF BROILER FATNESS

The main problem in research aimed at total and abdominal fat deposition in broilers is that a direct determination of total body fat is destructive and extremely laborious. Parameters correlated with fatness could therefore be useful.

The correlation needed between fatness and a parameter to estimate fatness depends on the purpose for which it is used. Feed conversion can be an interesting parameter to use in selection for a leaner broiler. However the weak correlation between fatness and feed conversion makes it a useless estimator of fatness. There have been several attempts to find parameters which can be measured with less labour or are non-destructive but do correlate well with fat deposition.

Destructive methods

As percentage of water and percentage of fat in the carcass are highly correlated (*Table 10.4*), water content is a good estimator of total fat content.

The composition of fat-free tissue in animals is, at a given age, relatively constant (Lin, 1981). The broiler can be seen as consisting of two components: fat and fat-free tissue. The specific weight of fat is lower than that of fat-free tissue, indicating that a fat broiler will have a lower specific weight than a lean broiler. Barton, Fletcher and Edwards (1981) found a correlation of −0.90 between specific weight of the eviscerated carcass and its fat content. However, Moran, Summers and Orr (1968) found only a correlation of −0.11 to −0.17 between specific weight and carcass fat score. Spencer *et al.* (1978) found an intermediate value of −0.36 to −0.69 for the correlation between specific weight and carcass fat content.

Another group of methods to estimate total fat content is to use correlations with the weight or fat content of parts of the carcass. Well definable fat depots, like abdominal and sartorial fat, do give reasonable estimates of total fat content of a carcass. The correlation between abdominal and total fat is higher (between 0.41 and 0.98) than the correlation between weight of sartorial depot fat and total carcass fat (between 0.35 and 0.92; Burgener, Cherry and Siegel, 1981).

As the skin is also an important fat depot in poultry, the weight of the total skin (Dansky and Hill, 1952) or the percentage of fat in parts of the skin have been examined as estimators of total fat content (Moran, Summers and Orr, 1968; Bartov, Bornstein and Lipstein, 1974; Bartov and Bornstein, 1976; Ehinger, 1976, 1977; Becker *et al.*, 1979, 1981).

Correlation coefficients mentioned vary between 0.5 and 0.9. The most promising estimator for total carcass fat of the ones mentioned above is abdominal fat.

Non-destructive methods

Non-destructive methods, which permit estimation of the fat content in a living bird, have enormous advantages for selection and nutrition research. Preferably such a method has to be fast and applicable to a large number of birds.

Among non-destructive methods examined are various skinfold measurements (Mirosh *et al.*, 1980) and ultrasonic measurements (Leenstra, 1980). Both types of measurement do not correlate well enough with total or abdominal fat content (r less than 0.2) to be useful. More promising is the measurement of the thickness of the abdominal fat pad by means of a pair of calipers (Pym and Thompson, 1980; Mirosh and Becker, 1982; Whitehead and Griffin, 1982; Rose and Michie, 1983), although correlation coefficients of 0.3–0.8 between caliper measurement and percentage of abdominal fat are too variable to produce a reliable technique.

Farrell (1974) used tritium labelled water to estimate total water content of live broilers. He could estimate both chemical water content ($r = 0.95$) and fat content ($r = -0.81$) rather accurately, but the method is not easy to apply to large numbers of birds. At the Poultry Research Centre a technique was developed to measure triglyceride concentrations in blood plasma (Griffin, Whitehead and Broadbent, 1982). The reliability of this method is strongly dependent on the type of diet fed and the nutritional status of the stock examined (Whitehead and Griffin, 1982; Griffin and Whitehead, 1982a). Within groups a correlation with body fat of 0.6–0.7 could be realized with this method.

Another method under investigation at Spelderholt at this moment is the plethysmometric method developed by Gundlach, Nijkrake and Hautvast (1980). With this method, the volume of a live animal can be measured rapidly and cleanly. For human beings the method correlated very well ($r = 0.99$) with underwater determination of specific weight. However results of this method for determining fat content in poultry are not yet available.

It can be concluded that a practical method for determining the fat content in large numbers of live broilers is not yet available.

Nutrition and fat deposition

It is not intended here to review the effects of nutrition on carcass composition, but only to indicate the complexity of the relation between nutrition and carcass composition.

The value of a broiler feed under practical circumstances is expressed as return in live body weight per unit feed cost and not as return in animal protein or energy per unit of protein or energy consumed. This difference between economic and biological efficiency hampers interpretation of nutritional research.

Nutritional factors do have significant effects on body composition in broilers (Lin, Friars and Moran, 1980; McLeod, 1982). In general, high energy diets do have advantages if body weight gain and feed conversion are considered. If carcass quality is taken into account, the benefits of high energy diets are less (Freeman, 1983). Fraps (1943) found, 40 years ago, that diets with a narrow energy to protein ratio caused less fat deposition than diets with a wide energy to protein ratio. This has been confirmed more recently by several authors (Bartov, Bornstein and Lipstein, 1974; Farrell, 1974; Kirchgessner, Roth-Maier and Müller, 1979).

Although much has been learned about the effects of dietary factors on carcass composition, it is still not possible to predict carcass composition accurately from a knowledge of dietary factors. This is mainly due to the fact that numerous factors and their interactions influence carcass composition. Factors as widely differing as amino acids, crude fibre, fat or salt content of the diet have significant effects on carcass fat deposition (Lipstein, Bornstein and Bartov, 1975; Ten Have and Scheele, 1981; Marks and Washburn, 1983). In addition there is a significant effect on fat deposition of the interaction between form of the feed (pellets, crumbs or mash) and energy content (Pesti, Whiting and Jensen, 1983), between sex of the broiler and feed composition (Summers, Slinger and Ashton, 1965; Holsheimer, 1975; Lipstein, Bornstein and Bartov, 1975; Salmon, Classen and McMillan, 1983), between broiler age and feed composition (Holsheimer, 1980; Scheele, Van Schagen and Ten Have, 1981; Salmon, Classen and McMillan, 1983) and between genotype and feed composition.

An illustration of these interactions is the influence of the ratio between energy and lysine on the effect of dietary fat on fat deposition (Ten Have and Scheele, 1981). Dietary fat had significantly more effect on fat deposition at low levels of dietary lysine than at high lysine levels. Guillaume and Summers (1970) and Jackson, Summers and Leeson (1982) found that the degree to which dietary energy was utilized was dependent on the energy:protein ratio of the diet. In diets relatively high in protein, energy is utilized less efficiently than in diets low in protein. These findings could explain the differences in energetic value of dietary fat that are reported (Rinehart, Green and Williamson, 1975; Summers, Slinger and Ashton, 1965; Griffiths, Leeson and Summers, 1977).

Another complicating factor in research towards the effects of nutrition on fat deposition is that most dietary factors do not have the same effect on the deposition of abdominal fat and the deposition of fat in the rest of the body; the effects of diet on abdominal fat content are greater than the effects of diet on total carcass fat (Elwinger, 1980; Jackson, Summers and Leeson, 1982). Scheele, Van Schagen and Ten Have (1981) found that regression formulas to predict abdominal fat from dietary factors differed significantly from those to predict total fat minus abdominal fat. Energy and crude fibre content of the diet were important in explaining percentage of abdominal fat, while lysine content was the main factor in predicting total fat minus abdominal fat. The equation to predict total fat minus abdominal fat at six weeks of age did not differ from the one at eight weeks of age. Abdominal fat at six weeks of age was however more dependent on dietary energy content than abdominal fat at eight weeks of age.

Such effects can be due to differences in the rate of deposition of abdominal fat relative to total fat. At older ages abdominal fat becomes a greater part of the amount of fat deposited (Griffiths, Leeson and Summers, 1978).

If more information on interactions between dietary factors (restricted feeding included) and between dietary factors and genotype is available, it should be possible to manipulate the total amount of carcass fat and the ratio between abdominal and total fat by dietary means.

Genetic aspects of fat deposition

Differences between breeds (Edwards and Denman, 1975) and strains (Ricard, 1975; Van Middelkoop, Kuit and Zegwaard, 1977; Cherry, Siegel and Beane, 1978; Twining, Thomas and Bossard, 1978; Neupert and Hartfiel, 1978; Griffiths, Summers and Leeson, 1978) in fat deposition indicate clearly the importance of genetic factors in fat deposition.

Genetic variation within strains in total fat content was found by Friars *et al.* (1983). They found a degree of heritability of 0.48. More estimates are available for the heritability of the amount of abdominal fat relative to body weight. Estimates vary between 0.3 and more than 1 (Ricard and Rouvier, 1967, 1969; Leclercq, Blum and Boyer, 1980; Becker *et al.*, 1981; Leenstra, 1982; Friars *et al.*, 1983). Leclercq, Blum and Boyer (1980) and Leenstra (1982) found that selection against abdominal fat is possible via selection on the performance of sibs, while Whitehead and Griffin (1982) used triglyceride concentrations in blood plasma to select broilers directly on their own performance.

Leclercq, Blum and Boyer (1980) found that selection against abdominal fat in males did not change abdominal fat in females to the same extent, thus indicating a different genetic regulation of fat deposition in males and females. The response to selection was also not symmetrical: selection for a high percentage of abdominal fat was more effective than selection for a low percentage of abdominal fat. Whitehead and Griffin (1982), however, found that their selection on plasmatriglyceride content produced exactly the same response in abdominal and total fat when selection occurred in both a positive and a negative direction.

The consequences of genetic changes in fat deposition on other parameters can be predicted by genetic correlations, which are not necessarily equal to phenotypic correlations, and examined directly from the results of selection experiments. The genetic correlation between percentage abdominal fat and body weight seems to be close to zero, while the phenotypic correlation is positive. Friars *et al.* (1983) estimated the genetic correlation to be −0.06, while Whitehead and Griffin (1982) and Leenstra (1982) did not find any differences in body weight between birds selected on fat deposition parameters and unselected birds. Leclercq, Blum and Boyer (1980) even found a higher body weight in the lean line compared to the fat line. These findings indicate that selection against fat deposition is possible without direct negative effects on body weight.

As already mentioned the phenotypic correlation between feed conversion and percentage fat is positive, indicating a lower feed conversion for the leaner bird. The genetic correlation between these parameters appears to be of the same magnitude. Friars *et al.* (1983) estimated this correlation to be 0.52, while Leenstra (1982) found a value of 0.13. Both Leclercq, Blum and Boyer (1980) and Whitehead and Griffin (1982) found changes in feed conversion in the desired direction as a consequence of selection against fat deposition. Pym and Solvyns

(1979) and Washburn, Guill and Edwards (1975) found that a consequence of selection for a favourable feed conversion was a lower fat content.

Due to changes in abdominal fat content with selection changes in total fat content can also be expected. Friars *et al.* (1983) estimated the genetic correlation between percentage abdominal fat and percentage total fat to be 0.72. Leclercq, Blum and Boyer (1980) found that for each gram of change in abdominal fat there was a change of 1.7 g in the same direction for fat in the rest of the body. As in pigs (Wood *et al.*, 1983) there are in broilers no indications that decreasing the fat content in one part of the body increases the fat content in other parts of the body.

The danger that selection against fat deposition in the abdominal region might produce a broiler which is too lean is small. Griffin and Whitehead (1982b) concluded that if the percentage of abdominal fat approaches zero, the total carcass still contains 10–11% fat.

The heritability, the genetic variation present and genetic correlations between abdominal fat content and other carcass parameters make the prospects for direct or indirect selection against abdominal fat very real.

Effects of genotype × nutrition interactions on fat deposition

With respect to fat deposition there are indications of significant interactions between genotype and nutritional environment. Cherry, Siegel and Beane (1978) and Ten Have and Scheele (1981) found that the influence of dietary factors on fat deposition was not the same for different commercial broiler strains. In particular the effect of dietary energy on fat deposition showed considerable strain differences.

Sørensen (1980) selected two lines on body weight; one line was selected after feeding a normal diet, the other after feeding a diet low in protein. For abdominal fat a line × diet interaction was clearly present: if both lines were fed a normal diet the line selected on the low protein diet had slightly more abdominal fat; if both lines were fed a low protein diet the line selected on the low protein diet had nearly twice as much abdominal fat. These indications of genotype × nutrition interactions should be considered in choosing suitable selection criteria and the diets to be used during breeding programmes.

Consequences for selection programmes

Before a selection programme is started, the breeding goal should be well defined. For broilers, the economic breeding goal at the moment is maximizing growth rate per unit of feed consumed, irrespective of body composition, even though breeders, feed manufacturers and processors all believe that the present broiler is too fat. Maximizing lean growth rate per unit of protein consumed is in biological terms the most appropriate breeding goal, more appropriate for instance than maximizing the lean to fat ratio, which is considered so intensively in current research. Maximizing lean growth rate per unit of protein consumed could lead to a leaner broiler if today's feeds are used, and possibly also with the introduction of diets with less protein and/or more energy.

A change towards a biological optimal breeding goal will only be possible if carcass quality is evaluated, or if it can be achieved without losses in progress in growth rate per unit of feed consumed. The latter is rather improbable since adding another selection parameter generally signifies a lowering in the selection intensity of the original parameter(s). Only if genetic progress in the added parameter has real economic advantages, which counterbalance the diminished progress in the original one(s), will a new parameter be added to a selection programme.

Maximizing growth rate per unit of feed consumed is achieved almost exclusively by selection for body weight at a fixed age using *ad libitum* feeding of a practical broiler ration. Due to the law of diminishing returns, however, selection on body weight alone will become less efficient. In the near future addition of another selection parameter might therefore be advantageous. This alternative selection parameter should meet several requirements. It should

(1) contribute to the economic breeding goal,
(2) contribute to the biological breeding goal,
(3) be measurable on large numbers of birds.

On the basis of variation, genetic influences and correlations found, several parameters can be taken into consideration. Feed conversion does contribute directly to the economic breeding goal and indirectly to the biological breeding goal since a favourable feed conversion is accompanied by less fat and more lean deposition. It can be measured on an individual basis, though not on a scale comparable with live body weight. Progeny testing of sires pre-selected on other criteria seems more practical, but Pym and James (1979) found more economic response from individual selection compared to progeny testing. Direct selection for lean to fat ratios or lean growth rate is, due to the laborious determination of these parameters, not yet possible. Selection against abdominal fat relative to body weight is possible via sib or progeny testing. Both will lead to leaner broilers (Leclercq, Blum and Boyer, 1980; Becker *et al.*, 1982). Caliper measurements of the abdominal fat content can, depending on the operator, perhaps be used to directly select an animal on its own performance against abdominal fat. Selection on the basis of plasma triglycerides appears also to hold some promise to achieve a less fat broiler.

All of these methods for determining fat deposition are laborious, and in a practical selection programme are only suitable as the second parameter in a tandem selection to find the leaner ones among pre-selected birds.

The effects of direct selection against fat deposition on the economic and biological breeding goal are not clear, as less fat deposition can be the result of less appetite (Whittemore, 1978).

Sørensen (1980) found that selection for body weight on a diet with a sub-optimal protein content resulted in more fat deposition compared to selection for body weight on a normal diet. It could be interesting to examine results of selection for body weight on a high protein diet. Theoretically broilers, storing relatively more protein, grow better on a high protein diet than broilers that have to metabolize the protein they cannot store.

Another possibility is to select broilers on body weight after restricted feeding. Under restricted feeding conditions (compared to *ad libitum* feeding conditions) appetite is less important in determining body weight, while efficient growth is more important. Comparisons between family means for amount of abdominal fat relative to body weight after *ad libitum* feeding, with family means for body weight after restricted feeding, revealed a negative, though weak, correlation ($r = -0.18$; Leenstra, 1982). This indicates that birds with a relatively high growth rate under restricted feeding conditions have a relatively low percentage of abdominal fat under *ad libitum* feeding conditions.

In 1984, at Spelderholt, results of selection on body weight after restricted feeding will be compared with results of individual selection on feed conversion, sib selection against relative amount of abdominal fat and individual selection on body weight after *ad libitum* feeding. Body weight, feed conversion and fat deposition of the four lines will be measured after feeding a normal broiler ration (Leenstra, 1982).

It is not yet possible to predict the contribution of alternative selection methods to the economic and the biological breeding goal. More information is needed on the genetic relationships between carcass parameters and their interactions with feed composition.

References

BARTON, A.G., FLETCHER, D.L. and EDWARDS Jr., H.M. (1981). *Poult. Sci.*, **53**, 574–576

BARTOV, I. and BORNSTEIN, S. (1976). *Br. Poult. Sci.*, **17**, 17–27

BARTOV, I., BORNSTEIN, S. and LIPSTEIN, B. (1974). *Br. Poult. Sci.*, **15**, 107–117

BECKER, W.A., SPENCER, J.V., MIROSH, L.W. and VERSTRATE, J.A. (1979). *Poult. Sci.*, **58**, 835–842

BECKER, W.A., SPENCER, J.V., MIROSH, L.W. and VERSTRATE, J.A. (1981). *Poult. Sci.*, **60**, 693–697

BECKER, W.A., SPENCER, J.V., MIROSH, L.W. and VERSTRATE, J.A. (1982). *Poult. Sci.*, **61**, 1415

BOUGON, M., JACQUET, J. and LECUYER, T. (1983). In *Quality of Poultry Meat*, pp. 69–77. Eds Lahellec, C., Ricard, F.H. and Colin, P. Ploufragan; France

BRODY, S. (1935). *Ann. Rev. Biochem.*, **4**, 383–412

BURGENER, J.A., CHERRY, J.A. and SIEGEL, P.B. (1981). *Poult. Sci.*, **60**, 54–62

CHERRY, J.A., SIEGEL, P.B. and BEANE, W.L. (1978). *Poult. Sci.*, **57**, 1482–1487

DANSKY, L.M. and HILL, F.W. (1952). *Poult. Sci.*, **31**, 912

DEATON, J.W., KUBENA, L.F., CHEN, T.C. and REECE, F.N. (1974). *Poult. Sci.*, **53**, 574–576

EDWARDS, Jr., H.M. and DENMAN, F. (1975). *Poult. Sci.*, **54**, 1230–1238

EHINGER, F. (1976). Methoden zur Bestimmung des Verfettungsgrades von Broilern. Dissertation, University Hohenheim; Stuttgart, Germany

EHINGER, F. (1977). *Archiv für Geflügelkunde*, **41**, 35–37

ELWINGER, K. (1980). In *Proceedings 6th European Poultry Conference*, Vol. III, pp. 256–263. Hamburg, Germany

FARRELL, D.J. (1974). *Br. Poult. Sci.*, **15**, 25–41

FRAPS, G.S. (1943). *Poult. Sci.*, **22**, 421–424

FREEMAN, C.P. (1983). *Proc. Nutr. Soc.*, **42**., 351–359

FRIARS, G.W., LIN, C.Y., PATTERSON, D.L. and IRWIN, L.N. (1983). *Poult. Sci.*, **62**, 1425

GRIFFIN, H.D. and WHITEHEAD, C.C. (1982a). *Br. Poult. Sci.*, **23**, 307–313

GRIFFIN, H.D. and WHITEHEAD, C.C. (1982b). *24th Br. Poult. Breeders Roundtable Conference*. Edinburgh, UK

GRIFFIN, H.D., WHITEHEAD, C.C. and BROADBENT, L.A. (1982). *Br. Poult. Sci.*, **23**, 15–23

GRIFFITHS, L., LEESON, S. and SUMMERS, J.D. (1977). *Poult. Sci.*, **56**, 638–646

GRIFFITHS, L., LEESON, S. and SUMMERS, J.D. (1978). *Poult. Sci.*, **57**, 1198–1203

GUILLAUME, J. and SUMMERS, J.D. (1970). *Can. J. Anim. Sci.*, **50**, 355–362

GUNDLACH, B.L., NIJKRAKE, H.G.M. and HAUTVAST, J.G.A.J. (1980). *Human Biol.*, **52**, 23–33

HOLSHEIMER, J.P. (1975). In *Quality of Poultry Meat*, 45(1)–45(10). Ed. Erdtsieck, B. Oosterbeek; The Netherlands

HOLSHEIMER, J.P. (1980). *Poult. Sci.*, **59**, 2060–2064

JACKSON, S., SUMMERS, J.D. and LEESON, S. (1982). *Poult. Sci.*, **61**, 2224–2231

KIRCHGESSNER, M., ROTH-MAIER, D.A. and MÜLLER, H.K. (1979). *Zeitschrift für Tierphysiologie, Tierernährung und Futtermittelkunde*, **41**, 218–225

KUBENA, L.F., DEATON, J.W., CHEN, T.C. and REECE, F.N. (1974). *Poult. Sci.*, **53**, 211–214

LECLERCQ, B., BLUM, J.C. and BOYER, J.P. (1980). *Br. Poult. Sci.*, **21**, 107–113

LEENSTRA, F.R. (1980). *Spelderholt Mededeling 331*. Beekbergen; The Netherlands

LEENSTRA, F.R. (1982). *24th Br. Poult. Breeders Roundtable Conference*. Edinburgh, UK

LEENSTRA, F.R. (1983). *Spelderholt Mededeling 386*. Beekbergen; The Netherlands

LIN, C.Y. (1981). *Wld Poult. Sci. J.*, **37**, 106–110

LIN, C.Y., FRIARS, G.W. and MORAN, E.T. (1980). *Wld Poult. Sci. J.*, **36**, 103–111

LIPSTEIN, B., BORNSTEIN, S. and BARTOV, I. (1975). *Br. Poult. Sci.*, **16**, 627–635

LOHMAN, T.G. (1973). *J. Anim. Sci.*, **32**, 647–653

McLEOD, J.A. (1982). *Wld Poult. Sci. J.*, **38**, 194–200

MARKS, H.L. and WASHBURN, K.W. (1983). *Poult. Sci.*, **62**, 263–272

MIROSH, L.W. and BECKER, W.A. (1982). *Poult. Sci.*, **62**, 1–5

MIROSH, L.W., BECKER, W.A., SPENCER, J.V. and VERSTRATE, J.A. (1980). *Poult. Sci.*, **59**, 945–950

MIROSH, L.W., BECKER, W.A., SPENCER, J.V. and VERSTRATE, J.A. (1981). *Poult. Sci.*, **60**, 509–512

MORAN, Jr., E.T., SUMMERS, J.D. and ORR, H.L. (1968). *Fd Technol.*, **22**, 999–1038

NEUPERT, B. VON and HARTFIEL, W. (1978). *Archiv für Geflügelkunde*, **42**, 150–158

PESTI, G.M., WHITING, T.S. and JENSEN, L.S. (1983). *Poult. Sci.*, **62**, 490–494

PYM, R.A.E. and JAMES, J.W. (1979). *Br. Poult. Sci.*, **20**, 99–107

PYM, R.A.E. and SOLVYNS, A.J. (1979). *Br. Poult. Sci.*, **20**, 87–97

PYM, R.A.E. and THOMPSON, J.M. (1980). *Br. Poult. Sci.*, **21**, 281–286

RICARD, F.H. (1975). In *Quality of Poultry Meat*, 4(1)–4(16). Ed. Erdtsieck, B. Oosterbeek; The Netherlands

RICARD, F.H. (1983). In *Quality of Poultry Meat*. pp. 49–68. Eds Lahellec, C., Ricard, F.H. and Colin, P. Ploufragan; France

RICARD, F.H., LECLERCQ, B. and MARCHÉ, G. (1982). *Ann. Génét. Sélect. anim.*, **14**, 551–556

RICARD, F.H. and ROUVIER, R. (1967). *Ann. Zootech.*, **16**, 23–39

RICARD, F.H. and ROUVIER, R. (1969). *Ann. Génét. Sélect. anim.*, **1**, 151–165

RINEHART, K.E., GREEN, D.E. and WILLIAMSON, J.L. (1975). *Poult. Sci.*, **54**, 1809–1810

ROSE, S.P. and MICHIE, W. (1983). *J. Agric. Sci.*, **101**, 345–350

SALMON, R.E., CLASSEN, H.L. and McMILLAN, R.K. (1983). *Poult. Sci.*, **62**, 837–845

SCHEELE, C.W., VAN SCHAGEN, P.J.W. and TEN HAVE, H.G.M. (1981). In *Quality of Poultry Meat*. pp. 398–411. Eds Mulder, R.W.A.W., Scheele, C.W. and Veerkamp, C.H. Beekbergen; The Netherlands

SIMPSON, M.D. and GOODWIN, T.L. (1979). *Poult. Sci.*, **58**, 1400–1402

SØRENSEN, P. (1980). In *Proc. 6th Europ. Poult. Conf.*, Vol. II. pp. 64–71. Hamburg, Germany

SPENCER, J.V., BECKER, W.A., VERSTRATE, J.A. and MIROSH, L.W. (1978). *Poult. Sci.*, **57**, 1164

SUMMERS, J.D. and LEESON, S. (1979). *Poult. Sci.*, **58**, 536–542

SUMMERS, J.D., SLINGER, S.J. and ASHTON, G.C. (1965). *Poult. Sci.*, **44**, 501–509

TEN HAVE, H.G.M. and SCHEELE, C.W. (1981). In *Quality of Poultry Meat*. pp. 387–397. Eds Mulder, R.W.A.W., Scheele, C.W. and Veerkamp, C.H. Beekbergen; The Netherlands

THOMAS, C.H., GLAZENER, E.W. and BLOW, W.L. (1958). *Poult. Sci.*, **37**, 1177–1179

TWINING JR., P.V., THOMAS, O.P. and BOSSARD, E.H. (1978). *Poult. Sci.*, **57**, 492–497

VAN ES, A.J.H. (1981). In *World Poultry Production: Where and how?* pp. 39–54. Eds Scheele, C.W. and Veerkamp, C.H. Beekbergen; The Netherlands

VAN MIDDELKOOP, J.H., KUIT, A.R. and ZEGWAARD, A. (1977). In *Growth and Poultry Meat Production*. pp. 131–143. Eds Boorman, K.N. and Wilson, B.J. British Poultry Science; Edinburgh

VELU, J.G., BAKER, D.H. and SCOTT, H.M. (1972). *Poult. Sci.*, **51**, 698–699

WASHBURN, K.W., GUILL, R.A. and EDWARDS, JR., H.M. (1975). *J. Nutr.*, **105**, 1311–1317

WEBSTER, A.J.F. (1978). In *XX Br. Poult. Breeders Roundtable Conference*. pp. 45–55. Birmingham, UK

WHITEHEAD, C.C. and GRIFFIN, H.D. (1982). *Br. Poult. Sci.*, **23**, 299–305

WHITTEMORE, C.T. (1978). In *XX British Poultry Breeders Roundtable Conference*. pp. 57–70. Birmingham, UK

WOOD, J.D., WHELEHAN, O.P., ELLIS, M., SMITH, W.C. and LAIRD, R. (1983). *Anim. Prod.*, **36**, 389–397

11

MEAT QUALITY IN BROILERS, WITH PARTICULAR REFERENCE TO PIGMENTATION

B. LIPSTEIN
Institute of Animal Science, Agricultural Research Organization, The Volcani Center, Israel

Meat quality is a term which should imply enjoyment of eating. In assessing meat quality, it is necessary to use parameters which have direct relevance to the producer, processor and consumer. The basic aim of both the producer and the processor is to satisfy the requirements of the consumer. The idea of a satisfied consumer, who enjoys eating a given product, is very relative and difficult to define. Taste, colour and texture are major components of meat quality. These characteristics are connected with the human organs of sense and, consequently, are affected by many factors, such as country, region, tradition and even state of health or feeling of hunger or satiety.

The quality of meat depends on numerous factors. It is influenced by genetic characteristics, physiological and nutritional factors, age, sex, environmental housing and management systems, and processing techniques. This chapter will review some recent research pertaining to the nutritional effects, without implying that other factors are of less importance.

Pigmentation of broilers

Pigmentation is a very important factor influencing meat quality. The broiler meat is usually sold as a complete carcass or as portions with intact skin; therefore skin colour is essential for acceptable eating quality in countries where consumers prefer a well-pigmented broiler.

Consumers of poultry meat have, for generations, looked at the colour and associated the yellow pigment with the nutritional value of the product. However, the yellow pigment has no nutritional value for the consumer, and its value lies in the aesthetic appeal. Its importance in poultry products (meat and eggs) is due only to consumer preference.

The degree of pigmentation of broilers in different areas of the world is variable and dependent on tradition. Some countries prefer white-skinned birds, some yellow and some orange. Consumer preference appears to be related to geographical areas, and sometimes in the same country it is also related to regional preferences. For example, consumers in Belgium and Italy prefer yellow-orange pigmented broilers, as do those in Australia, Mexico, and some countries in Latin America. However, in the UK, Switzerland and most of France, a less pigmented

First published in *Recent Advances in Animal Nutrition – 1984*

chicken is preferred. In Spain, some areas prefer the well-pigmented broiler, while in others no marked preference is shown. The northeastern region of the USA demands a highly pigmented broiler, whereas the southwest puts less emphasis on colour. Some traditions are easily explainable; for instance, the northeastern part of the USA is the origin of breeds such as Rhode Island Red and Plymouth Rock, breeds which readily deposit the yellow pigment in their skin and shanks. On the other hand, the English breeds (Sussex, Orpington, etc.) are white-skinned birds.

The most important factor affecting the yellow colour deposited in broilers is the amount of xanthophyll contained in the feed. This was reported by Palmer in 1915. Producing highly pigmented broilers may become an expensive factor, especially since much higher levels of dietary pigments are required for broiler pigmentation than for egg yolk pigmentation, perhaps because a much larger surface area has to be pigmented. The cost of adding pigment sources to the diet makes it imperative to obtain the maximum pigmentation and efficiency from the dietary source. The resulting pigmentation is modified by several factors, such as genetic characteristics, sex, environment, health of the broilers, etc.

The pigmentation sources used in poultry feeds are the natural xanthophylls and synthetic carotenoids. Computer-formulated rations based on least-cost linear programs have increased the pressure to use synthetic carotenoids in small amounts to provide the colour desired by the consumer.

Good pigmentation depends on the characteristics of the administered xanthophylls. These have to be: available, so that they can be biologically utilized; stable and thus insensible to oxidation; and effective, so that the colour of the broiler skin has the right tone.

EVALUATING BROILER PIGMENTATION

The methods generally used for evaluation of broiler pigmentation are as follows:

(1) Visual scoring by comparison with the colour fan of Hoffmann La Roche and the Heiman-Carver Colour-Reter.
(2) Direct instrument reading by using a reflectance colorimeter which is able to determine colour by measuring the three dimensions of colour from the direct values XYZ.
(3) Chemical–spectrophotometric determinations using methods approved by AOAC.

Generally, the chemical–spectrophotometric methods are used to determine relative pigmenting properties of feedstuffs or relative pigmentation levels, whereas the visual techniques and the reflectance colorimetry are used to evaluate final broiler pigmentation.

Visual evaluations are often complemented by chemical methods. It appears, generally, that there exists a good relationship between visual scoring of the legs and toe-web xanthophyll levels (Lipstein, Bornstein and Budowski, 1967).

Visual scoring has proved to be a sensitive criterion of pigmentation, but in the range of deep colour it does not always show the best repeatability. The broiler pigmentation response to dietary xanthophyll increments in terms of visual impressions will follow the law of diminishing returns in diets containing high levels of xanthophyll (Bartov and Bornstein, 1966).

There exists an optimal-yellow which must be appetizingly pretty as determined by the corruptible human eye. Consumer acceptability thus depends on visual impression, which is not always well correlated with chemical determinations. That is the reason why the human eye is not very sensitive to the darker shades of yellow, and the standard colorimetric method (AOAC) is not sensitive to reddish pigments, which deepen the visual colour. Therefore, the use of high levels of xanthophyll to improve the visually observed broiler pigmentation seems to be questionable from the economic point of view, if a reddish pigment is available to enhance the yellow colour as seen by the eyes.

Plasma xanthophyll level appears to reflect only the present state of xanthophyll absorption, irrespective of previous dietary history, and hence is least correlated with actual pigmentation.

Recently, Middendorf, Childs and Cravens (1980a) used a new method to determine availability of xanthophyll by comparing the xanthophyll content in the diet with that in the serum following intubation with equal amounts of xanthophyll from various sources. A comparison of results concerning the availability of xanthophyll pigments in a variety of feed ingredients, obtained by Fletcher (1983) with those obtained by Middendorf, Childs and Craven (1980) and according to our findings (Lipstein, Bornstein and Budowski, 1967; Lipstein and Talpaz, 1984), raises doubts as to the suitability of blood xanthophyll as a practical criterion for pigment evaluation.

Most of the methods employed for determining both xanthophyll content in certain feed ingredients and diets as well as broiler pigmentation are not satisfactory. They do not differentiate among the carotenoid patterns nor reflect changes in their proportions.

SOURCES OF XANTHOPHYLLS

The pigments deposited in the skin, shank and foot pad are not synthesized by the birds themselves, but must be derived from their feed. Palmer (1915) and Palmer and Kempster (1919) concluded that the xanthophyll in body tissue was physiologically identical to that in the feed. In contrast, Smith and Perdue (1966) reported that the xanthophyll in broiler skin differed significantly from that in the diet. These pigments are carotenoids of various colours, from yellow to red, and are natural pigments structurally related to vitamin A. Only a few carotenoids are utilized by the broilers for their pigmentation. Pigmenting carotenoids are mainly found in those members of the group which have no vitamin A activity at all and are transferred unchanged to the skin, shank and foot pad. Carotenoids may be regarded as oxidation products of the carotenes. The hydroxycarotenoids are known as oxycarotenoids or xanthophylls.

The most important pigment-containing materials used in poultry feeds are yellow maize, alfalfa and grass meals and various products made from these materials (Day and Williams, 1958; Ratcliff, Day and Hill, 1959; Waldroup *et al.*, 1960; Ratcliff *et al.*, 1962; Dua *et al.*, 1967; Kuzmicky *et al.*, 1968). The predominant xanthophylls in these are lutein and zeaxanthine. The xanthophyll contents of these common feed ingredients vary within exceptionally wide limits (*Table 11.1*).

Marigold meal has been shown to be an effective xanthophyll source (Brambila, Pino and Mendoza, 1963; Twinning *et al.*, 1971), as has florofil-blossom meal of *Tagetes erecta* (Leibetseder and Schweighardt, 1979) and dehydrated bluegrass

Table 11.1 AVERAGE XANTHOPHYLL CONTENT OF SOME NATURAL MATERIALS

Materials	Xanthophyll (mg/kg)
Yellow maize	20–25
Maize gluten meal (60% protein)	330
Maize gluten meal (41% protein)	90–180
Dehydrated alfalfa meal (17% protein)	185–350
Dehydrated alfalfa leaf meal (20% protein)	400–550
Marigold petal meal (*Tagetes erecta*)	6000–10 000
Coastal Bermuda grass	185–350
Mexico pollen	345
New York State pollen	440
Broccoli leaf meal	670
Acidulated soyabean soapstock	168–260
Common algae meal	2200[a]

(Data from Patrick and Schaible, 1980)

[a]Assays at our laboratory showed that the xanthophyll concentration of sewage-grown algae ranged from 250 to 1150 mg/kg

(Halloran, 1974). A by-product of the refining of soyabean oil, acidulated soyabean soapstock, was tested by Lipstein, Bornstein and Budowski (1967). Results obtained from this study indicated that acidulated soapstock containing 168–260 mg/kg xanthophyll may serve as a pigmenter for broilers.

In search of less expensive sources of pigment-containing feedstuffs, dehydrated turf grass was tested. According to Willis and Baker (1980), the latter contains at least four times more available xanthophyll than maize gluten meal.

Several other feed ingredients, including algae meal, have been shown to be effective pigmenting agents (Perdue and Smith, 1962; Smith and Perdue, 1966; Marusich and Bauernfeind, 1970). The last mentioned reported that the responses to xanthophylls from two algae meal samples (*Chlorella* and *Spongiococcum*) and alfalfa meal were about equal. Sewage-grown algae meal (*Chlorella* and *Micractinium*) contain large amounts of xanthophylls and can be used effectively as pigment sources in broiler diets (Grau and Klein, 1957; Lipstein and Hurwitz, 1980, 1981; Lipstein and Talpaz, 1984).

Synthetic xanthophylls such as apo-ethyl ester, canthaxanthin and zeaxanthine have been widely tested (Marusich and Bauernfeind, 1970; Marusich *et al.*, 1976). The manner in which the chemically produced xanthophylls were used, their ratio to each other, and to the natural mixed xanthophylls was thought to be important to achieve optimal pigmentation.

AVAILABILITY OF XANTHOPHYLLS FROM DIFFERENT SOURCES

Efficiency of utilization of equal quantities of xanthophyll from different sources varies from case to case, and they do not necessarily produce identical broiler pigmentation. Whether a product with a large amount of xanthophyll can be a good pigmenter source depends not only on the amount of xanthophyll but also on its pigmenting capacity, and in this respect there exist differences between broiler and yolk pigmentation. The intensity of pigmentation in broilers is related mainly to the total amount of xanthophylls consumed and their availability. The availability depends on many factors, such as the carotenoid pattern and the relative degree of deposition of each individual carotenoid, its origin and processing and storage conditions.

The literature on broiler pigmentation shows considerable variation regarding the relative pigmentation utilization of alfalfa and yellow maize. Quackenbush *et al.* (1963) reported that the total xanthophyll content and the composition differ markedly among yellow maize inbreds. They analysed 125 varieties of hybrid maize and found that the lutein content ranged from 2.0 to 33.1 µg/g; for zeaxanthin this range was 0.6 to 27.4; moreover, the lutein:zeaxanthin ratio varied from 6.0:1.0 to 1.0:2.7. A high lutein–low zeaxanthin maize has been shown to be a much more effective pigmenter than a high zeaxanthin-low lutein maize (Quackenbush *et al.*, 1965). Bartov and Bornstein (1967) reported variations of up to 25% in the xanthophyll utilization from maize of different origins and storage conditions and a difference in their xanthophyll patterns.

Middendorf, Childs and Cravens (1980b) reported that when the availability of apo-ethyl ester was given a relative value of 100, the availability of xanthophyll from maize gluten meal ranged from 48 to 89%, and that from dehydrated alfalfa ranged from 35 to 65%.

Processing and storage conditions of dehydrated alfalfa meal affect both the total xanthophyll content and the xanthophyll composition. Livingston *et al.* (1969) observed greater availability of xanthophyll from extracts of dehydrated alfalfa and maize gluten meal than from the original product, even though the individual xanthophyll compositions were similar. Reports concerning the individual xanthophyll composition of dehydrated alfalfa meals reveal considerable variation, from 46% lutein and 6% zeaxanthin (Bickoff *et al.*, 1954) up to 76% lutein and 4% zeaxanthin (Livingston *et al.*, 1969). The same amounts of xanthophyll from dehydrated alfalfa and fresh alfalfa caused reduced utilization of the former as compared with that of the latter, and this points to the influence of the process of dehydration (Tortuero and Centeno, 1976). The xanthophylls (450–950 mg/kg) from high-protein alfalfa leaf concentrate (X-PRO) showed xanthophyll availability equivalent to that from dehydrated alfalfa meal (Halloran, 1974). Kuzmicky *et al.* (1977) concluded that high protein alfalfa leaf concentrate, Pro-Xan, could be prepared commercially with an xanthophyll availability of about 1.7 times that of dehydrated alfalfa or maize gluten meal, and about three times that of marigold meal. This high availability of Pro-Xan xanthophyll could have a marked effect on its economic value as a feed ingredient.

It appears, therefore, that the relative utilization of alfalfa meal varies from 31 to 75% (Day and Williams, 1958; Ratcliff, Day and Hill, 1959; Ratcliff *et al.*, 1962; Dua *et al.*, 1967) and from 84 to 100% (Lipstein, Bornstein and Budowski, 1967) of that of yellow maize. However, Kuzmicky *et al.* (1968) reported that the xanthophylls in alfalfa, yellow maize and maize gluten meals produce equal broiler skin pigmentation. The availability of marigold meal ranged from 38 to 65% of that of maize gluten meal xanthophyll (Twinning *et al.*, 1971).

The relative utilization of soyabean oil soapstock was 40–46% in comparison with the pigmentation efficiency of maize xanthophylls (Lipstein, Bornstein and Budowski, 1967). The lutein makes up 13–16% of the soyabean oil soapstock-derived xanthophylls and 55–65% of the maize xanthophylls (Bartov and Bornstein, 1967). These results may be construed to indicate that lutein plays a more important role than other xanthophyll fractions in broiler pigmentation. The same amount of dietary xanthophyll from three different sources produced different results in broiler pigmentation (*Table 11.2*), which were caused by the differences in

Table 11.2 THE INFLUENCE OF ACIDULATED SOYABEAN SOAPSTOCK (ASS)[a], ALFALFA MEAL[b] AND YELLOW MAIZE[c] ON BROILER PIGMENTATION. (AVERAGE OF 12 CHICKS)

Diets		*Xanthophylls*		*Xanthophylls in*		*Visual score*[d] *(%)*	*Relative utilization*[e]
Xanthophyll supplements (%)		*In diet (μg/g)*	*Intake (mg)*	*Plasma (μg/100 ml)*	*Toe-web (μg/100 cm²)*		
—	—	1.3	2.2	22	25	0.2	
ASS	1.9	5.2	8.6	71	37	0.4	
	3.8	9.3	15.2	107	61	1.2	40–46
	5.7	13.9	22.5	161	69	1.7	
Alfalfa	1.4	5.3	8.6	93	49	0.6	
meal	2.7	8.2	13.5	181	76	1.7	84–100
	4.0	11.6	18.8	260	110	2.7	
Yellow	17.0	5.6	8.5	104	53	1.1	
maize	34.0	8.7	13.6	207	93	2.5	100
	51.0	12.7	20.8	282	131	3.5	

(Data from Lipstein, Bornstein and Budowski, 1967)
[a]Xanthophyll content 209.3 μg/g
[b]Xanthophyll content 294.0 μg/g
[c]Xanthophyll content 23.3 μg/g
[d]Hoffmann–La Roche colour units
[e]Relative to yellow maize which was considered to be 100%

Table 11.3 THE INFLUENCE OF YELLOW MAIZE, ALFALFA MEAL, APO-ETHYL ESTER AND CANTHAXANTHIN ON SHANK PIGMENTATION

Xanthophyll source	*Xanthophyll level (g/tonne feed)*			*Shank visual score (Fan 1–15)*
	Natural	*Synthetic*	*Total*	
Yellow maize (YM)[a]	10	—	10	3.9
YM + 2.3% AM[b]	20	—	20	6.3
YM + 4.6% AM	30	—	30	7.5
YM + 6.9% AM	40	—	40	8.5
YM + apo-ethyl ester[c]	10	10	20	7.3
YM + apo-ethyl ester	10	20	30	9.2
YM + apo-ethyl ester	10	30	40	10.2
YM + 2.3% AM + canthaxanthin[d]	20	2.5	22.5	8.2
YM + 2.3% AM + canthaxanthin	20	4.0	24.0	9.9
YM + 4.6% AM + canthaxanthin	30	2.5	32.5	9.1
YM + 4.6% AM + canthaxanthin	30	4.0	34.0	10.1

(Data from Marusich and Bauernfeind, 1970)
[a]60% Yellow maize (YM) ration provided 10.0 g xanthophyll/tonne
[b]Alfalfa meal (AM) contained 440 g xanthophyll/tonne
[c]Stabilized beadlets (115 g/kg)
[d]Stabilized beadlets (104 g/kg)

relative xanthophyll utilization among the three xanthophyll supplements (Lipstein, Bornstein and Budowski, 1967).

Marusich and Bauernfeind (1970) reported that the response of the xanthophylls from the natural ingredients was lower than the synthetic xanthophylls. *Table 11.3* shows that apo-ethyl ester provided for higher shank visual scores than the equivalent levels of xanthophylls from alfalfa. The addition of canthaxanthin at 2.5

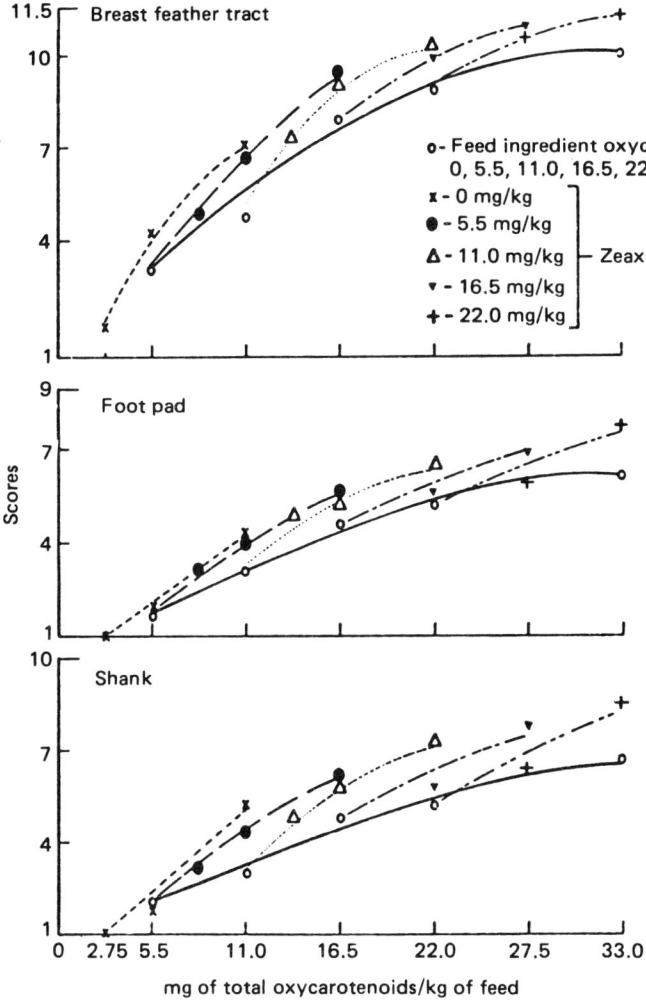

Figure 11.1 Visual pigmentation scores of broilers fed graded concentrations of feed ingredient for seven weeks and zeaxanthin for only the last three weeks. (Data from Marusich *et al.*, 1976)

or 4.0 mg/kg to a ration containing from 20–30 mg/kg of xanthophylls from maize and alfalfa resulted in still higher visual scores, all in the acceptable colour range.

Marusich *et al.* (1976) reported that synthetic zeaxanthin is a very effective broiler pigmenter producing a highly acceptable yellow to yellow-orange colour. Visual scores for the shank, foot pad and breast skin (*Figure 11.1*) all increased as the dietary level of feed ingredient xanthophyll increased and as the dietary concentration of zeaxanthin increased. Marusich *et al.* (1976) also reported that the response to the xanthophylls from natural ingredients was lower than to synthetic zeaxanthin: 1 g of zeaxanthin can replace up to 2 g of mixed xanthophylls from feed ingredients.

Precise information on a wide spectrum of significant numbers of representative xanthophyll-available sources will enable the nutritionist to supply the degree of pigmentation required by a specific market area with the consistency it demands. It is not advisable for a feed manufacturer to rely solely on natural pigment sources, since their effects on the colour of shanks cannot be assessed reliably in advance. In commercial feed manufacture it is usually not practical to attempt to analyse continuously feed ingredient materials such as maize and alfalfa meals in order to check their xanthophyll contents.

The advantage of the synthetic carotenoids lies in their stability, which enables the feed manufacturer to guarantee diets with a declared pigment content to meet a demand for a specific degree of pigmentation. Moreover, the use of the synthetic carotenoids makes possible the formulation of rations containing variable, and even very high, concentrations of pigments without affecting dietary composition or choice of ingredients as far as other nutrient requirements are concerned.

DIETARY FACTORS AFFECTING BROILER PIGMENTATION

Xanthophylls from natural ingredients are not stable components and their concentration in feeds is reduced by oxidation. The addition of vitamin E or synthetic antioxidants to the feed was found to increase the pigment deposition in broilers (Potter *et al.*, 1956; Fritz and Wharton, 1957; Elrod *et al.*, 1958; Waldroup *et al.*, 1960). However, there seems to be considerable variation in the action of different antioxidants. The effect of ethoxyquin (EQ) was shown to be more effective than butylated hydroxytoluene (BHT) in improving broiler pigmentation (Ratcliff, Day and Hill, 1961). Bartov and Bornstein (1966) reported that the effect of ethyoxyquin is restricted to the stabilization of dietary xanthophylls and to reducing their destruction in the feed.

Dietary fat supplementation tends to improve xanthophyll absorption. Some authors (Day and Williams, 1958; Health and Shaffner, 1972) have reported a positive correlation between dietary fat and pigment deposition. However, Ratcliff, Day and Hill (1959) and Herrick *et al.* (1970) found that added fat had either no effect or a negative one on pigmentation.

The conflicting results may be explained by the fact that the effect of dietary fats on broiler pigmentation depends on several factors, such as the amount and composition of the fat supplement, the presence of antioxidants, the quality of the added xanthophylls, and the presence of unsaponifiable materials present in the diets, such as sterols and vitamins.

EFFECTS OF LENGTH OF FEEDING PERIOD

The intensity of pigmentation in broilers is related mainly to the total amount of xanthophylls consumed. This amount is the product of the xanthophyll concentration in the diet, the daily feed intake and the length of the feeding period.

The investigations concerning the minimum feeding period required for satisfactory pigmentation have resulted in contradictory data. Fritz, Wharton and Classen (1957), Mitchell, Bletner and Tugwell (1961) and Combs and Nicholson (1963) reported that desirable pigmentation can be obtained by feeding xanthophyll-containing diets for the last four weeks of the growing period, after chicks had been depleted of their initial xanthophyll reserves. According to Couch, Farr and Camp (1963), Day and Williams (1958) and Braunlich (1974), the accumulation of pigmenting compounds in the skin and shanks of broilers can be accomplished even in two to three weeks, and hence it is not necessary to use feed high in xanthophylls except during the finishing period.

Broilers raised up to seven weeks of age on a diet with a low level of xanthophylls and then fed a finisher diet containing a high level of xanthophylls for four weeks, did not show desirable pigmentation, whereas the same finisher diet fed to non-depleted chicks produced deeply pigmented shanks (Bartov and Bornstein, 1969).

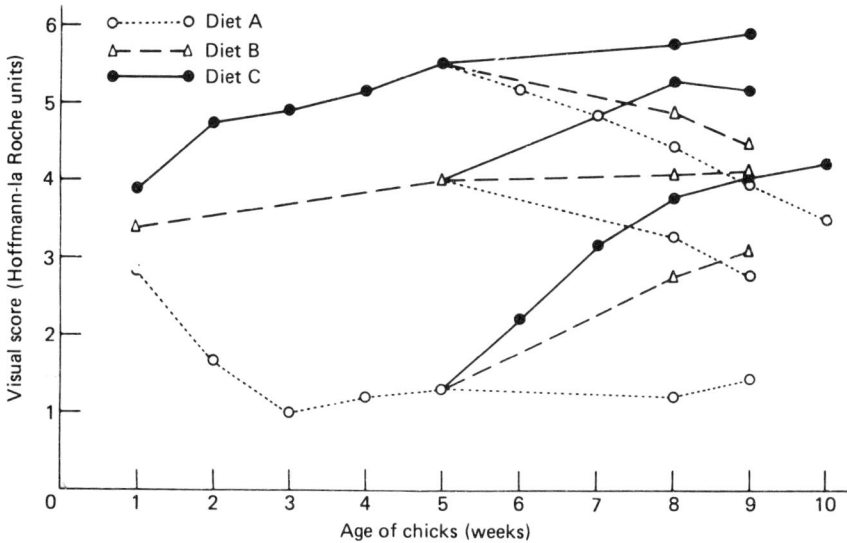

Figure 11.2 Effect of low, medium and high levels of xanthophylls in starter (6.7, 16.9 and 25.4 μg/g in diets A, B and C, respectively) and finisher (6.4, 15.0 and 24.9 μg/g in diets A, B and C, respectively) diets in a factorial arrangement on the development of shank pigmentation. (Data from Bartov and Bornstein, 1969)

There exists a close interdependence between the xanthophyll status of the young chick (up to five weeks of age) and that of the eight- to ten-week old broiler. Chicks fed low levels of xanthophylls in their starter diets were not able to achieve a desirable degree of pigmentation at eight to ten weeks of age, in spite of the relatively high levels of xanthophylls included in their finisher diets. Chicks which received the lowest xanthophyll level in their starter diet and the highest level in their finisher diet, exhibited less shank pigmentation than those broilers which had been fed the reverse combination (*Figure 11.2*). The level of xanthophylls in the starter diets seems to have a pronounced effect on final pigmentation. A similar trend involving synthetic carotenoids was reported by Marusich and Bauernfeind (1970), who compared the feeding of xanthophylls for the entire eight to nine week period with having them present in the diet for the last three to five weeks only. They found that about 50% more total xanthophyll must be present if it is fed for the last three to five weeks only.

NON-DIETARY FACTORS AFFECTING BROILER PIGMENTATION

Environment

Besides dietary factors affecting broiler pigmentation, the environment may have considerable influence on shank colour (Collins, Thayer and Skoglund, 1955). Bartov and Bornstein (1967) reported that exposure of maize to the sun reduced its total xanthophyll content and the pigmenting ability of the diets containing this maize. The individual xanthophylls are not equally affected by adverse storage conditions, thereby altering the relative ratios of the individual xanthophylls and thus modifying the resulting colour. Fletcher *et al.* (1977) found that broilers raised in conventional open-sided pens pigmented better than broilers raised in window-less low-light environmentally modified houses. They suggested that higher light intensity could improve pigmentation.

Goodwin (1971) discussed several studies in which it was shown that xanthophylls such as zeaxanthin and lutein could be isomerized through reversible epoxidation cycles requiring oxygen and light. There is evidence that the effect of light on broiler pigmentation could be due to the light altering the metabolism of the xanthophylls in the bird, by

(1) causing selective oxidation and degradation of only some of the xanthophylls present in the diet, or
(2) triggering the alternation of pigments in the feed.

Fletcher (1981) reported that the influence of light intensity on broiler pigmentation is due at least in part to the effect of light on the xanthophyll pigments within the feed. Although heat and light reduce the total xanthophyll concentration, light by itself may also cause the structural alteration of isomerization discussed previously. Thus, the total concentration of xanthophyll may decrease, but the resulting colour from the altered xanthophyll may be increased. Other factors that can affect the final colour or pigmentation are the effects of light on the bird's metabolism or absorption and metabolism. The problem of decreased pigmentation as a result of housing birds under reduced light housing can be resolved by the dietary inclusion of synthetically produced red xanthophyll pigments or feedstuffs such as maize gluten meal (Janky, 1983), which contain relatively larger concentrations of red xanthophyll pigments, since the addition of red pigment to yellow pigment enhances the latter more than further addition of yellow pigment, as explained previously.

Pathological conditions

Coccidiosis, in common with other diseases affecting the digestive tract of birds, may interfere with normal intestinal absorption processes and lead to a poorly pigmented broiler (Littlefield, Bletner and Shirly, 1970; Kowalski and Reid, 1970; Marusich *et al.*, 1972). However, Marusich *et al.* (1973) reported that shank and breast skin visual scores, and plasma and toe-web xanthophylls were not depressed as a result of coccidial infection in diets containing canthaxanthin. These results indicate the effectiveness of canthaxanthin as a pigmenter adjuvant when added to diets containing xanthophylls from natural feed ingredients.

The presence of the mycotoxins, ochratoxin (Huff and Hamilton, 1975) as well as aflatoxin (Tung and Hamilton, 1973) decreases plasma xanthophylls, which can result in an underpigmented broiler. Both aflatoxin and ochratoxin caused a general inhibition of lipid transport, which, in turn, tended to impair xanthophyll absorption.

Sex, age, strains and processing of birds

Xanthophyll concentration in the plasma of male birds was significantly higher than in female birds (Collins, Thayer and Skoglund, 1955; Grabowski and Szuman, 1963; Lipstein, Bornstein and Budowski, 1967). In contrast, Fry, Harms and Moeller (1976) found that females generally have more pigmentation than males when fed a common diet.

Lipstein, Bornstein and Budowski (1967) showed that age had a significant effect on the level of xanthophyll retention and plasma xanthophyll. Xanthophylls derived from alfalfa meal were retained significantly better at 58 days than at 33 days of age. In a parallel trend, Delpech, Dumont and Nefzaoui (1983), reported that colour was more intense in older chickens than in younger ones.

Significant differences in broiler pigmentation exist among different strains and crosses of broilers. Collins, Thayer and Skoglund (1955) reported that shank colour of New Hampshire chickens was significantly darker than that of White Plymouth Rock. Harms, Fry and McPherson (1977) and Scholtyssek (1978) reported that male lines of broiler chicks have a decided effect on the degree of pigmentation obtained. The strain with the least amount of pigmentation would require approximately 4.1 g more xanthophyll per tonne to achieve a pigmentation equal to the strain with the highest pigmentation. This indicates that the degree of pigmentation may be a heritable characteristic.

Improper processing can reduce the intensity of pigmentation of broilers. If the temperature and length of scald are increased too much the degree of colouring may be reduced. Much of the pigmentation is lost if the cuticle is removed due to excessive scalding or maladjusted picking machines (Health and Thomas, 1973, 1974). The pH of the scald water resulting from an optimum level of additive for feather removal also retained the most yellow colour in the skin (Health and Wabeck, 1975).

Carcass fatness

The quantity and quality of carcass fat and meat are important aspects of the texture. Texture is one of the most important characteristics of broiler meat quality. It seems to be very difficult to find an appropriate definition for the attributes of the term 'texture'. The consumer hopes that certain sensory criteria— tenderness, firmness, juiciness, fatness, flavour and taste—will be met, and these are the qualities that are included in the concept of texture (Scholtyssek, 1980). The above mentioned terms are interrelated, influence each other, and are all dependent on the main feed ingredients. By adjusting dietary constituents it is possible to produce broilers with widely varying amounts of body fat (Fraps, 1943).

Carcass composition changes occur mainly in the moisture-to-fat ratio, with the protein level remaining relatively constant. Fat is deposited into tissue at the expense of water and is closely related to tenderness (Maw, 1935). The determination of body water provides a simple but reliable means of estimating the concentration of body fat (Velu, Scott and Baker, 1972). A high correlation was observed between the level of fat and dry matter in the back skin, and between dry matter in the back skin and the visual scoring of the degree of fatness, and between the latter and the determined quantity of abdominal fat (Bartov, Bornstein and Lipstein, 1974; Becker *et al.*, 1979). Moisture and fat contents of meat are affected by sex and age (Evans, Goodwin and Andrews, 1976; Singh and Essary, 1974; Grey *et al.*, 1983). This applies also to fat content in the skin (Suderman and Cunningham, 1980).

The development of visceral fat during the finishing period is rapid. From four to eight weeks of age carcass fat content was found to increase some 12% while visceral fat content increased by 40%, with marked differences between the sexes (Summers and Leeson, 1979). Tenderness of eight-week old broilers is less than that of either six- or seven-week old birds. The highest tenderness value was obtained from seven-week old broilers, but females of this age have a higher fat content (Evans, Goodwin and Andrews, 1976; Ricard, 1983). The marked difference between males and females emphasizes the importance of differentiating between slaughtering ages of females and males. Early slaughtering of females is of utmost significance.

Differences among commercial broiler strains in abdominal and carcass fat provided partial evidence that these traits are influenced by heredity (Goodwin, Andrews and Webb, 1969; Moran, Orr and Larmond, 1970; Edwards and Denman, 1975; Cherry, Siegal and Beame, 1978; Markley *et al.*, 1980). However, Summers and Leeson (1979) and Becker *et al.* (1981) did not find any statistically significant differences in visceral and abdominal fat among tested strains.

At present, the widely accepted theory is that modern broiler strains have been bred with greater ability to grow fast in a short time by increasing their capacity for feed consumption without a parallel improvement in the birds' capability to convert feed into lean tissue (Summers and Leeson, 1979; Lin, Friars and Moran, 1980). Therefore, the selection for rapid growth also produces broilers with the propensity to deposit large quantities of fat. Since excessively fat broilers have become of acute and worldwide concern, it is important to detail the various aspects of the problem.

Broilers containing excessive abdominal fat are often less desirable for:

(1) a producer, because the abdominal fat is associated with a waste of dietary energy;
(2) a poultry-processing plant, because the removed fat increases waste disposal problems and the processor has a reduced yield problem; and
(3) a consumer, when the fat comprises part of the marketable product and the consumer has increased cooking loss.

The consumer, the processor and the producer all consider this excessive fat a waste product.

NUTRITIONAL FACTORS AFFECTING CARCASS FATNESS

Fattening of broilers occurs when the amount of energy consumed exceeds the bird's requirements for maintenance and growth. Decreasing dietary energy concentration but maintaining the same protein level reduces carcass fat content (Hill and Dansky, 1954; Essary *et al.*, 1960; Carew and Hill, 1964). High dietary energy level alone is not a major factor responsible for the excess energy consumption (Bartov, Bornstein and Lipstein, 1974; Edwards and Hart, 1971; Bartov and Bornstein, 1976a; Griffiths, Leeson and Summers, 1976b; Coon, Beker and Spencer, 1981).

The ratio of dietary energy to protein (E:P) or energy to balanced amino acids is a more important regulator of carcass fat content. Broilers fed diets deficient in protein, with a wide E:P ratio, consume more energy than is required for the relatively low growth rate. Under slight protein or amino acid deficiencies, growth will not necessarily be impaired, since the broilers are able to satisfy their respective needs by increasing their feed intake. In both cases the extra energy consumed results in increased fat deposition.

Reduction in the rate of hepatic lipogenesis is observed in broilers fed diets containing increased protein levels (Yeh and Leveille, 1969). This effect cannot be explained adequately on the basis of reduction of dietary carbohydrates that accompanies the increased protein level (Leveille *et al.*, 1975). Increasing dietary protein levels elevates uric acid concentration in the body of birds (Okumura and Tasaki, 1969). Considerable energy is required for synthesis of this compound (Buttery and Boorman, 1976). Part of the effect of a high-protein diet in decreasing carcass fat content may be due to the need for the broiler to expend more energy in order to eliminate excess nitrogen from the body.

The effect of dietary protein level or E:P ratio on the amount of carcass fat is rapid and reversible (Thomas and Twinning, 1971; Bartov, Bornstein and Lipstein, 1974; Jackson, Summers and Leeson, 1982; Salmon, Classen and McMillan, 1983). *Figure 11.3* shows that a relatively low E:P ratio prevents the accumulation of dermal and visceral fat. Transferring the bird from a diet with a low E:P ratio to

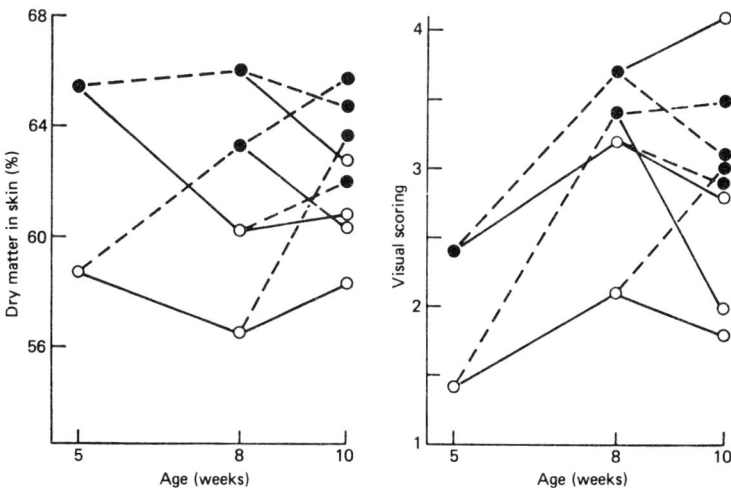

Figure 11.3 Effect of feeding diets containing low (——O) and higher (---●) E:P ratios during different stages of the growth period on the percentage of dry matter in the back skin (left) and on visual scoring of the amount of visceral fat (right). The scores ranged from 0.5 for no visible fat to 5.0 for extreme obesity. (Data from Bartov, Bornstein and Lipstein, 1974)

one with a high E:P ratio or *vice versa* results in a rapid and marked change in the degree of fatness within a two-week period.

The effect of specific amino acid deficiency on feed intake and body composition is different from that of a low-protein diet or a diet containing a wide E:P ratio. While both of these latter diets retard growth, broilers fed a diet with normal protein concentration but severely deficient in a single amino acid fail to show any over-consumption of feed and concomitant increase in body fat (Lipstein, Bornstein and Bartov, 1975; Griffiths, Leeson and Summers, 1977a; Velu, Baker and Scott, 1971).

Addition of methionine and lysine to well-balanced finisher diets, which were lowered in protein by replacing soyabean meal with sorghum, decreased fat deposition (Lipstein, Bornstein and Bartov, 1975). Increasing the level of glutamic acid in a diet containing a low level of an essential amino acid mixture reduces carcass fat (Velu, Scott and Baker, 1972). Narrowing the E:P ratio from a level considered optimum for growth by the addition of feather meal (poor quality protein) to the diet is equally as effective in reducing abdominal fat as the addition of high-quality protein (Griffiths, Leeson and Summers, 1977a). Mabray and Waldroup (1981) reported that through dietary manipulations it was possible to produce carcasses of widely varying abdominal fat content. *Table 11.4* shows that the

Table 11.4 ABDOMINAL FAT PAD WEIGHT (g) OF BROILERS FED DIETS CONTAINING DIFFERENT CONCENTRATIONS OF METABOLIZABLE ENERGY (ME) AND ESSENTIAL AMINO ACIDS

ME	*Percent NRC amino acid requirements*				
(MJ/kg)	*70*	*80*	*90*	*100*	*120*
12.43	41.5[d]	36.2[c]	32.2[b]	31.2[b]	27.2[a]
13.35	46.0[e]	44.7[de]	36.7[c]	33.7[bc]	32.0[b]
14.27	56.5[f]	45.0[c]	47.6[c]	45.0[e]	31.1[b]

(Data from Mabray and Waldroup, 1981)
a,b,c,d,e,f Values with different superscripts are significantly different ($P < 0.05$)

degree of fatness could be reduced by increasing the amino acid level of the diet within the same energy level, or decreasing the energy concentration within the same essential amino acids level and pattern.

Attempts to control the fat content of the broiler carcass by nutrient restriction have yielded varying results. Griffiths, Leeson and Summers (1977b) found that restricting the energy intake of chicks in the birth to three weeks of age group resulted in no significant decrease in abdominal fat deposition at eight weeks of age. March and Hansen (1977), however, demonstrated that the restriction of nutrient intake by dietary dilution of a basal diet from birth until six weeks of age decreased growth rate and lipid accumulation in the tissues of broilers. This is in agreement with Simon *et al.* (1978) who found that restricted feed intake from six to eight weeks reduced carcass fat content and body weight at the end of this period. Arafa *et al.* (1983) found that energy restriction during the ten day finishing period showed promise as a practical means of reducing abdominal fat without a loss in feed efficiency, even though body weight was reduced.

Fasting, even for short periods of time, depresses lipogenesis, whereas re-feeding following the fast increases it (Leveille *et al.*, 1975). Over-consumption of feed due to force-feeding increases the rate of hepatic lipogenesis and carcass fat (Shapira, Nir and Budowski, 1978).

TYPE OF CARCASS FAT

The importance of type of carcass fat for meat quality is due to the fact that a part of it appears as intramuscular fat. The latter has a great influence on the tenderness of meat and its stability.

The composition of carcass fat is affected mainly by dietary fat and the degree of lipogenesis. Including even relatively low levels of unsaturated fat in the diet markedly reduces the degree of saturation of carcass lipids.

The most saturated carcass fat was obtained by broilers fed diets containing tallow and diets without any fat supplementation, and the least saturated carcass fat by feeding diets with acidulated soyabean soapstock (Bartov, Lipstein and Bornstein, 1974). Acidulated cottonseed soapstock resulted in more saturated fat, apparently due to interference of the cyclopropenoid acids with the metabolic desaturation of stearic acid in the body (Evans *et al.*, 1962; Frampton *et al.*, 1966; Raju and Reiser, 1967; Lipstein, Bornstein and Budowski, 1970). Fattening of broilers due to enhanced consumption of diets relatively low in protein and lacking added fat, is mainly the result of increased biosynthesis of fat from surplus carbohydrates. The process of converting feed carbohydrates into body fat will cause a predominance of palmitic, palmitoleic, oleic and stearic acids. Linoleic acid, the main fatty acid in most cereal grain oils and in most of the fat supplements of vegetable origin, and linoleic acid are not synthesized by birds (Reiser, 1950). However, these fatty acids are readily absorbed and deposited in the body at the expense of palmitic, palmitoleic and oleic acids (Marion and Woodroof, 1966; Edwards *et al.*, 1973; Bartov and Bornstein, 1976a).

The graded effect of diets containing higher levels of unsaturated fat on the relative degree of unsaturation of adipose tissue in broilers being fed relatively low protein diets (*Table 11.5*) may be construed as evidence that synthesis of endogenous fat is not depressed in the presence of unsaturated dietary fatty acids. The above relationship between dietary E:P ratio and the composition of carcass fat may compound the effects of factors such as type or proportion of added fat supplements on the composition of carcass fat.

Feeding broilers diets containing low E:P ratio consistently produced lean carcasses (Bartov and Bornstein, 1976b). This phenomenon was accompanied by a greater susceptibility of the abdominal fat of lean broilers to oxidation and a decrease in the stability of their meat. These data on the relationship between E:P ratio and meat stability support results reported by Marion and Woodroof (1966) and Marion, Boggess and Woodroof (1967), that increasing dietary protein content decreases the stability of broiler meat. The lower stability of meat and abdominal fat from lean broilers seems to be due to the decreased saturation of their carcass fat (due to the reduced lipogenesis of body fat from dietary carbohydrates) which in turn decreases its stability (Klose *et al.*, 1951; Salmon and O'Neil, 1973; Bartov, Lipstein and Bornstein, 1974; Bartov and Bornstein, 1976b).

The chemical and physical properties of carcass fat can be substantially improved by omitting or replacing the unsaturated dietary fat supplement during the last

Table 11.5 EFFECT OF UNSATURATED FAT (ASS)[a] AND ENERGY TO PROTEIN (E:P) RATIOS ON THE INDEX OF FATTY ACID SYNTHESIS[b], RELATIVE DEGREE OF UNSATURATION[c] AND COEFFICIENT OF VARIATION (CV) OF IODINE VALUE (IV) OF ABDOMINAL FAT AND DRY MATTER (DM) IN SHIN

Unsaturated fat content (%)	*E:P ratio (MJ/kg per % protein)*	*Index of fatty acid synthesis*	*Relative unsaturation (%)*	*CV*	
				IV	*DM*
0	0.536	5.11	100	4.4	7.7
	0.774	7.75	96	4.0	6.8
	0.536	2.21	100	3.1	5.4
2	0.632	2.63	99	4.6	5.1
	0.791	3.23	93	2.0	2.0
	0.531	1.07	100	2.9	8.7
5	0.628	1.50	94	4.0	3.9
	0.791	1.85	89	3.6	5.1

(Data from Bartov and Bornstein, 1976a)
[a]Acidulated soyabean soapstock
[b]Percentages of 16:0 + 16:1 + 18:0 + 18:1 divided by percentages of 18:2 + 18:3
[c]Iodine value of broiler fed the narrowest E:P ratio = 100%

three or four weeks before slaughter (Lipstein and Bornstein, 1973; Bartov, Lipstein and Bornstein, 1974). It appears, therefore, that from the point of view of carcass quality—namely, of preventing liquid carcass fat and excess water uptake in broiler processing plants, or improving shelf-life—unsaturated dietary fat supplements could and should be withdrawn from the ration or replaced by a more saturated fat during the last few weeks of the growing period.

EXTENDING SHELF-LIFE BY STABILIZING CARCASS FAT

In order to avoid any misunderstanding, it should be emphasized that in the following paragraphs vitamin E is considered in its non-specific role as a natural antioxidant, rather than as a specific vitamin, corresponding to the synthetic antioxidants butylated hydroxytoluene (BHT) and ethoxyquin (EQ). The latter are added routinely to poultry rations in order to protect their fat-soluble vitamins against oxidation. Vitamin E is distributed over almost all the adipose tissue and in the intracellular fat of other tissues.

Unsaturated carcass fat is relatively less stable to oxidation and this affects meat stability. Vitamin E supplementation of broiler diets usually leads to its storage in body fat (Mecchi *et al.*, 1956; Bartov and Bornstein, 1976b) and muscle (Opstvedt, 1973; Marusich *et al.*, 1975), and prevents oxidative deterioration. Bartov and Bornstein (1977a) investigated the effect of vitamin E supplementation of diets which contained no added fat or fats of various sources, on the stability of the abdominal fat and the thigh muscle of broilers. *Table 11.6* shows that dietary vitamin E improved significantly the stability of abdominal fat and meat in broilers having relatively saturated carcass fat, whereas its beneficial effect was rather limited as the degree of unsaturation of carcass fat increased.

Synthetic antioxidants, such as BHT and EQ, are commonly used in poultry diets. These compounds act in a similar manner in protecting dietary lipid components against oxidation, as well as in improving performance of broilers (Bartov and Bornstein, 1972). Furthermore, synthetic antioxidants can partly or

Table 11.6 EFFECT OF DIETARY FAT SUPPLEMENTS AND VITAMIN E CONCENTRATIONS ON THE IODINE VALUE OF ABDOMINAL FAT, ITS STABILITY AND THE STABILITY OF THIGH MUSCLES OF NINE-WEEK OLD BROILERS

Dietary variable		Iodine value	Peroxide value (mEq/kg)	Thigh muscle TBA[a] value
Type of fat (4%)	Vitamin E (mg/kg)			
None	0	68.5[b]	43.2[b]	144[c]
	20	67.3	5.4	75
	40	67.6	0	46
Tallow	0	66.0	38.7	144
	20	66.5	16.9	62
	40	65.9	0	38
Acidulated soyabean soapstock	0	85.0	42.4	155
	20	86.7	32.8	94
	40	88.0	31.9	87
Refined soyabean oil	0	95.6	69.4	160
	20	97.7	58.2	132
	40	95.5	53.7	145

(Data from Bartov and Bornstein, 1977a)
[a]Expressed as µg sodium salt of malonaldehyde-bis-bisulphite/g of tissue
[b]Averages of three pooled samples of two broilers each
[c]Average of six broilers

completely prevent signs of vitamin E deficiency in chicks, EQ being more active than BHT (Machlin, Gordon and Meisky, 1959; Krishnamurthy and Bieri, 1962; Machlin and Gordon, 1962; Bartov and Bornstein, 1972). EQ and BHT improve the stability of relatively saturated, as well as unsaturated, abdominal fat but only EQ improves muscle stability (Bartov and Bornstein, 1976b). The stability of abdominal fat and meat increased with the combination of EQ and vitamin E in broilers fed diets containing saturated and unsaturated fats, beyond the effect of each of these materials fed alone. The combination of BHT with vitamin E increased the stability of fat and meat only in broilers fed diets containing saturated fat (Bartov and Bornstein, 1981).

High-level supplementation of vitamin E or EQ is rather expensive, and the question arises whether or not this non-routine addition might be limited to the period during which most of the body fat is deposited. Bartov and Bornstein (1978) concluded that any protective effect that dietary antioxidant exerts is similar whether they are supplied for the last four weeks before marketing or for the entire growth period.

Flavour and taste

Flavour components of poultry are becoming increasingly important as more and more poultry products are marketed frozen after extended freezer storage. Also, the increased emphasis on further processing and the development of new products utilizing all parts of poultry meat makes preservation of acceptable fresh flavour and retardation of off-flavour of importance.

Flavour, odour, off-flavour and taint have been investigated extensively as major components of eating quality of poultry meat. For recent reviews *see* the papers of Land and Hobson-Frohock (1977) and Land (1980).

EFFECT OF FATNESS ON FLAVOUR

Conflicting results have been reported on the relationship between fatness and eating quality (Patterson, 1975), although it is clear that the amount of intramuscular fat is closely related to juiciness (Williams, 1968).

In broiler chickens, the amount of abdominal fat is generally highly correlated with age (Yamashita *et al.*, 1976). These authors also noted that younger birds were preferred, both on the basis of flavour and of greater tenderness. Another report, however, while confirming the greater tenderness of the younger birds, found that flavour was not a function of age for male chickens, and actually improved with age for females (Larmond and Morgun, 1969). Basker, Angel and Bartov (1983) concluded that meat of medium fatness had the highest sensory quality, followed by the fat meat product and last by the lean meat product.

EFFECT OF FEED INGREDIENTS

Diet-associated fishy flavours have arisen from fish meal and products derived from fish oil which are readily susceptible to oxidation (Carlson *et al.*, 1957; Hardin, Milligan and Sidwell, 1964; Fry *et al.*, 1965; Holdas and May, 1966; Miller *et al.*, 1967a; Dean *et al.*, 1969; Miller and Robish, 1969; Opstvedt, 1971; Lipstein and Bornstein, 1973). A dietary level of up to 2% fish oil was more than enough to render the birds unfit for consumption (Lipstein and Bornstein, 1973). In spite of generally high nutritional value, fish oils have no application in practical feed programmes because of the fishy flavour they impart to the broiler carcass.

Table 11.7 EFFECT OF FAT SUPPLEMENT (FISH OIL (FO) AND SOYABEAN OIL (SO) SOAPSTOCKS), FED AT TWO CONCENTRATIONS, ON FLAVOUR SCORE[a] OF BROILERS

Level of fat (%)	2		4		
Type of fat	SO	FO	SO	FO	SO+FO
Thigh meat flavour score	8.3	1.9	8.3	1.3	3.7
Breast meat flavour score	9.2	4.6	9.2	2.6	7.7

(Data from Lipstein and Bornstein, 1973)
[a]Averages of four birds scored by six panel members each, the scores ranging from 0 (strong fishy flavour) to 10 (normal, fully-acceptable flavour)

Fish flavour has often been associated with the special long-carbon polyunsaturated fatty acids (LC-PUFA) of fish lipids (Opstvedt, 1971). It is usually assumed that the fishy flavour of broilers fed small amounts of fish oil or considerable quantities of fish meal is the result of the accumulation of LC-PUFA in the broiler carcass, chiefly in the liver, less in white meat, still less in dark meat, and in negligible amounts in abdominal fat (Leong *et al.*, 1964; Miller, Leong and Smith, 1969). On the other hand, *Table 11.7* presents data indicating that the off-flavour was always more prominent in the dark than in the white meat. The thigh meat contains two and a half to three times as much total lipid as breast meat (Marion, 1965; Marion and Woodroof, 1965; Katz, Dugan and Dawson, 1966).

The assumption that LC-PUFA are responsible for a fishy flavour is strongly supported by the fact that animals which were not fed any marine products may exhibit typically fishy flavours. It may be assumed that the fishy flavour is not due to LC-PUFA alone, but is the result of some volatile water-soluble degradation product, since the off-flavour was largely transferred from the meat to the broth during cooking (Carlson *et al.*, 1957). The mechanisms of the reactions and the nature of the chemical compounds responsible for fishy flavour and odour are still not adequately understood.

According to organoleptic tests, a three-week withdrawal period of fish oil soapstock (Lipstein and Bornstein, 1973) and its substitution by soyabean oil soapstock resulted in a relatively rapid improvement in the flavour score.

Vitamin E supplementation significantly reduced fishy flavour in the meat of broilers fed fish oils and fish meals; however, vitamin E did not affect the uptake and deposition of LC-PUFA linolenates by the bird (Crawford *et al.*, 1974, 1975).

In a review on rapeseed meal and its use in poultry diets, Fenwick and Curtis (1980) mentioned that there have been several circumstantial reports of off-flavour in broiler carcasses from diets containing rapeseed meal, but none of the cases was supported by convincing evidence. Petersen (1969) fed chickens on diets including sorghum supplemented with tannin, known to be present in a high concentration in rapeseed meal, and reported an off-flavour in the meat from birds receiving the tannin. No adverse effects on flavour and texture were reported when broilers were fed diets containing 5% expeller-processed rapeseed meal, but concentrations above 5% could produce off-flavour in broiler meat (Yule and McBride, 1976). Steedman *et al.* (1979a,b) found that feeding 15% Span rapeseed meal with 5% herring meal to broilers adversely affected the eating quality of the meat (odour, flavour, overall acceptability of light and dark meat). Inclusion of 15% Span rapeseed meal in the ratio resulted in significantly poorer flavour, tenderness and overall acceptability of dark meat, than in meat from broilers fed soyabean meal. There were no significant differences in juiciness or water-holding capacity attributable to the ration.

The influence of various grains on the flavour of broiler meat was investigated by Jensen (1964) and by Petersen (1969). The former found that broilers fed the sorghum diet had the poorest flavour. Petersen (1969) reported that broilers fed on oats, barley and maize were graded higher for flavour than the birds which had received sorghum or wheat. The sorghum plus tannin diet seemed to give the meat an off-flavour.

Light and dark meat from chickens fed a standard diet had more flavour than meat from chickens fed a low-fat purified diet (Lewis *et al.*, 1956). Leong *et al.* (1964) found that differences in fatty acid components of the diet could result in flavour differences.

EFFECT OF SPECIFIC ADDITIONS

Various additives to the diet of broilers have been used in attempts to influence flavour, such as garlic (Newman, Schaible and Dawson, 1958), curry, chili and black pepper (Williams and Kienholz, 1974) all without success.

Wabeck and Health (1982) tested the addition of broth or fat-based ingredients ('En-hance') to the diet ten days prior to slaughter in an effort to improve flavour and aroma of broiler meat. The results showed that flavour, aroma and moistness were improved significantly through the use of 'En-hance'. Objective tests showed that tenderness was not improved, but that the moisture content of muscles was significantly higher.

Broilers were sometimes processed without removing viscera at the time of slaughtering. The birds processed in this manner developed a characteristic flavour resembling a 'visceral taint' or 'gamey' flavour. It was suggested by Shrimpton (1966) to be associated with the metabolic activities of intestinal micro-organisms, with the subsequent absorption of microbial metabolites by skeletal tissue. Sheldon and Essary (1982) found that dietary antibiotics (streptomycin, penicillin, flavomycin) significantly modified the intestinal microflora and metabolites, and thus resulted in a significant change in broiler meat broth flavour.

Despite the advantages of methyl bromide fumigation for the destruction of undesirable organisms in broiler diets, the process produced deleterious effects on the flavour of the roasted bird (Griffiths *et al.*, 1978).

PROLONGING SHELF-LIFE

Off-flavour, tainted odour and other repulsive changes are the results of two very different processes: chemical oxidation and microbiological multiplication, each one requiring different preventative treatments besides freezing or adequate refrigeration.

Years of research and development of food preservation by irradiation has shown that this technology requires less energy than other preservation methods, eliminates pathogenic micro-organisms efficiently and can replace or drastically minimize the use of food additives. The worldwide interest in food irradiation technology is clearly shown by the activities on this subject in many countries.

Radicidation is defined as a treatment with ionizing radiation to adequately reduce the number of viable non-sporing pathogenic micro-organisms in feed (Franken, 1981). The use of irradiation to extend shelf-life of broiler meat had no adverse effect on flavour.

ORGANOLEPTIC TESTS

Organoleptic properties of meat are perceived directly by the consumer when buying or eating the meat. These are mainly colour, tenderness, juiciness and flavour. According to Touraille (1983), there are two kinds of methods that may be used to measure organoleptic characteristics: sensory and instrumental. Sensory evaluation is done by taste panels. A taste panel measures all the organoleptic characteristics, whereas different physiochemical methods must depend on the following criteria: composition of meat, tenderness evaluation, water-binding capacity, influence of thermal processes. Selection and training of panel members, use of different tests according to the purpose of the experiment, and comparing results obtained are very difficult. An effort should, therefore, be made to standardize the procedures. The literature demonstrates the complexity of the problem and the difficulties in interpretation of sensory data, and thus the need for standardization in the area (Dumont, Delpech and Arrive, 1983).

References

ARAFA, A.S., BOONE, M.A., JANKY, D.M., WILSON, H.R., MILES, R.D. and HARMS, R.H. (1983). *Poult. Sci.*, **62**, 314–320
BARTOV, I. and BORNSTEIN, S. (1966). *Poult. Sci.*, **45**, 297–305
BARTOV, I. and BORNSTEIN, S. (1967). *Poult. Sci.*, **46**, 795–805
BARTOV, I. and BORNSTEIN, S. (1969). *Poult. Sci.*, **48**, 495–504
BARTOV, I. and BORNSTEIN, S. (1972). *Poult. Sci.*, **51**, 859–868
BARTOV, I. and BORNSTEIN, S. (1976a). *Br. Poult. Sci.*, **17**, 29–38
BARTOV, I. and BORNSTEIN, S. (1976b). *Br. Poult. Sci.*, **17**, 17–27
BARTOV, I. and BORNSTEIN, S. (1977a). *Br. Poult. Sci.*, **18**, 59–68
BARTOV, I. and BORNSTEIN, S. (1977b). *Br. Poult. Sci.*, **18**, 47–57
BARTOV, I. and BORNSTEIN, S. (1978). *Br. Poult. Sci.*, **19**, 129–135
BARTOV, I. and BORNSTEIN, S. (1981). *Poult. Sci.*, **60**, 1840–1845
BARTOV, I., BORNSTEIN, S. and LIPSTEIN, B. (1974). *Br. Poult. Sci.*, **15**, 107–117
BARTOV, I., LIPSTEIN, B. and BORNSTEIN, S. (1974). *Poult. Sci.*, **53**, 115–124
BASKER, D., ANGEL, S. and BARTOV, I. (1983). *J. Fd Qual.* (in press)
BECKER, W.A., SPENCER, J.V., MIROSH, L.W. and VERSTRATE, J.A. (1979). *Poult. Sci.*, **58**, 835–842
BECKER, W.A., SPENCER, J.V., MIROSH, L.W. and VERSTRATE, J.A. (1981). *Poult. Sci.*, **59**, 693–697
BICKOFF, E.M., LIVINGSTON, A.L., BAILEY, G.F. and THOMPSON, C.R. (1954). *J. Agr. Fd Chem.*, **2**, 563–566
BRAMBILA, S.J., PINO, J.A. and MENDOZA, C. (1963). *Poult. Sci.*, **42**, 294–300
BRAUNLICH, K. (1974). *15th Wld Poult. Cong.* pp. 7–18. New Orleans
BUTTERY, P.J. and BOORMAN, K.N. (1976). *Protein Metabolism and Nutrition.* pp. 197–204. Ed. Cole, J.A. Butterworths; London
CAREW, L.B. and HILL, F.W. (1964). *J. Nutr.*, **83**, 293–299
CARLSON, D., POTTER, L.M., MATTERSON, L.D., SINGSEN, E.P., GILPIN, G.L., RESTRAM, R.A. and DAWSON, E.H. (1957). *Fd Techn., Champaign.*, **11**, 615–620
CHERRY, J.A., SIEGAL, P.B. and BEANE, W.L. (1978). *Poult. Sci.*, **57**, 1482–1487
COLLINS, W.M., THAYER, S.C. and SKOGLUND, W.C. (1955). *Poult. Sci.*, **34**, 223–228
COMBS, G.F. and NICHOLSON, J.L. (1963). *Feedstuffs*, **35**(1), 36–38
COON, C.N., BECKER, W.A. and SPENCER, J.V. (1981). *Poult. Sci.*, **60**, 1264–1271
COUCH, J.R., FARR, F.M. and CAMP, A.A. (1963). *Proc. Conf. Distillers Feed Res. Council*, pp. 31–39
CRAWFORD, L., PETERSON, D.W., KRETSCH, M.J., LILYBLADE, A.L. and OLCOTT, H.S. (1974). *Fish Bull.*, **72**, 1032–1036
CRAWFORD, L., KRETSCH, M.J., PETERSON, D.W. and LILYBLADE, A.L. (1975). *J. Fd Sci.*, **40**, 751
DAY, E.J. and WILLIAMS, Jr., W.P. (1958). *Poult. Sci.*, **37**, 1373–1381
DEAN, P., LAMOREUX, W.F., AITKEN, J.R. and PROUDFOOT, F.G. (1969). *Can. J. Anim. Sci.*, **49**, 11–15
DEATON, J.W., McNAUGHTON, J.L., REECE, F.N. and LOTT, B.D. (1981). *Poult. Sci.*, **60**, 1250–1253
DELPECH, P., DUMONT, B.L. and NEFZAOUI, A. (1983). *Quality of Poultry Meat.* pp. 21–27. Eds Lahellec, C., Ricard, F.H. and Colin, P. Ploufragan; France
DUA, P.N., DAY, E.J., HILL, J.E. and SROGEN, C.O. (1967). *J. Agric. Fd Chem.*, **15**, 324–328

DUMONT, B.L., DELPECH, P. and ARRIVE, J. (1983). *Proc. 6th Eur. Symp.* pp. 495–504. Eds Lahellec, C., Ricard, F.H. and Colin, P. Ploufragan; France

EDWARDS, H.M., DENMAN, F., AGO-ASHOUR, A. and NUGARA, D. (1973). *Poult. Sci.*, **52**, 934–948

EDWARDS, H.M. and DENMAN, F. (1975). *Poult. Sci.*, **54**, 1230–1238

EDWARDS, H.M. and HART, P. (1971). *J. Nutr.*, **101**, 989–996

ELROD, R.C., ROBAJDEK, E.S., GLEDHILL, R.H., WITZ, W.N., DISER, G.M. and HAYWARD, J.W. (1958). *Feedstuffs*, **30**(33), 26–32

ESSARY,E.O.,DAWSON,L.E.,WISMAN,E.L.and HOLMES,C.E.(1960).*Poult.Sci.*,**39**,1249

EVANS, R.J., BANDEMER, S.L., ANDERSON, M. and DAVIDSON, J.A. (1962). *J. Nutr.*, **76**, 314–319

EVANS, D.G., GOODWIN, T.L. and ANDREWS, L.D. (1976). *Poult. Sci.*, **55**, 748–755

FENWICK, G.R. and CURTIS, R.F. (1980). *Anim. Fd Sci. Technol.*, **5**, 255–298

FLETCHER, D.L. (1981). *Poult. Sci.*, **60**, 68–75

FLETCHER, D.L. (1983). *Feedstuffs*, **55**(23), 11–12

FLETCHER, D.L., JANKY, D.M., VOITLE, R.A. and HARMS, R.H. (1977). *Poult. Sci.*, **56**, 953–956

FRAMPTON, V.L., KUCK, J.E., PEPPERMAN, A.B. Jr., PONS, W.A. Jr., Watts, H.B. and JOHNSTON, C. (1966). *Poult. Sci.*, **45**, 527–535

FRANKEN, E. (1981). *Quality of Poultry Meat.* pp. 480–488. Eds Mulder, R.W., Scheele, C.W., Simons, P.C. and Verkamp, C.H. The Netherlands

FRAPS, G.S. (1943). *Poult. Sci.*, **22**, 421–424

FRITZ, J.C. and WHARTON, F.D. (1957). *Poult. Sci.*, **36**, 1118

FRITZ, J.C., WHARTON Jr., F.D. and CLASSEN, L.J. (1957). *Feedstuffs*, **29**(43), 18–24

FRY, J.L., HARMS, R.H. and MOELLER, M.W. (1976). *Poult. Sci.*, **55**, 744–748

FRY, J.L., VAN WALLEGHEN, P., WALDROUP, P.W. and HARMS, R.H. (1965). *Poult. Sci.*, **44**, 1016–1019

GOODWIN, T.L., ANDREWS, L.D. and WEBB, J.E. (1969). *Poult. Sci.*, **48**, 548–552

GOODWIN, T.L. (Ed.) (1971). *Carotenoids. VII Biosynthesis.* Birkhauser Verlag; Basel

GRABOWSKI, I. and SZUMAN, J. (1963). *Rocz. Nauk. Roln.* **83**-B-2, 227–245

GRAU, C.R. and KLEIN, N.W. (1957). *Poult. Sci.*, **36**, 1046–1051

GREY, T.C., ROBINSON, D., JONES, J.M., STOCK, S.W. and THOMAS, N.L. (1983). *Br. Poult. Sci.*, **24**, 219–231

GRIFFITHS, L., LEESON, S. and SUMMERS, J.D. (1977a). *Poult. Sci.*, **56**, 638–646

GRIFFITHS, L., LEESON, S. and SUMMERS, J.D. (1977b). *Poult. Sci.*, **56**, 1018–1026

GRIFFITHS, N.M., HOBSON-FROHOCK, A., LAND, D.G., LEVETT, J.M., COOPER, D.M. and ROWELL, J.G. (1978). *Br. Poult. Sci.*, **19**, 529–535

HALLORAN, H.R. (1974). *15th Wld Poult. Cong.*, pp. 19–28. New Orleans

HARDIN, J.O., MILLIGAN, J.L. and SIDWELL, V.D. (1964). *Poult. Sci.*, **43**, 858–860

HARMS, R.H., FRY, J.L. and McPHERSON, B.N. (1977). *Poult. Sci.*, **56**, 86–90

HEALTH, J.L. and SHAFFNER, C.S. (1972). *Poult. Sci.*, **51**, 502–506

HEALTH, J.L. and THOMAS, O.P. (1973). *Poult. Sci.*, **52**, 967–971

HEALTH, J.L. and THOMAS, O.P. (1974). *Poult. Sci.*, **53**, 291–295

HEALTH, J.L. and WABECK, C.J. (1975). *Poult. Sci.*, **54**, 1288–1292

HERRICK, G.H., FRY, J.L., DAMRON, B.L. and HARMS, R.H. (1970). *Poult. Sci.*, **49**, 222–225

HILL, F.W. and DANSKY, L.M. (1954). *Poult. Sci.*, **33**, 112–119

HOLDAS, A. and MAY, K.N. (1966). *Poult. Sci.*, **45**, 1405–1407

HUFF, W.E. and HAMILTON, P.B. (1975). *Poult. Sci.*, **54**, 1308–1310

JACKSON, S., SUMMERS, J.D. and LEESON, S. (1982). *Nutr. Rep. Int.*, **25**, 601–612
JANKY, D.M. (1983). *Broiler Ind.*, **2**, 50–52
JENSEN, J.F. (1964). *Forsgslakoratioriets Arbog*, 352
KATZ, M.A., DUGAN, L.R. and DAWSON, L.E. (1966). *J. Fd Sci.*, **31**, 717–720
KLOSE, A.A., MECCHI, E.P., HANSON, H.L. and LINEWEAVER, H. (1951). *J. Am. Oil Chem. Soc.*, **28**, 126–164
KOWALSKI, L.M. and REID, W.M. (1970). *Poult. Sci.*, **49**, 1405
KRISHNAMURTHY, S. and BIERI, Y.G. (1962). *J. Nutr.*, **77**, 245–252
KUZMICKY, D.D., KOHLER, G.O., LIVINGSTON, A.L., KNOWLES, R.E. and NELSON, J.W. (1968). *Poult. Sci.*, **47**, 389–397
KUZMICKY, D.D., LIVINGSTON, A.L., KNOWLES, R.E., KOHLER, G.O., GUENTHNER, E., OLSON, O.E. and CARLSON, C.W. (1977). *Poult. Sci.*, **56**, 1504–1509
LAND, D.G. (1980). *Meat Quality in Poultry and Game Birds*. pp. 17–30. Eds. Mead, G.C. and Freeman, B.M. Longman Group; Edinburgh
LAND, D.G. and HOBSON-FROHOCK, A. (1977). *Growth and Poultry Meat Production*. pp. 301–334. Eds Boorman, K.N. and Wilson, B.J. British Poultry Science Ltd; Edinburgh
LARMOND, E. and MORGAN, E.T. (1969). *Can. Inst. Fd Tech.*, **2**, 185–187
LEIBETSEDER, J. and SCHWEIGHARDT, H. (1979). *Wld Poult. Sci. J.*, **35**, 236–243
LEONG, K.C., KNOBL, C.M., SNYDER, D.G. and GRUGER, E.M. (1964). *Poult. Sci.*, **43**, 1235–1240
LEONG, K.C., SUNDE, M.L., BIRD, H.R. and WECKEL, K.G. (1958). *Poult. Sci.*, **37**, 1170–1172
LEVEILLE, G.A., ROMSOS, D.R., YEH, Y.Y. and O'HEA, E.K. (1975). *Poult. Sci.*, **54**, 1075–1093
LEWIS, R.W., SANFORD, P.E., ERICSON, A.T., HARRISON, D.L. and CLEGG, R.E. (1956). *Poult. Sci.*, **35**, 251–253
LIN, C.Y., FRIARS, G.W. and MORAN, E.T. (1980). *Wld Poult. Sci. J.*, **36**, 103–111
LIPSTEIN, B. and BORNSTEIN, S. (1973). *Br. Poult. Sci.*, **14**, 279–289
LIPSTEIN, B. and BORNSTEIN, S. (1975). *Br. Poult. Sci.*, **16**, 189–200
LIPSTEIN, B., BORNSTEIN, S. and BUDOWSKI, P. (1967). *Poult. Sci.*, **46**, 626–638
LIPSTEIN, B., BORNSTEIN, S. and BUDOWSKI, P. (1970). *Poult. Sci.*, **49**, 1631–1638
LIPSTEIN, B., BORNSTEIN, S. and BARTOV, J. (1975). *Br. Poult. Sci.*, **16**, 627–635
LIPSTEIN, B. and HURWITZ, S. (1980). *Br. Poult. Sci.*, **21**, 9–21
LIPSTEIN, B. and HURWITZ, S. (1981). *Poult. Sci.*, **60**, 2628–2638
LIPSTEIN, B. and TALPAZ, H. (1984). *Br. Poult. Sci.*, **25** (in press)
LITTLEFIELD, L.H., BLETNER, J.K. and SHIRLY, H.V. (1970). *Poult. Sci.*, **49**, 1407
LIVINGSTON, A.L., KUZMICKY, D.D., KNOWLES, R.E. and KOHLER, G.O. (1969). *Poult. Sci.*, **48**, 1678–1683
MABRAY, C.J. and WALDROUP, P.W. (1981). *Poult. Sci.*, **60**, 151–159
MACHLIN, L.J., GORDON, R.S. and MEISKY, K.H. (1959). *J. Nutr.*, **67**, 333–343
MACHLIN, L.J. and GORDON, R.S. (1962). *Poult. Sci.*, **41**, 473–477
MARCH, B.E. and HANSEN, C. (1977). *Poult. Sci.*, **56**, 886–894
MARKLEY, J.W., WEINLAND, B.T., MALONE, G.W. and CHALOUPKA, G.V. (1980). *Poult. Sci.*, **59**, 1755–1760
MARION, J.E. (1965). *J. Nutr.*, **85**, 38–44
MARION, J.E., BOGGESS Jr., T.S. and WOODROOF, J.G. (1967). *J. Fd Sci.*, **32**, 426–429
MARION, J.E. and WOODROOF, J.G. (1965). *J. Fd Sci.*, **30**, 38–43
MARION, J.E. and WOODROOF, J.G. (1966). *Poult. Sci.*, **45**, 241–247
MARUSICH, W.L. and BAUERNFEIND, J.C. (1970). *Poult. Sci.*, **49**, 1566–1579

MARUSICH, W.L., DeRITTER, E., OGRINZ, E.F., KEATING, J., MITROVIC, M. and BUN-NELL, R.H. (1973). *Br. Poult. Sci.*, **14**, 541–546

MARUSICH, W.L., DeRITTER, E., OGRINZ, E.F., KEATING, J., MITROVIC, M. and BUN-NELL, R.H. (1975). *Poult. Sci.*, **54**, 831–844

MARUSICH, W.L., OGRINZ, E.F., CAMERLEUGO, N., McCAMBLEY, J. and MITROVIC, M. (1976). *Poult. Sci.*, **55**, 1486–1494

MAW, W.A. (1935). *US Egg Poult. Mag.*, **41**, 32

MECCHI, E.P., POOL, M.F., BEHMAN, G.A., HAMACHI, M. and KLOSE, A.A. (1956). *Poult. Sci.*, **35**, 1238–1246

MIDDENDORF, D.F., CHILDS, G.R. and CRAVENS, W.W. (1980a). *Poult. Sci.*, **59**, 1442–1450

MIDDENDORF, D.F., CHILDS, G.R. and CRAVENS, W.W. (1980b). *Poult. Sci.*, **59**, 1460–1470

MILLER, D., GRUGER, E.H., LEONG, K.C. and KNOBL, G.M. (1967a). *J. Fd Sci.*, **32**, 342–345

MILLER, D., GRUGER, E.H., LEONG, K.C. and KNOBL, G.M. (1967b). *Poult. Sci.*, **46**, 438–444

MILLER, D., LEONG, K.C. and SMITH, P. (1969). *J. Fd Sci.*, **34**, 136–141

MILLER, D. and ROBISH, P. (1969). *Poult. Sci.*, **48**, 2146–2157

MITCHELL, Jr. R.P., BLETNER, J.K. and TUGWELL, R.L. (1961). *Poult. Sci.*, **40**, 1432

MORAN, E.T., ORR, H.L. and LARMOND, E. (1970). *Fd Tech.*, **24**, 73–78

NEWMAN, P.S., SCHAIBLE, P.J. and DAWSON, L.E. (1958). *Bull. Mich. Agric. Coll. Exp. Stn*, **40**, 747

OKUMURA, J. and TASAKI, J. (1969). *J. Nutr.*, **97**, 316–320

OPSTVEDT, J. (1971). *University of Nottingham Nutritional Conference for Feed Manufacturers 5*, pp. 70–93. Eds Swan, H. and Lewis, D. Edinburgh

OPSTVEDT, J. (1973). *Acta Agric. Scand. (Suppl.)*, **19**, 64–71

PALMER, L.S. (1915). *J. Biol. Chem.*, **23**, 261–279

PALMER, L.S. and KEMPSTER, H.L. (1919). *J. Biol. Chem.*, **39**, 331–337

PATRICK, H. and SCHAIBLE, P. (1980). *Poult. Fds Nutr.* p. 221. 2nd edn. Westport, Connecticut

PATTERSON, R.L.S. (1975). In *Meat*. Eds Cole, D.J.A. and Lawrie, R.A. Butterworths; London

PERDUE, H.S. and SMITH, J.D. (1962). *Proc. Texas Nutr. Conf.*, 77–84

PETERSEN, V.E. (1969). *Poult. Sci.*, **48**, 2006–2013

POTTER, L.M., BUNNELL, R.H., MATTERSON, L.D. and SINGSEN, E.P. (1956). *Poult. Sci.*, **35**, 452–456

QUACKENBUSH, F.W., FIRCH, J.G., BRUNSON, A.M. and HOUSE, L.R. (1963). *Cereal Chem.*, **40**, 250–259

QUACKENBUSH, F.W., KRAKOVSKY, S., HOOVER, T. and ROGLER, J.C. (1965). *J. Ass. Offic. Agr. Chem.*, **48**, 1241–1244

RAJU, P.K. and REISER, R. (1967). *J. Biol. Chem.*, **242**, 379–384

RATCLIFF, R.G., DAY, E.J., GROGAN, C.O. and HILL, J.E. (1962). *Poult. Sci.*, **41**, 1529–1532

RATCLIFF, R.G., DAY, E.J. and HILL, J.E. (1959). *Poult. Sci.*, **38**, 1039–1048

RATCLIFF, R.G., DAY, E.J. and HILL, J.E. (1961). *Poult. Sci.*, **40**, 716–720

REISER, R. (1950). *J. Nutr.*, **42**, 325–336

RICARD, R.H. (1983). *Proc. 6th Eur. Symp.*, pp. 49–65. Ed. Lahellec, C., Ricard, F.H. and Colin, P. Ploufragan; France

SALMON, R.E., CLASSEN, H.L. and McMILLAN, R.K. (1983). *Poult. Sci.*, **62**, 837–845

SALMON, R.E. and O'NEIL, J.B. (1973). *Poult. Sci.*, **52**, 314–317

SCHOLTYSSEK, S. (1978). *Wld Poult. Sci. J.*, **34**, 222–228

SCHOLTYSSEK, S. (1980). *Meat Quality in Poultry and Game Birds*. pp. 51–58. Eds Mead, G.C. and Freeman, B.M. Longman Group; Edinburgh

SHAPIRA, N., NIR, J. and BUDOWSKI, P. (1978). *J. Nutr.*, **108**, 490–496

SHELDON, B.W. and ESSARY, E.O. (1982). *Poult. Sci.*, **61**, 280–287

SHRIMPTON, D.H. (1966). *J. Appl. Bact.*, **29**, 222–230

SIMON, P.J., ZYBKO, A., GUILLAUME, J. and BLUM, J.C. (1978). *Arch. Gefluegelk.*, **42**, 6–9

SINGH, S.P. and ESSARY, E.O. (1974). *Poult. Sci.*, **53**, 2143–2147

SMITH, J.D. and PERDUE, H.S. (1966). *Poult. Sci.*, **45**, 577–581

STEEDMAN, C.D., HAWRYSK, Z.J., HARDIN, R.T. and ROBLE, A.R. (1979a). *Poult. Sci.*, **58**, 148–155

STEEDMAN, C.D., HAWRYSK, Z.J., HARDIN, R.T. and ROBLE, A.R. (1979b). *Poult. Sci.*, **58**, 337–340

SUDERMAN, D.R. and CUNNINGHAM, F.E. (1980). *Poult. Sci.*, **59**, 2247–2249

SUMMERS, J.D. and LEESON, S. (1979). *Poult. Sci.*, **58**, 536–542

SUMMERS, S.J., SLINGER, S.J. and ASHTON, G.C. (1965). *Poult. Sci.*, **44**, 501–509

THOMAS, O.P. and TWINNING, P.V. (1971). *Proc. Md. Nutr. Conf.*, 87–90

TORTUERO, F. and CENTENO, C. (1976). *Br. Poult. Sci.*, **17**, 245–248

TOURAILLE, C. (1983). *Proc. 6th Eur. Symp.* pp. 469–492. Eds Lahellec, C., Ricard, F.H. and Colin, P. Ploufragan; France

TUNG, H.T. and HAMILTON, P.B. (1973). *Poult. Sci.*, **52**, 80–82

TWINING, P.V., BOSSARD, Jr. E.H., LAND, P.G. and THOMAS, O.P. (1971). *Proc. Univ. Maryland Nutr. Conf.*, 91–95

VELU, J.G., BAKER, D.H. and SCOTT, H.M. (1971). *J. Nutr.*, **101**, 1249–1256

VELU, J.G., SCOTT, H.M. and BAKER, D.H. (1972). *J. Nutr.*, **102**, 741–748

WABECK, C.J. and HEALTH, J.L. (1982). *Poult. Sci.*, **61**, 719–725

WALDROUP, P.W., DOUGLAS, C.R. McCALL, J.T. and HARMS, R.H. (1960). *Poult. Sci.*, **39**, 1313–1317

WILLIAMS, E.F. (1968). In *Quality Control in the Feed Industry*, **2**, 251–301. Ed. Herschdoerfer, S.M. Academic Press; London

WILLIS, G.M. and BAKER, D.H. (1980). *Poult. Sci.*, **59**, 404–411

WILLIAMS, N. and KIENHOLZ, E.W. (1974). *Poult. Sci.*, **53**, 2233–2234

YAMASHITA, C., ISHIMOTO, Y., MEKADA, H., EBISAWA, S., MARI, T. and NONAKA, S. (1976). *Jap. Poult. Sci.*, **13**, 14–19

YEH, Y.Y. and LEVEILLE, G.A. (1969). *J. Nutr.*, **98**, 356–366

YULE, W.J. and McBRIDE, R.L. (1976). *Br. Poult. Sci.*, **17**, 231–239

12

FEEDING THE REPLACEMENT PULLET

S. LEESON and J.D. SUMMERS
Department of Animal and Poultry Science, University of Guelph, Guelph, Ontario, Canada

Of all the factors affecting profitability in the laying house, none is more important than the health and condition of the pullet at point-of-lay. Environment, management system, health and nutrition are confounding factors that are largely responsible for pullet condition within any genetical line. This chapter will deal with only one of these factors, namely nutrition. However it is essential to realize the potential confounding effects of the other parameters.

The nutritonal requirements of the growing pullet are not well defined. This is due partly to the fact that goals are not clearly defined. Everyone would agree that laying performance is the ultimate standard: thus, within the growing period, it is necessary to define guidelines that will ultimately influence laying performance. For this reason, such factors as mature body weight, status of sexual maturity, uniformity of body weight and state of feathering, are most often considered in evaluating a feeding programme.

Considering the fact that amino acid balance of diets has improved in recent years, there have been few changes either in the recommended nutrient requirements of the pullet in the last 20 years (National Research Council, 1960, 1977), or in feeding systems. With the advent of complete diets, rearing programmes were arbitrarily divided into starter (0–6 or 0–8 weeks) and a subsequent grower or rearer period. Thus, relatively high-protein starter diets, used to ensure that the chicks had a 'good start' in life, were necessarily followed by lower-protein grower diets in order to reduce growth and delay sexual maturity to avoid small egg size and physical disorders of the reproductive tract associated with early maturing pullets. With this traditional approach, the type of growth curve shown in *Figure 12.1* is noticed. Early rapid growth (A) is associated with high-protein, high-nutrient density starter diets, while nutrient restriction through physical or dietary manipulation (B) is necessary in order to achieve the breeders' recommended target weight at point of lay (POL). Both early rapid growth and subsequent nutrient restriction can result in stress and thus a more ideal situation is that depicted by C (*Figure 12.1*), where constant steady growth occurs from day 1 of age throughout the rearing period. This situation has not yet been achieved with conventional feeding systems.

As an initial step in achieving the above goal, Summers, Pepper and Moran (1972) advocated the use of low-protein starter diets for chicks. A 14 per cent rather than a 20 per cent CP starter diet from 0–8 weeks of age resulted in a

First published in *Recent Advances in Animal Nutrition – 1980*

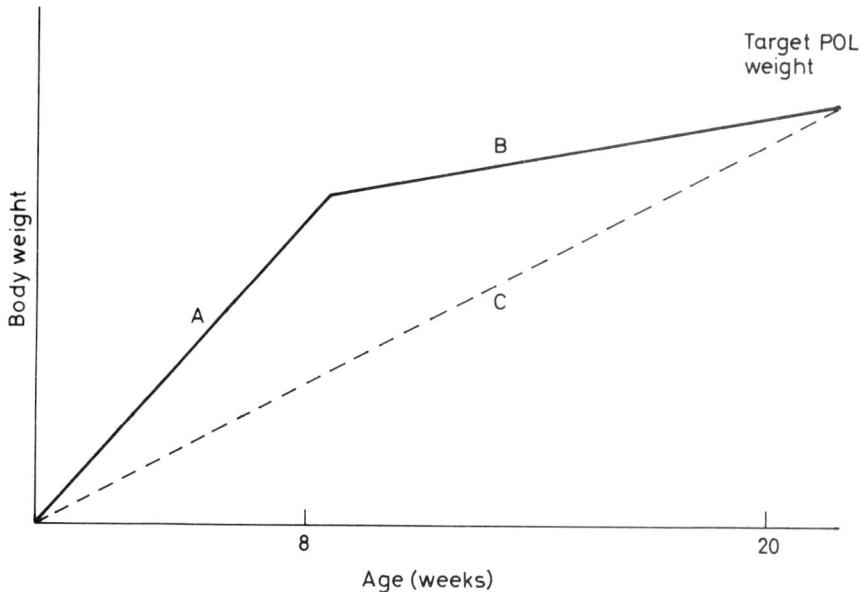

Figure 12.1 Alternative growth curves of pullets from hatching to point of lay (POL). ———,
conventional growth curve (A: high-protein starter diet; B; reduced-protein grower diet.
————, ideal growth curve (C)

20 per cent reduction in feed intake and a 25 per cent decrease in body weight. During the subsequent grower phase, when all birds received comparable diets, some control of body weight was still maintained, although there was an indication that birds restricted in growth during the starter period consume large amounts of energy when given the opportunity to compensate.

Because the chick seems well able to select a balance of dietary nutrients when presented with an array of ingredients (Dove, 1935), it was decided to investigate the nutrient intake and growth pattern of birds allowed a degree of diet self-selection. As described by Summers and Leeson (1978), the diet selection was based on maize versus soya bean meal, each being balanced with respect to minerals and vitamins and with the levels of methionine and lysine adjusted to ⩾ 2 per cent and ⩾ 5 per cent of CP, respectively (*Table 12.1*).

These 'split-diets', together with a conventional rearing diet, were offered *ad libitum* to floor-reared White Leghorn pullets from 4 to 20 weeks of age.

Pullets consuming these split-diets were consistently smaller in body size to 20 weeks of age, as indicated by the data in *Table 12.2*.

If the body weight data for the split-diet fed birds is presented in graphical form, an almost perfect straight line is achieved with a correlation coefficient (r^2) of 0.99. This type of growth is identical to that suggested in *Figure 12.1*.

Up to 11 weeks of age, birds fed split-diets consumed less total feed than did control birds. This effect was also true in terms of energy and protein consumption (*Table 12.3*).

An interesting feature of these data is that, from 11 weeks onwards, those birds practising self-selection consumed significantly more protein: thus their energy—protein intake ratio was significantly lower than that of the control birds. A consideration of the physiological changes occurring in the developing

Table 12.1 COMPOSITION OF FREE–CHOICE DIETS (g/kg) FED TO PULLETS FROM 4 TO 20 WEEKS OF AGE

| Ingredients | Experiment 1 | | |
| | Split-diets | | Control |
	1	2	3
Ground maize	–	945.4	360.0
Soya bean meal	944.5	–	140.0
Oatmeal	–	–	446.5
Wheatmeal	–	–	–
Animal–vegetable blend fat	10.0	10.0	10.0
Calcium phosphate (20% P)	20.0	20.0	20.0
Limestone	12.5	12.5	12.5
Iodized salt (0.015% KI)	2.5	2.5	2.5
Mineral mix[a]	2.5	2.5	2.5
Vitamin mix[a]	5.0	5.0	5.0
DL-methionine	3.0	–	0.4
L-lysine HCl	–	2.1	0.6
Calculated metabolizable energy (MJ/kg)	10.30	13.32	11.83
Determined crude protein (g/kg)	463	80	152

[a] To NRC (1977) specifications.

Table 12.2 BODY WEIGHT (g) OF WHITE LEGHORN PULLETS AT DIFFERENT STAGES OF GROWTH WHEN FED EITHER THE CONTROL OR SPLIT DIETS

| Group | Age (weeks) | | | | | |
	4	8	11	14	17	20
Control	313	701	891	1081	1230	1465
Split-diets	313	580	766	949	1136	1338
	NS	**	**	**	NS	**

NS: $P > 0.05$; ** $P < 0.01$

Table 12.3 FOOD, ENERGY AND PROTEIN INTAKE OF WHITE LEGHORN PULLETS DURING DIFFERENT GROWTH STAGES WHEN FED EITHER CONTROL OR SPLIT DIETS

| Food, energy and protein intake | Age (weeks) | | | | |
	4–8	8–11	11–14	14–17	17–20
Food intake (g/bird d)					
Control	43.0	53.5	58.1	61.8	78.0
Split-diets	35.6	45.4	62.7	62.6	73.0
	*	*	*	NS	NS
Energy intake (kJ ME/bird d)					
Control	510	632	686	732	924
Split-diets	464	586	787	778	904
	NS	*	*	NS	NS
Protein intake (g/bird d)					
Control	6.5	8.1	8.8	9.4	11.9
Split-diets	4.1	5.1	10.7	11.9	14.1
	**	**	*	**	*

NS: $P > 0.05$; * $P < 0.05$; ** $P < 0.01$

pullet may help to explain this apparently 'erratic' self-determined intake of protein. Up to approximately 14 weeks of age, the growing pullet requires protein only for feather development and a relatively slow, genetically predetermined rate of muscle deposition. However, at this time initial development of the ovaries and their subsequent ovum occurs, together with a fairly rapid increase in the growth of the oviduct (Gilbert, 1971). It is postulated that this activity may initiate an increased appetite for protein. If this is correct, then it would appear that we are feeding our pullets 'back-to-front', because at around 14 weeks of age, it is usually low- rather than high-protein diets that are recommended (Scott, Nesheim and Young, 1976), while conventional restricted-feeding programmes at this age further limit protein intake. It is postulated that birds fed according to the split-diet programme may reach the laying house with a more physically mature body structure, and hence be better suited to the rigours of restricted feeding in the laying period (Balnave, 1976).

From a consideration of total feed, energy and protein intakes, the rather unorthodox pattern of nutrient intake is clearly shown in calculation of the (conventional) dietary equivalent consumed by these pullets (*Table 12.4*).

Table 12.4 EQUIVALENT DIET CONSUMED BY PULLETS FED SPLIT-DIETS

| Diet specification | Age (weeks) | | | | |
	4–8	8–11	11–14	14–17	17–20
ME (MJ/kg)	12.89	13.01	12.47	12.34	12.38
CP (g/kg)	114	113	170	189	193

In order to test the theory of increased protein needs with advancement of pullet age, birds were fed a series of diets based on the specifications shown in *Table 12.4* (Leeson and Summers, 1979a); these diets are subsequently referred to as step-up protein diets (*Table 12.5*).

These individual step-up protein and control diets were fed *ad libitum* to caged White Leghorn type pullets according to the time schedule given in *Table*

Table 12.5 COMPOSITION OF REARING DIETS (g/kg) FED TO WHITE LEGHORN TYPE PULLETS

| Ingredient | Step-up protein | | | Control | | |
	1	2	3	1	2	3
Yellow maize	797.0	620.0	660.0	685.1	647.0	543.0
Barley	57.5	140.0	16.9	–	137.2	302.3
Soya bean meal	93.0	187.1	270.0	250.0	163.0	102.0
Animal–vegetable fat	10.0	10.0	10.0	22.0	10.0	10.0
Limestone	12.5	12.5	12.5	12.5	12.5	12.5
Calcium phosphate	20.0	20.0	20.0	20.0	20.0	20.0
Iodized salt (0.015% KI)	2.5	2.5	2.5	2.5	2.5	2.5
Mineral-vitamins	7.5	7.5	7.5	7.5	7.5	7.5
DL-methionine	–	0.4	0.6	0.4	0.3	0.2
Calculated analysis						
Crude protein (g/kg)	120.0	160.0	190.0	180.0	150.0	130.0
ME (MJ/kg)	12.89	12.43	12.43	12.76	12.52	12.35
Feeding period (weeks)	0–12	12–16	16–20	0–8	8–12	12–20

12.5. Twenty-four replicate cages of 10 birds each were assigned to each treatment, with the birds being maintained on constant bright light from 0–3 d, the intensity subsequently being gradually reduced to 10 lux and the photoperiod to 8 h/d. From 8 to 20 weeks of age, pullets fed the step-up protein diets were smaller in body size than the control birds, although uniformity of body size, as measured by the proportion of birds being within 10 per cent of flock mean weight, was not adversely affected. Comparison with breeder specifications showed these test birds to be exactly on POL target weight, while the control birds were about 200 g overweight (*Table 12.6*).

Table 12.6 BODY WEIGHT (g) OF PULLETS OF WHITE LEGHORN TYPE AT DIFFERENT STAGES DURING THE GROWING PHASE: COMPARISON OF CONTROL AND STEP-UP PROTEIN DIETS

Diet treatment			*Body weight* (g) *at age:*		*20-week*
	1d	*8 w*	*12 w*	*20 w*	*uniformity*
Control	39.9	633	1043	1552	89.1%
Step-up protein	39.8	432	773	1347	83.3%

During the early stages of growth, it was observed that birds fed the initial low-protein diet of the step-up protein series, were not as attractively feathered as were control birds: feathering was adequate, although a little rough. As the high-protein diets in this series were introduced, feather appearance improved, and there were no visible differences at POL. One of the disadvantages mentioned by poultry producers with respect to low-protein starter diets, is inadequate early feathering which necessitates higher brooding temperatures for a longer time; the saving in protein feed cost is thus offset by high fuel costs. Feathering was not retarded with these diets, possibly because of maintenance of an adequate amino acid balance with these low-protein starter diets (*Table 12.5*).

The feed and nutrient intake pattern of these pullets reflects the dietary specifications (*Table 12.7*).

Table 12.7 FEED AND NUTRIENT INTAKE OF PULLETS OF WHITE LEGHORN TYPE ON STEP-UP PROTEIN AND CONTROL DIETS

Feed, energy and protein		*Age (weeks)*		
intake	*0–8*	*8–12*	*12–16*	*16–20*
Feed intake (g)	**	**	**	NS
Control	1891	2031	2090	1883
Step-up protein	1534	1715	2003	1847
Energy intake (MJ)	**	**	**	NS
Control	24.2	25.4	25.8	23.3
Step-up protein	19.8	22.1	24.9	23.0
Protein intake (g)	**	**	**	**
Control	340	305	272	245
Step-up protein	184	206	320	351

The overall pattern of protein and energy intake is of interest in terms of economics. Although it may appear that the step-up protein programme is expensive, because high-protein diets are used at the time when the birds are consuming the most feed, calculations show that to 20 weeks of age, these birds consumed about 100 g protein, and 8.4 MJ ME/bird, less than did the control-fed birds.

It was originally stated that the ultimate goal of a pullet rearing programme, is performance in the laying house. To this end, these birds were transferred to individual laying cages at 20 weeks of age and their performance monitored for a 40-week laying cycle. All birds were offered a maize-soya bean meal mash laying diet providing 11.7 MJ ME/kg and containing either 130, 150, 170 or 190 g crude protein/kg diet. Various levels of protein were included to test the assumption that these step-up protein reared birds would be more sexually mature at 20 weeks, and thus better suited to low-protein diets throughout lay. Rearing treatment had no effect on egg production, although step-up protein birds consumed less feed while producing smaller eggs; there was no interaction with protein level of the laying diet (*Table 12.8*).

Table 12.8 PERFORMANCE OF LAYERS FROM 20 TO 60 WEEKS OF AGE FOLLOWING CONTROL OR STEP-UP PROTEIN DIETS DURING THE REARING PERIOD

Regime	Feed intake (g)	Egg production (%)†	Egg weight (g)	Egg shell deformation (μm)
Rearing	*	NS	**	*
Control	112.8	78.7	57.8	22.7
Step-up protein	109.1	80.1	56.4	23.4
Laying diet CP (g/kg)	*	NS	*	NS
130	114.1[b]	77.7	56.4[a]	22.6
150	107.9[a]	80.3	56.7[ab]	23.3
170	108.7[a]	78.4	57.8[b]	23.4
190	111.0[ab]	80.9	57.5[ab]	23.1

a vs b: $P < 0.05$
† Calculated on hen bird-day basis

The smaller body size of step-up protein reared birds recorded at 20 weeks of age (*Table 12.6*) was maintained throughout the laying period (1986 vs 1881 g), while mortality throughout lay was not influenced by rearing treatment.

It was originally proposed that step-up protein diets would result in a more physically mature bird at 20 weeks of age, because of the effect of the higher protein diets in the latter half of the rearing period. Although body weight at point-of-lay was controlled, the initial rate of egg production suggested that these birds were more immature than the controls. Thus, from 20 to 24 weeks of age, production by control and test birds was 50.7 and 37.6 per cent respectively, (age at 50 per cent production was 157 and 165 days for control and test-reared birds, respectively). However, after this time, production by the 'step-up protein' reared birds was consistently higher.

While the reduction in egg size of the step-up protein reared birds is likely to be a result of the reduced body size at POL, the reason for the loss in egg shell quality (as measured by deformation) is not readily apparent. It was surprising that no interaction between type of rearing and the protein content of the laying diet was recorded. Immature birds at 20 weeks of age would be expected to show a marked improvement in performance when offered the laying diets of higher protein content. Analysis of individual four-week data did not demon-strate this effect. As these birds showed no compensatory growth during the 40-week laying cycle, it is likely that the step-up protein reared birds were small but well-fleshed at point-of-lay. As suggested by Krueger *et al.* (1976), such a high level of production would probably not have been achieved had fleshing been inadequate.

The step-up protein rearing system seems ideally suited for growing replace-ment pullets to conventional standards. Thus, control of body weight is achieved without recourse to the difficulties often encountered with restricted feeding programmes, and seems to be economical from a consideration of nutrient intake. Body weight control at point-of-lay did not sacrifice uniformity. This is an important consideration: our own observations suggest this to be in contrast to results obtained with manual feed restriction which suggests that variability in intake (associated with peck order etc.) is involved, rather than variability in the metabolic adaptation of birds to restricted quantities of nutrients.

At this stage, it may be useful to question some of our conventional standards for pullet growing.

Field reports indicate ever-increasing problems of controlling sexual maturity in pullets. Although it is generally thought that light-induced early maturity results in the production of small eggs (Card and Nesheim, 1972), there are no recent reports on the effect of inducing early maturity with modern hybrid strains. Economically, it seems detrimental to delay pullet maturity, because recent estimates show additional costs of approximately 0.5 p/pullet/day for birds maintained in the growing-house after 19 weeks of age (Strong, 1978). Leeson and Summers (1979b) have recently conducted a trial to investigate the effect of early light-induced maturity with commercial strain White Leghorn pullets.

Cage-reared birds were offered all mash diets, comparable in specifications to those indicated as the control diets in *Table 12.5*. From 3 d of age, birds were maintained on 8 h of low-intensity light. At 15, 18 or 21 weeks of age, birds were transferred to individual laying cages, being immediately subjected to 14 h con-stant bright light. Birds housed at the younger age came into production earlier although after 23 weeks of age, rearing treatment had no effect on egg produc-tion (*Table 12.9*).

Table 12.9 PULLET MATURITY AND EGG PRODUCTION IN WHITE LEGHORN PULLETS TRANSFERRED TO LAYING CAGES AT DIFFERENT AGES

Age at housing (wk)	Egg production at:			Mean to 65 wk
	18–20 wk	*21–23 wk*	*24–26 wk*	
15	29.8[c]	87.7[c]	91.8	72.9
18	12.1[b]	70.0[b]	91.5	71.8
21	0.0[a]	51.5[a]	90.5	71.3

a vs b vs c: $P < 0.05$

Delaying the move to the laying house resulted in a dramatic increase through-out the experiment in the number of large-grade, and a decrease in the number of small-grade, eggs (*Table 12.10*).

In terms of overall economics, this loss in egg-grading must be considered in the light of the extra production attained at an early age: thus, when comparing

Table 12.10 EGG GRADING (%) OF WHITE LEGHORN POULTRY TRANSFERRED TO LAYING CAGES AT DIFFERENT AGES

Age at housing (wk)	30-week pullets			48-week pullets			63-week pullets		
	L[a]	M	S	L	M	S	L	M	S
15	17	64	18	37	48	14	44	36	17
18	21	67	10	41	42	13	65	22	12
21	37	57	5	56	35	6	69	17	8

[a] Egg weights; L: 56.7–63.8 g; M: 49.6–56.6 g; S: 42.5–49.5 g

the 15-week and 21-week housed groups, from 18 to 23 weeks of age, about 14 extra eggs were produced by the former group. Although birds were comparable in body size by 30 weeks of age, reduced egg grading is probably associated with reduced body size at point-of-lay (1180, 1300 and 1575 g, for 15, 18 and 21 weeks respectively), confirming previous observations (Owings *et al.*, 1973; Krueger, Creger and Bradley, 1974). However, in terms of subsequent production, liveability and the non-occurrence of prolapse and blow-outs, modern hybrids seem well able to withstand the 'stress' of early lighting. However, in order to shorten the growing period, rearing treatments will have to be modified in such a way that a more mature body weight is obtained at an earlier age.

Harms (1978) concluded that point-of-lay pullets do not consume enough feed to meet their energy requirements in hot weather, and that birds with large fat reserves are required in this environment. It may be that this situation is comparable to that arising with early induced maturity in a normal environment, and we may have to rethink our ideas on carcass composition of the pullet, if early induced maturity is required. From observations with step-up protein reared birds (Leeson and Summers, 1979a), from attempts at early induced maturity (Leeson and Summers, 1979b) and from results of other workers (Payne, 1966; McNaughton *et al.*, 1977), it seems that the key to shortening the growing period is attainment of mature body size and/or composition at an earlier age than that dictated by today's conventions.

As an initial step in studying this problem, the pullets' response to high-nutrient density diets throughout a 15-week growing period has been monitored (*Table 12.11*).

All diets were offered *ad libitum* to cage-reared birds, subject to 8 h low light intensity to 15 weeks of age, and thereafter maintained on 14 h light in laying cages. Birds fed broiler-type diets throughout rearing were heavier at 8 weeks of

Table 12.11 DIET TREATMENTS AND SPECIFICATIONS FOR THREE DIFFERENT REARING TREATMENTS

Treatment		
A	Chick starter 0–8 weeks → chick grower 8–15 weeks	
B	Chick starter 0–8 weeks → broiler starter 8–15 weeks	
C	Broiler starter 0–8 weeks → broiler grower 8–15 weeks.	

Specification		g CP/kg diet	MJ ME/kg
(1)	Chick starter	180	12.76
(2)	Chick grower	140	12.34
(3)	Broiler starter	240	13.00
(4)	Broiler grower	200	13.40

age than conventionally fed birds, although the difference was not maintained at 15 weeks of age (*Table 12.12*). The protein and energy intake of these birds is given in *Table 12.13*.

Table 12.12 PULLET BODY WEIGHT (g) FOLLOWING THREE DIFFERENT REARING TREATMENTS

Rearing treatment	8 weeks	15 weeks
A	704[a]	1272
B	702[a]	1267
C	730[b]	1291

a vs b: $P<0.05$

Table 12.13 NUTRIENT INTAKE OF PULLETS ON THREE DIFFERENT REARING TREATMENTS (TOTAL INTAKES FOR PERIOD)

Rearing treatment	Energy (MJ)		Protein (g)	
	0–8 wk	8–15 wk	0–8 wk	8–15 wk
A	18.8	40.9	265	464
B	18.4	38.9	260	718
C	23.2	39.8	428	599

Up to 8 weeks of age therefore, increased weight gain is correlated with increased intake of both energy and protein. However, during the latter half of the rearing period the birds appeared to consume comparable quantities of energy, irrespective of diet energy concentration, which is probably the reason for similarity in weight at 15 weeks of age. Use of the high nutrient density diets did, however, result in improved early egg production and egg size, as shown in *Table 12.14*.

Table 12.14 EGG PRODUCTION AND EGG WEIGHT OF PULLETS SUBJECTED TO THREE DIFFERENT REARING TREATMENTS

Results	16–19 wk	20–23 wk	24–27 wk
Egg production (%)			
Rearing A	2.8[a]	64.2	91.0
B	4.7[b]	65.0	87.2
C	4.1[b]	68.5	90.2
Egg weight (g)			
Rearing A	40.8[a]	49.2	53.4
B	41.5[b]	49.2	52.7
C	44.5[c]	49.8	52.5

a vs b vs c: $P<0.05$

Production was very low in this first period: thus, the practical or economical implications of these rearing treatments may be questioned. Egg size was substantially improved by the high-nutrient density diets, but this, of course, relates to a relatively small number of eggs, and hence the economics of this effect relative to the price situation of the rearing diets is debatable. Again, these results indicate the durability of modern hybrid strains when brought into production at an early age. Peak production was achieved around 22 weeks of age, and hence this represents a substantial saving in maintenance requirement during rearing. The trial is still under way, and further results should be of interest.

We were surprised at the constancy of energy intake of these pullets, regardless of the energy level of the diet, reinforcing the concept of regulatory mechanisms of feed intake relative to energy 'needs'. This facet may cause problems in attaining the desired goal of mature body structure at earlier ages, and will be the topic of more intensive research.

Our studies indicate the durability of the modern laying hen to withstand the rigours of early egg production, and it is our goal to develop feeding systems which will allow such management practices within the confines of economic feasibility.

References

BALNAVE, D. (1976). *Br. Poult. Sci.,* **17**, 145–150

CARD, L.C. and NESHEIM, M.C. (1972). *Poultry Production,* 11th Edn. Philadelphia; Lea and Febiger

DOVE, W.F. (1935). *American Naturalist,* **69**, 469–544

GILBERT. A.B. (1971). In *Physiology and Biochemistry of the Domestic Fowl,* Vol. 3., pp. 1163–1208. Eds D.J. Bell and B.M. Freeman. London; Academic Press

HARMS, R.H. (1978). In *Proc. Florida Nutr. Conf. St. Petersburg, Florida. Jan. 12–13, 1978,* pp. 11– 17.

KRUEGER, W.F., CREGER, C.R. and BRADLEY, J.W. (1974). *Poult. Sci.,* **53**, 1639

KRUEGER, W.F., BRADLEY, J.W., CREGER, C.R. and THOMPSON, D. (1976). *Poult. Sci.,* **55**, 1598

LEESON, S. and SUMMERS, J.D. (1979a). *Poult. Sci.,* **58**, 681–686

LEESON, S. and SUMMERS, J.D. (1979b). *Poult. Sci.,* **59** (in press.)

McNAUGHTON, J.L., KUBENA, L.F., DEATON, J.W. and REECE, F.N. (1977). *Poult. Sci.,* **56**, 1391–1398

NATIONAL RESEARCH COUNCIL (1960). *Nutrient Requirements of Poultry.* Washington, DC; National Academy of Sciences

NATIONAL RESEARCH COUNCIL (1977). *Nutrient Requirements of Poultry,* 7th revised edition. Washington, DC; National Academy of Sciences

OWINGS, W.J., MULLER, R.D., BALLOUN, S.L. and TRAKULCHANG, N. (1973). *Poultry Sci.,* **52**, 2071

PAYNE, C.G. (1966). In *Proc. 13th Wlds. Poult. Congr. Kiev. Russia,* pp. 480–485.

SCOTT, M.L., NESHEIM, M.C. and YOUNG, R.J. (1976). *Nutrition of the Chicken,* 2nd Edn. Ithaca, NY; M.L. Scott and Assoc.

STRONG, C.F. (1978). *Poultry Tips, Georgia Co-operative Extension Service,* **78**, (10)

SUMMERS, J.D. and LEESON, S. (1978). *Br. Poult. Sci.,* **19**, 425–430

SUMMERS, J.D., PEPPER, W.F. and MORAN, E.T. (1972). *Can. J. Anim. Sci.,* **52**, 761–766

13

PULLET FEEDING SYSTEMS DURING REARING IN RELATION TO SUBSEQUENT LAYING PERFORMANCE

R.G. WELLS
Harper Adams Poultry Husbandry Experimental Unit, Edgmond, Newport, Shropshire

Introduction

There is a growing tendency in the egg industry to place importance on body weight as an indicator of replacement pullet value. Breeders commonly publish target weights for their stock which, it is claimed, are related to optimum laying performance. Most attention in practice is focused on body weight at point-of-lay, that is, at the age of about 18 weeks when pullets are normally moved from rearing to laying quarters.

The recommended target body weights are usually somewhat below the expected weights of birds fed conventional chick starter and pullet grower diets *ad libitum*. This implies some form of restricted feeding, which is in keeping with reports that this can give improved efficiency of food conversion in the laying stage as well as savings in food cost during the growing period (Balnave, 1973). However, there is very little information on precisely how growth should be controlled in order to meet the target weights.

The experiments in this study were undertaken to examine the short- and long-term effects of qualitative and/or quantitative dietary restriction of replacement pullets, designed to provide variations in the pattern of growth, but similar body weights at point-of-lay. The objective was to identify the economically desirable growth patterns of pullets and thereby to establish a basis for future recommendations on the nature, timing and degree of restricted feeding.

Experimentation

Two experiments were undertaken, the first involving light-weight Shaver Starcross '288' pullets and the second involving medium-weight Warren Studler 'Sex-Sal-Link' pullets.

REARING STAGE (DAY-OLD TO 18 WEEKS OF AGE)

Both stocks were reared from day-old to 18 weeks of age on deep litter in two windowless, insulated buildings, each of which was divided into 12 (6.1 m × 6.1 m)

First published in *Recent Advances in Animal Nutrition – 1980*

pens. Details of routine management of the birds during the rearing stage, in terms of lighting, heating, ventilation and vaccination programmes, are reported elsewhere (Oldale, Coggins and Wells, 1977; Coggins and Wells, 1979).

Experiment 1

A total of 9600 white Shaver '288' pullets was housed at day-old, with 400 birds in each of the 24 pens. The pens were divided into three blocks and eight dietary treatments were randomly allocated to the pens within each block.

The compositions and calculated nutrient densities of the diets used are shown in *Table 13.1*. The eight dietary treatments, which comprised a full-fed control and seven qualitatively and/or quantitatively restricted feeding programmes, are outlined in *Table 13.2*. These were designed to give a variety of growth curves (patterns), but similar body weights in the restricted groups at 18 weeks of age.

Table 13.1 COMPOSITION, NUTRIENT DENSITY AND METABOLIZABLE ENERGY CONTENT OF THE REARING DIETS

| Ingredient[c] | Chick starter | | Pullet grower (CS)[a] | Pullet grower | Pullet developer |
	Exp. 1	Exp. 2	Exp. 1	Exps. 1 and 2	Exp. 1
Wheat, ground	400	400	400	400	500
Maize, ground	300	300	300	300	250
Wheat middlings	100	100	137.5	150	150
Soya bean meal	75	75	87.5	87.5	50
Herring meal	50	—	—	—	—
White fish meal	—	75	12.5	12.5	—
Meat and bone meal	50	37.5	25	25	25
Limestone flour	—	—	12.5	12.5	12.5
Vitamin/mineral supplement	25	12.5	25	12.5	12.5
Protein (g/kg)[b]	182	185	153	151	133
Methionine (g/kg)[b]	3.3	3.6	2.6	2.6	2.2
Lysine (g/kg)[b]	9.2	10.0	7.3	7.2	5.7
Calcium (g/kg)[b]	8.8	10.0	11.1	11.3	10.4
Phosphorus (g/kg)[b]	7.5	8.1	6.3	5.9	5.5
Metabolizable energy (MJ/kg)	12.5	12.4	12.2	12.1	12.1

[a] Pullet grower (CS) contained the same vitamin/mineral supplement as the chick starter diet.
[b] Calculated from figures provided by Bolton and Blair (1974).
[c] kg/tonne, except where stated.

All changes from one diet to another were performed over a three-day period. The specified degrees of quantitative food restriction (rationing) for appropriate treatment groups were based on the intake of their full-fed counterparts during the previous week, treatment 1 being the full-fed counterpart of treatments 5, 6 and 7, and treatment 3 constituting the full-fed counterpart of treatment 8. The weekly rations were calculated on a surviving-bird basis and were supplied daily, one-seventh each day.

Table 13.2 DIETARY TREATMENTS APPLIED IN EXPERIMENT 1

Treatment	Feeding programme			
	Type of restriction	*Diets (see Table 13.1)*		
1	Unrestricted (control)	Starter	:	0–8 weeks
		Grower	:	8–18 weeks
2	Qualitative	Starter	:	0–6 weeks
		Developer	:	6–18 weeks
3	Qualitative	Grower (CS)	:	0–6 weeks
		Grower	:	6–18 weeks
4	Qualitative	Grower (CS)	:	0–6 weeks
		Developer	:	6–15 weeks
		Starter	:	15–18 weeks
5	Quantitative	Starter	:	0–8 weeks
	10% : 6–9 weeks	Grower	:	8–18 weeks
	20% : 9–18 weeks			
6	Quantitative	Starter	:	0–8 weeks
	30/40% : 15–18 weeks	Grower	:	8–18 weeks
7	Quantitative	Starter	:	0–8 weeks
	15% : 6–9 weeks	Grower	:	8–18 weeks
	30% : 9–15 weeks			
8	Qualitative and quantitative	Grower (CS)	:	0–8 weeks
	10% : 6–9 weeks	Grower	:	8–15 weeks
	20% : 9–15 weeks	Starter	:	15–18 weeks

Individual body weights of a sample of 50 birds from each replicate (pen) were recorded at three weeks of age and thereafter at weekly intervals for one block and at three-week intervals for the other two blocks. At six, 12 and 18 weeks the same 50-bird samples were also scored for feather coverage according to the following scale: very good (5); good (4); average (3); poor (2) and very poor (1).

Food intake for each pen of birds was recorded on a weekly basis. Mortality was recorded daily, birds dying in the first 24 h being replaced.

Experiment 2

A total of 7584 brown Warren 'SSL' pullets was housed at day-old, with 316 birds in each of the 24 pens. The latter were divided into four blocks and six dietary treatments were randomly allocated to the pens within each block.

All birds followed a two-stage feeding programme using chick starter diet to eight weeks of age, followed by pullet grower diet to 20 weeks. The grower diet was common to both experiments, while the two starter diets used were of very similar composition (*Table 13.1*). The six dietary treatments, which are set out in *Table 13.3*, involved a comparison of full-feeding with five programmes of restricted feeding. The latter were all quantitative in nature and applied some-time between six and 18 weeks of age. They were designed to achieve varying

Table 13.3 DIETARY TREATMENTS APPLIED IN EXPERIMENT 2

| Treatment | Timing and degree of food rationing | | | | |
	0–6 weeks	6–9 weeks	9–12 weeks	12–15 weeks	15–18 weeks
1	None	None	None	None	None
2	None	None	None	None	40%
3	None	None	None	20%	20%
4	None	10%	20%	20%	10%
5	None	15%	30%	30%	None
6	None	20%	40%	None	None

growth patterns but similar body weights, relative to the full-fed control at 18 weeks of age.

As in Experiment 1, the specified food rationing in each block was based on the intake of the control group during the previous week and was calculated on a surviving-bird basis. However, the weekly rations were not supplied daily in Experiment 2, but on three days/week; two-sevenths on Mondays and on Wednesdays and the remaining three-sevenths on Fridays.

Individual body weights of a sample of 30 birds from each pen were recorded weekly from six weeks of age. At six, 12 and 18 weeks of age the same 30-bird samples were also scored for feather coverage as described under Experiment 1.

Food intake for each pen of birds was recorded on a weekly basis. Mortality was recorded daily, all birds dying in the first six weeks being replaced.

LAYING STAGE (18–80 WEEKS OF AGE)

Birds were moved from the rearing units to controlled-environment battery houses at 18 weeks of age. Management of the birds during the laying stage complied with normal commercial practice in the United Kingdom. Food was supplied *ad libitum*. All the layer diets used were calculated to contain approximately 11.7 MJ/kg metabolizable energy. In Experiment 1, the diets fed before and after 40 weeks of age contained about 166 and 160 g/kg protein, respectively. Corresponding figures for the protein contents of layer diets used in Experiment 2 were 175 and 155 g/kg.

A random sample of 870 birds/treatment (6960 in all) was involved in the laying stage of Experiment 1, each treatment group housed as three 290-bird replicates in a randomized block design using tiers of cages as blocks. Corresponding figures for Experiment 2 were 960 birds/treatment (5760 in all), accommodated as six 160-bird replicates.

STATISTICAL ANALYSES

Analyses of variance were performed on the replicate data for all traits except egg grading, for which a chi-square test was used to determine the significance of differences between treatment means. Where appropriate (for example, with peak egg production and mortality), data were transformed to arcsin $\sqrt{\%}$ before being subjected to analyses.

Table 13.4 BIRD LIVE WEIGHT, CUMULATIVE FOOD INTAKE, MORTALITY AND FEATHER SCORE IN THE REARING STAGE: EXPERIMENT 1

	Age (weeks)	Treatment (see Table 13.3)							
		1	2	3	4	5	6	7	8
Live weight (g)	6	409b	405b	342a	352a	400b	411b	407b	359a
	9	691e	634d	595bc	555a	610cd	687e	610cd	564ab
	12	905c	831b	833b	762a	809b	891c	777a	757a
	15	1099d	1018c	1037c	962b	961b	1087d	918a	917a
	18	1224e	1187d	1183d	1143b	1089a	1130b	1150bc	1173cd
Cumulative food intake (kg/ surviving bird)	6	1.12ab	1.17bc	1.18bc	1.09a	1.12ab	1.20c	1.15abc	1.17bc
	9	2.22bc	2.35d	2.30cd	2.17b	2.04a	2.35d	2.03a	2.07a
	12	3.58d	3.66de	3.64de	3.42c	3.17b	3.75e	3.00a	3.12ab
	15	5.07d	5.11de	5.08de	4.83c	4.36b	5.22e	4.04a	4.27b
	18	6.48c	5.56c	6.56c	6.25b	5.56a	6.18b	5.57a	5.70a
Cumulative mortality (%)	6	1.67	2.58	1.92	2.25	1.42	1.25	1.33	2.50
	9	2.42	3.42	3.58	2.58	2.17	1.92	1.92	3.17
	12	2.75	3.50	3.92	3.33	2.17	2.00	2.25	3.42
	15	2.83	3.67	3.92	3.33	2.25	2.00	2.42	3.75
	18	2.83	3.67	4.08	3.08	2.25	2.17	2.50	3.83
Feather score	6	3.1	3.0	2.8	2.9	3.0	2.8	3.0	2.9
	12	3.3bcd	3.1abc	3.0ab	3.5cd	3.5cd	2.9a	3.2abcd	3.4cd
	18	3.2	2.8	3.0	3.1	3.2	3.0	3.0	3.1

185

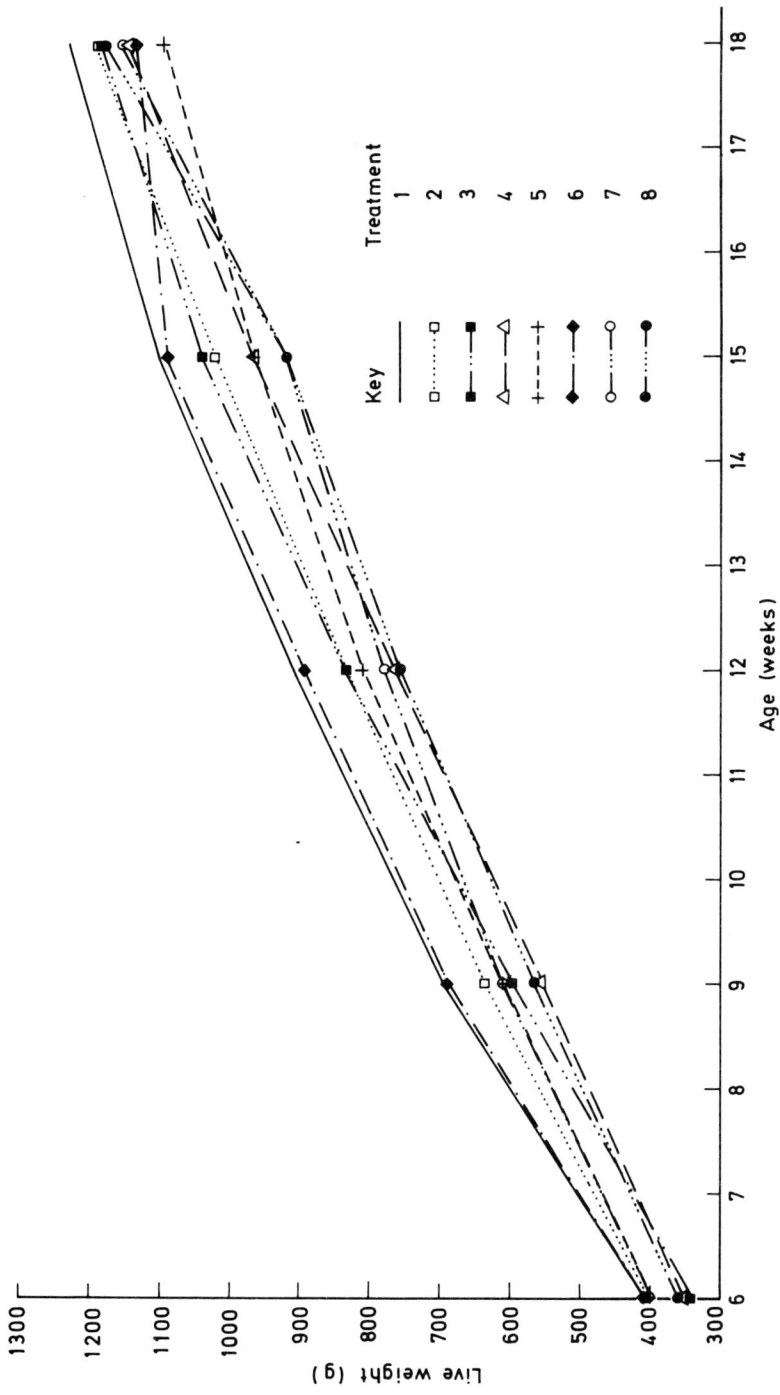

Figure 13.1 Growth curves of pullets (Shaver '288') during the rearing stage when subjected to the different dietary regimes used in Experiment 1 (Details of dietary treatments are given in Table 13.2)

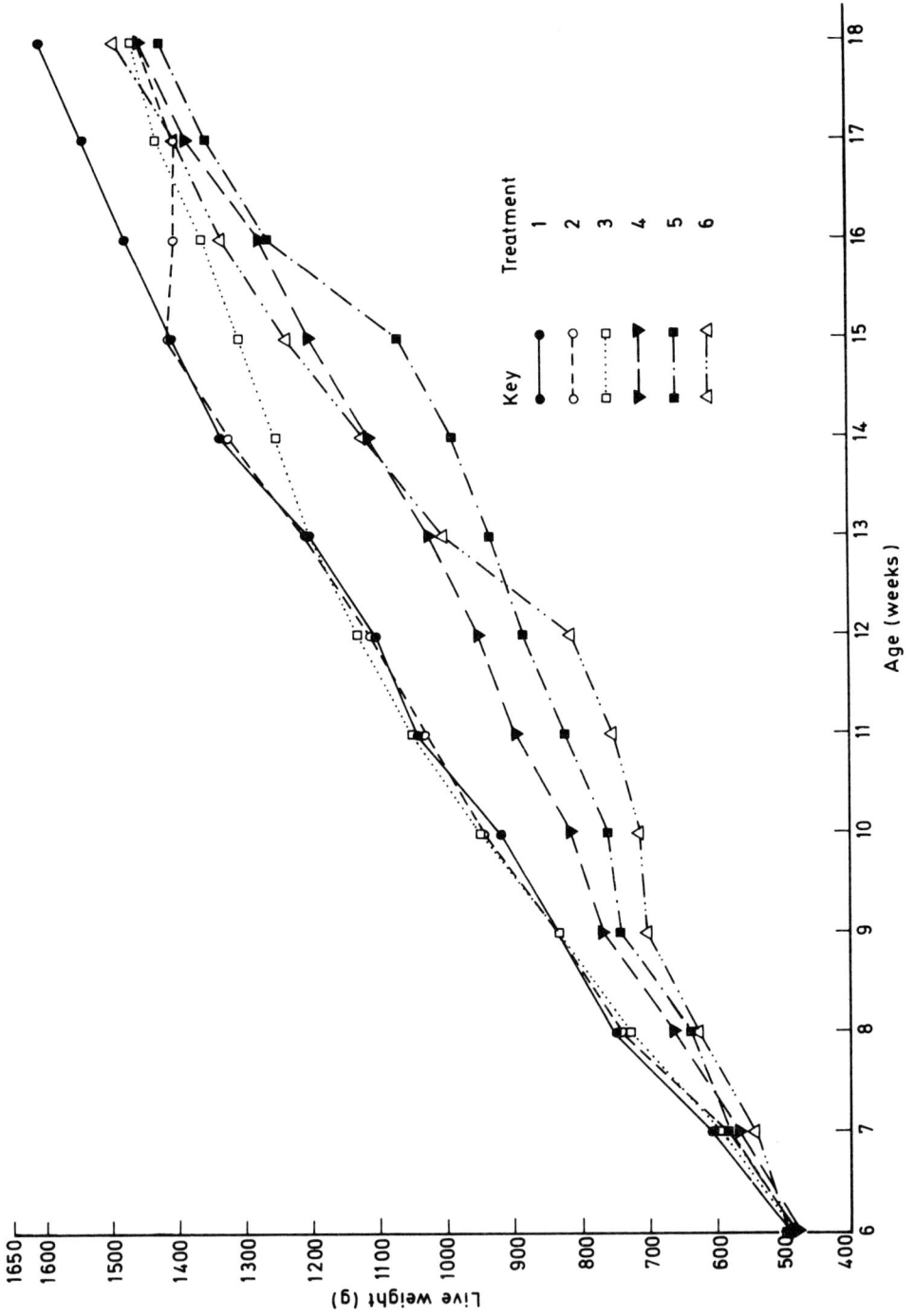

Figure 13.2 Growth curves of pullets (Warren 'SSL') during the rearing stage when subjected to the different dietary regimes used in Experiment 2. (Details

Results of the analyses are shown in the tables by appropriate lettering of the data. Values for treatment means which are not followed by the same letter are significantly different (*P*<0.05). The absence of lettering indicates that the overall variation between treatments was non-significant (*P*>0.05).

Results and discussion

RELATIONSHIP BETWEEN GROWTH AND FOOD INTAKE DURING REARING

Graphs plotting live weight against age during the rearing stage are presented in *Figures 13.1* and *13.2*. Mean treatment data for body weight and food intake at selected stages in the rearing period are shown in *Tables 13.4* and *13.5*.

Table 13.5 BIRD LIVE WEIGHT, CUMULATIVE FOOD INTAKE, MORTALITY AND FEATHER SCORE IN THE REARING STAGE: EXPERIMENT 2

	Age (weeks)	Treatment (see Table 13.3)					
		1	*2*	*3*	*4*	*5*	*6*
Live weight (g)	6	489	483	490	476	484	493
	9	835a	833a	831a	768b	742b	705c
	12	1104a	1113a	1132a	952b	885c	813d
	15	1409a	1414a	1309b	1202d	1071e	1238c
	18	1612a	1461bc	1471bc	1461bc	1428c	1495b
Cumulative food	6	1.20	1.21	1.17	1.17	1.18	1.20
intake (kg/	9	2.56a	2.54a	2.51a	2.25b	2.20b	2.17b
surviving bird)	12	4.12a	4.14a	4.13a	3.46b	3.27c	3.08d
	15	5.82a	5.87a	5.55b	4.81c	4.43d	4.91c
	18	7.67a	7.13b	7.03b	6.52d	6.45d	6.77c
Mortality (%)	0–6	0.71	0.71	0.32	0.32	0.63	0.40
	6–9	0.24	–	0.16	0.40	0.16	0.16
	6–12	0.32	0.08	0.32	0.40	0.40	0.32
	6–15	0.40	0.08	0.32	0.47	0.40	0.32
	6–18	0.47	0.08	0.40	0.47	0.40	0.32
Feather score	6	3.04	3.03	3.08	3.07	3.11	3.11
	12	3.12	3.25	3.28	3.33	3.19	3.25
	18	3.17a	3.12ab	3.11abc	3.06bcd	3.02cd	3.01d

A wide range of growth patterns was achieved in both experiments, although in each case the body weights attained at 18 weeks of age by the several groups of restricted fed birds were all within the breeders' recommended target weight ranges (1.11–1.22 kg/bird for the Shaver '288's in Experiment 1, and 1.41–1.55 kg/bird for the Warren 'SSL's in Experiment 2).

With qualitative restriction of diet, as applied through treatments 2–4 in Experiment 1, growth rate was effectively reduced, but food intake tended to be maintained at a level similar to that of the non-restricted control birds. The savings in rearing food cost with these treatments were therefore small, being limited to those resulting from the lower costs/unit weight of the diets of reduced quality compared with the control diets (*see also Table 13.12*).

Quantitative restriction of food, as applied by treatments 5–8 in Experiment 1 and treatments 2–6 in Experiment 2, always reduced growth significantly and gave substantial savings in food cost compared with the control feeding programmes (*see also Tables 13.12* and *13.13*). With these quantitatively restricted regimens, the relationship between total food intake in the rearing period and body weight attained at 18 weeks of age was positive, but it was not exact. For example, the birds subjected to treatments 6 and 7 in Experiment 1 had similar body weights at 18 weeks of age, but significantly different food intakes. Moreover, in Experiment 2, birds under treatment 6 attained a higher body weight than those under treatment 2 at 18 weeks, but had a significantly lower food intake. Those treatments, like treatment 7 in Experiment 1 and treatment 6 in Experiment 2, which were the most effective in reducing food intake, and hence food cost, while achieving a given 'target' weight at 18 weeks of age involved a comparatively high degree of rationing, with a marked depression of growth rate in the middle of the rearing period and allowed compensatory growth to take place in the later stages. Treatments like treatment 6 in Experiment 1 and treatment 2 in Experiment 2, which were the least effective in this respect, involved severe rationing with near suspension of growth in the final stages of rearing.

The evenness of growth within groups, measured by the variance of body weights about the mean, was generally unaffected by qualitative or quantitative dietary restriction. The only significant difference in body weight variation arose among birds of treatment 6 in Experiment 2 at 12 weeks of age. At this stage, the birds in question showed a significantly smaller body weight range and a lower mean weight than any other treatment group. Thus, they were uniformly light and the severe rationing at the 12-week stage had, if anything, improved evenness of growth, rather than the reverse.

EFFECT OF RESTRICTED FEEDING DURING REARING ON MORTALITY AND FEATHER SCORE

Relevant data for the rearing period are shown in *Tables 13.4* and *13.5*, while those for the laying stage are provided in *Tables 13.6* and *13.7*.

Mortality in the rearing and laying stages of both experiments was low. The only significant treatment differences in mortality arose between 20 and 40 weeks in Experiment 1, when the groups which had been quantitatively restricted during rearing experienced slightly fewer losses than those which had been full-fed. This indicated that food rationing imposed little stress on the birds.

Scoring the birds for feather coverage gave an indication of the incidence of feather pecking within the treatment groups, as well as providing a measure of the birds' appearance. There were some differences between treatments for feather score which reached statistical significance, but no consistent or important trends were noticeable. In Experiment 1, differences in feathering which were apparent mid-way through rearing had disappeared by 18 weeks of age (the important stage from the standpoint of birds' appearance and the effect this might have on a purchaser's judgement of their quality). In Experiment 2, significant differences between treatments in feather score did occur at the important final stage of the rearing period, when there was an apparent downward trend from treatment 1 to treatment 6. However, feathering throughout

Table 13.6 BIRD LIVE WEIGHT, FOOD INTAKE, MORTALITY AND FEATHER SCORE IN THE LAYING STAGE: EXPERIMENT 1

				Treatment (see Table 13.2)				
	1	2	3	4	5	6	7	8
Live weight (kg)								
20 weeks	1.32c	1.28bc	1.29c	1.23ab	1.21a	1.30c	1.23ab	1.23ab
40 weeks	1.72b	1.75b	1.75b	1.69ab	1.62a	1.73b	1.67ab	1.69ab
60 weeks	1.74ab	1.79b	1.77ab	1.74ab	1.68a	1.75ab	1.72ab	1.72ab
80 weeks	1.86b	1.85b	1.88b	1.78ab	1.73a	1.82ab	1.78ab	1.80ab
Food intake †† (kg bird-day basis)								
20–40 weeks	14.3bc	14.5c	14.3c	14.2bc	13.9ab	14.4c	13.9a	14.2bc
40–80 weeks	31.6ab	32.8c	32.3bc	31.7ab	31.2a	32.3bc	31.2a	31.6ab
20–80 weeks	45.9ab	47.3c	46.6bc	45.9ab	45.1a	46.7bc	45.1a	45.8ab
Food intake† (kg/bird housed)								
20–40 weeks	14.1abc	14.4c	14.3bc	14.1abc	13.9ab	14.4c	13.8a	14.2bc
40–80 weeks	30.6abc	31.9d	31.2bcd	30.5abc	30.3ab	31.6cd	30.0a	30.7abc
20–80 weeks	44.7abc	46.3d	45.5bcd	44.6ab	44.2a	46.0cd	43.8a	44.9abc
Mortality (%)								
20–40 weeks	1.49bc	1.03abc	1.26bc	1.72c	0.46a	0.57ab	0.80ab	0.92abc
40–80 weeks	4.25	4.25	4.83	4.71	5.06	3.91	4.37	4.83
20–80 weeks	5.74	5.28	6.09	6.43	5.52	4.48	5.17	5.75
Feather score								
20 weeks	3.3	3.0	3.1	3.2	3.2	3.1	3.1	3.2
40 weeks	3.1	2.9	3.0	3.2	3.0	3.0	3.1	3.0
60 weeks	1.5	1.3	1.4	1.6	1.5	1.4	1.6	1.7
80 weeks	2.1	2.0	2.1	2.1	2.0	2.1	2.1	2.4

† Calculated as (total food intake for group ÷ number of birds in group at start of the laying period).
†† An allowance has been made in this figure for mortality during the laying period for each group.

Table 13.7 BIRD LIVE WEIGHT, FOOD INTAKE, MORTALITY AND FEATHER
SCORE IN THE LAYING STAGE: EXPERIMENT 2

	Treatment (see Table 13.3)					
	1	2	3	4	5	6
Live weight (kg)						
20 weeks	1.79a	1.74b	1.68c	1.64cd	1.63d	1.63d
40 weeks	2.26a	2.26a	2.17b	2.16b	2.20ab	2.21ab
60 weeks	2.33	2.33	2.26	2.28	2.29	2.31
80 weeks	2.35	2.33	2.31	2.29	2.30	2.31
Food intake†† (kg bird-day basis)						
20−40 weeks	17.0	17.0	16.8	16.9	17.0	16.9
40−80 weeks	36.0	35.9	35.5	35.3	35.2	35.2
20−80 weeks	53.0	52.8	52.3	52.2	52.1	52.1
Food intake† (kg/bird housed)						
20−40 weeks	16.9	16.9	16.8	16.8	16.9	16.9
40−80 weeks	35.1	34.6	34.5	34.3	34.6	34.5
20−80 weeks	52.1	51.5	51.2	51.0	51.5	51.4
Mortality (%)						
20−40 weeks	0.94	1.46	0.73	0.94	0.63	0.73
40−80 weeks	3.75	5.21	5.10	4.79	2.81	3.65
20−80 weeks	4.69	6.67	5.83	5.73	3.44	4.38
Feather score						
20 weeks	4.0a	3.9a	3.8ab	3.7bc	3.6c	3.9a
40 weeks	3.2	3.4	3.4	3.3	3.3	3.5
60 weeks	2.0	2.1	2.1	2.0	2.1	2.1
80 weeks	1.4	1.4	1.5	1.3	1.4	1.4

† Calculated as (total food intake for group ÷ number of birds in group at start of the laying period).
†† An allowance has been made in this figure for mortality during the laying period for each
group.

was quite even, all scores were slightly above average, and any differences disappeared fairly early in the laying period. It should be noted that none of the
birds' beaks was trimmed.

RELATIONSHIP BETWEEN FEEDING PROGRAMME DURING REARING AND
SUBSEQUENT EGG PRODUCTION

Data related to egg production in the initial 20 weeks and final 40 weeks of
the laying period are given in *Tables 13.8* and *13.9*.
   The main influence of restricted feeding on the rate of sexual maturity, as
measured by age at 50 per cent egg production, and on egg yield in the early
stages of the laying period, was through its effect on body weight at 18−20
weeks of age. The heaviest birds at 18−20 weeks of age came into lay first and
produced the highest number of eggs between 20 and 40 weeks of age. Subsequent egg production, however, tended to be influenced more by the pattern
of growth during rearing than by the body weight attained at the end of the
rearing period. Thus, in Experiment 1 those treatment groups, apart from the
full-fed control, which laid comparatively poorly between 40 and 80 weeks of
age had experienced feeding programmes which retarded growth in the initial
(treatments 4 and 8) or final (treatment 6) stages of rearing, whereas those
groups which laid comparatively well over this period had been subjected to
most restriction of food, and thereby depression of growth, in the middle of the

Table 13.8 EGG PRODUCTION DATA: EXPERIMENT 1

	1	2	3	4	5	6	7	8
					Treatment (see Table 13.2)			
Age of 50% production (days)†	164a	166ab	166ab	169c	166ab	167bc	168bc	167bc
Peak egg production (%)††	93.2ab	95.1c	94.3bc	91.9a	93.6bc	92.9ab	94.1bc	93.8bc
Egg number (bird-day basis)								
20–40 weeks	106c	106c	106c	101a	104bc	104bc	104bc	104bc
40–80 weeks	207a	213b	211ab	209ab	213b	206a	213b	207a
20–80 weeks	313abc	319c	317bc	310a	317bc	310a	317bc	311ab
Egg number (per bird housed)								
20–40 weeks	105cd	105cd	105cd	100a	104bc	104bc	103b	104bc
40–80 weeks	201	207	204	201	206	202	204	201
20–80 weeks	306ab	312c	309bc	301a	310bc	306ab	307abc	305ab
Egg weight (g/egg)								
20–40 weeks	53.2de	53.3e	53.0cde	52.6bc	52.4ab	52.7bcd	52.5ab	52.0a
40–80 weeks	62.1cd	62.2d	61.8abcd	61.4ab	61.2a	62.0bcd	61.8abcd	61.5abc
20–80 weeks	59.2c	59.3c	58.9abc	58.5ab	58.3a	58.9bc	58.7abc	58.4ab
% large-grade eggs (>62 g)ƒ								
20–40 weeks	8.4bcd	9.3d	7.8abc	6.8a	7.3ab	9.0cd	7.4ab	7.3ab
40–80 weeks	43.3d	42.7cd	41.0bc	37.0a	37.0a	40.6b	40.4b	37.8a
20–80 weeks	31.1c	31.3c	29.6b	26.7a	26.8a	29.7b	29.2b	27.2a
% medium- and small-grade eggs (42–53 g)ƒ								
20–40 weeks	31.9bc	29.4a	32.4bc	33.0bcd	35.2d	30.9ab	33.4cd	35.0d
40–80 weeks	3.3abc	2.9a	3.6bcd	4.2de	4.4e	3.0ab	3.7cde	4.4e
20–80 weeks	13.3bc	12.0a	13.5bc	14.1cd	14.9de	12.7ab	13.8c	15.1e
% downgraded eggsƒ								
20–40 weeks	4.2	3.6	4.4	4.3	4.0	4.0	3.8	4.0
40–80 weeks	6.7	6.4	6.6	6.7	6.8	6.7	5.9	5.8
20–80 weeks	5.8	5.5	5.9	5.8	5.9	5.8	5.2	5.2

† Calculated from the second day of the first 2-day period during which egg production averaged at least 50% on a hen-housed basis.
†† Highest percentage egg production attained in any one week calculated on a hen-housed basis.
ƒ Eggs were graded according to the official UK standards operative at the time (Low, 1970).

Table 13.9 EGG PRODUCTION DATA: EXPERIMENT 2

			Treatment (see Table 13.3)			
	1	*2*	*3*	*4*	*5*	*6*
Age at 50% production (days)†						
	165a	168bc	167b	169c	171d	171d
Peak egg production (%)††						
	92.5	92.3	91.7	92.7	92.7	92.8
Egg number (bird-day basis)						
20–40 weeks	105a	103b	102bc	102b	100c	101c
40–80 weeks	196	197	194	197	197	196
20–80 weeks	301	300	296	299	297	297
Egg number (per bird housed)						
20–40 weeks	104a	102b	102b	101bc	100c	100c
40–80 weeks	191	189	188	191	194	192
20–80 weeks	295	292	290	292	294	293
Egg weight (g/egg)						
20–40 weeks	55.8	55.4	56.0	55.3	55.8	55.4
40–80 weeks	64.6	64.4	64.4	64.5	64.6	64.3
20–80 weeks	61.6	61.4	61.6	61.4	61.7	61.3
% large-grade eggs (> 62 g)ƒ						
20–40 weeks	22.3	19.7	20.8	20.3	23.4	21.5
% eggs in grade sizes 1 and 2 (65 g and over)ƒ						
20–80 weeks	35.7	34.1	36.1	36.3	35.5	34.5
% downgraded eggs						
20–40 weeks	2.78	2.99	2.74	2.61	2.79	2.92
40–80 weeks	5.35b	4.63a	4.68a	4.63a	5.36b	4.91a
20–80 weeks	4.47cd	4.07ab	4.01ab	3.94a	4.50d	4.24bc

† Calculated from the second day of the first 2-day period during which egg production averaged at least 50% on a hen-housed basis.
†† Highest percentage egg production attained in any one week calculated on a hen-housed basis.
ƒ At the 48-week stage in Experiment 2 the official grading standards were changed from the original UK weight grades (Low, 1970) to the EEC grade sizes in current use (Overfield, 1973). An estimate was made of the percentage eggs produced in each of the new weight grades to this stage of lay using accumulative basis conversion graphs (Eggs Authority, 1977)

rearing period (treatments 2, 5 and 7). In Experiment 2, however, differences between treatments in egg production during the latter two-thirds of the laying period were small and non-significant.

There were significant treatment differences in Experiment 1 for mean egg weight and egg-weight grades. Restricted feeding seemed to influence egg weight through its effect on body weight at point-of-lay and at subsequent stages in laying. The heaviest birds tended to produce the heaviest eggs. This trend was not apparent, however, in Experiment 2, where all differences between treatments in egg weight were non-significant.

The quality of eggs produced, as measured by the incidence of downgrading at the commercial packing station, did not appear to be affected by feeding treatment during rearing. The proportion of downgraded eggs did vary significantly in the latter stages (40–80 weeks of age) in Experiment 2, treatments 1 and 5 showing a higher proportion than other treatments, but this cannot be explained except in terms of a chance effect.

RELATIONSHIP BETWEEN FEEDING PROGRAMME DURING REARING AND
SUBSEQUENT BODY WEIGHT GAIN AND FOOD INTAKE

Data for body weight and food intake at selected stages in the laying period are
presented in *Tables 13.6* and *13.7*.

In both experiments, treatment groups tended to retain their body-weight
rankings from point-of-lay to end-of-lay, although the body-weight differentials
between treatments narrowed with time and in Experiment 2 failed to reach
statistical significance after the 40-week stage. Among the quantitatively restricted groups, those which were rationed in the final stage of rearing (treatment 6
in Experiment 1 and treatment 2 in Experiment 2) recouped body weight more
rapidly than other groups which were rationed at earlier ages.

There were overall positive relationships in both experiments, between body
weight and food intake during the laying period and between food intake in
rearing and in lay. There was not, therefore, compensation for reduced food
intake in the rearing stage by increased consumption during the laying period. It
follows from this, that quantitative restriction of food in the rearing period
effected a greater reduction in subsequent food intake than qualitative restriction
(compare treatments 5–8 with treatments 2–4 in Experiment 1). In particular,
food intake in lay was low following quantitatively restricted feeding programmes which gave higher growth rates than the control during the final three
weeks of the rearing stage, but much lower rates of growth relative to the control
between six and 15 weeks of age (treatments 5, 7 and 8 in Experiment 1 and
treatments 5 and 6 in Experiment 2).

Although in Experiment 2 there were no significant differences in cumulative
food intake/bird, in the 40–80 week period consumption for treatments 1 and
2 was considerably and significantly higher than that for treatments 4, 5 and 6.

RELATIONSHIP BETWEEN FEEDING PROGRAMME DURING REARING AND
SUBSEQUENT EFFICIENCY OF FOOD CONVERSION TO EGGS

Synopses of the egg output and food intake data showing the comparative
biological and economic efficiencies of conversion of food to eggs are provided
in *Tables 13.10* and *13.11*. In each experiment, as expected, there was a very
close relationship between the biological efficiency of conversion of food to
eggs and the economic margin of egg income over food cost. Treatments which
ranked first and second in both experiments for these measures of efficiency
of laying performance, did so largely because of lower food intakes.

It was, in fact, the same feeding programme which gave the most efficient
biological and economic conversion in each experiment: treatment 7 in Experiment 1 and treatment 5 in Experiment 2. Moreover, both second-ranking treatments (treatment 5 in Experiment 1 and treatment 4 in Experiment 2) were very
similar (*see Tables 13.2* and *13.3*). A common feature of all four of these high-
ranking treatments was that they involved comparatively heavy rationing of food
in the middle of the rearing period, which caused marked reductions in body
weight in comparison with the control birds between six and 15 weeks of age,
but allowed a rapid recovery in growth thereafter. Rates of growth in this middle
period in Experiment 1 were 74 per cent (treatment 7) and 81 per cent (treatment 5) relative to the full-fed control birds; the corresponding figures in

Table 13.10 SYNOPSIS OF EGG OUTPUT AND FOOD INTAKE DATA (20–80 WEEKS)† : EXPERIMENT 1

Treatment	Bird-day egg output†† (kg)	Bird-day food intake (kg)	Food conversion ratio (kg food/kg eggs)	Margin of egg income over food cost (£/bird)
1	18.53 (4)	45.9 (4=)	2.477 (3)	2.15 (3=)
2	18.92 (1)	47.3 (8)	2.500 (5)	2.15 (3=)
3	18.67 (2)	46.6 (6)	2.496 (4)	2.12 (5)
4	18.13 (8)	45.9 (4=)	2.532 (7)	1.99 (7)
5	18.48 (5)	45.1 (1=)	2.440 (2)	2.26 (2)
6	18.26 (6)	46.7 (7)	2.557 (8)	1.98 (8)
7	18.61 (3)	45.1 (1=)	2.423 (1)	2.28 (1)
8	18.16 (7)	45.8 (3)	2.522 (6)	2.06 (6)

† Figures in brackets indicate treatment rankings.
†† Bird-day egg number x mean egg weight.

Experiment 2 were 64 per cent (treatment 5) and 79 per cent (treatment 4).

Heavy rationing of food in the latter stages of the rearing period (treatment 6 in Experiment 1 and treatments 2 and 3 in Experiment 2) and qualitative dietary restriction in the first six weeks of rearing (treatments 3, 4 and 8 in Experiment 1) resulted in comparatively poor biological and economic performance.

It was noted in the introduction that, when judging the quality of pullets, egg producers often attach considerable importance to body weight at point of lay, that is, at about 18 weeks of age. However, it is apparent from the results of these experiments that body weight at 18 weeks is not a reliable indicator of subsequent laying performance, when considered in isolation from the pattern of growth leading to that weight. A comparison of treatment 4 and treatment 7 in Experiment 1, which showed respectively one of the least and one of the most efficient laying performances, affords the best illustration of this point. These two treatments produced very similar body weights at 18 weeks of age, but markedly different growth patterns. Treatment 4 showed a practically linear

Table 13.11 SYNOPSIS OF EGG OUTPUT AND FOOD INTAKE DATA (20–80 WEEKS)† : EXPERIMENT 2

Treatment	Bird-day egg output†† (kg)	Bird-day food intake (kg)	Food conversion ratio (kg food/kg eggs)	Margin of egg income over food cost (£/bird)
1	18.53 (1)	53.0 (6)	2.859 (3)	2.79 (4)
2	18.40 (2)	52.9 (5)	2.871 (6)	2.76 (5)
3	18.26 (5)	52.3 (4)	2.861 (4=)	2.75 (6)
4	18.35 (3)	52.2 (3)	2.843 (2)	2.85 (2)
5	18.34 (4)	52.1 (1=)	2.842 (1)	2.87 (1)
6	18.21 (6)	52.1 (1=)	2.861 (4=)	2.82 (3)

† Figures in brackets indicate treatment rankings.
†† Bird-day egg number x mean egg weight.

Table 13.12 MARGIN OF EGG INCOME OVER FOOD COST: EXPERIMENT 1

				Treatment (see Table 13.2)				
	1	2	3	4	5	6	7	8
Egg income†								
Average price per dozen eggs (p)	29.78	29.92	29.70	29.55	29.49	2.70	29.76	29.59
Egg income per bird housed (£)	7.59	7.78	7.65	7.41	7.62	7.59	7.61	7.52
Food cost per bird housed†† (£)								
Rearing (day-old –20 weeks)	0.78	0.77	0.76	0.74	0.67	0.75	0.67	0.69
Laying (20 –80 weeks)	5.44	5.63	5.53	5.42	5.36	5.61	5.33	5.46
Total (day-old –80 weeks)	6.22	6.40	6.29	6.16	6.03	6.36	6.00	6.15
Margin per bird housed (£)								
20 –80 weeks	2.15	2.15	2.12	1.99	2.26	1.98	2.28	2.06
Day-old –80 weeks	1.37	1.38	1.36	1.25	1.59	1.23	1.61	1.37

† Egg prices include the packing station bonuses.
†† Calculated from starter, grower and developer diets at £131.20, £116.50 and £114.10 per tonne, respectively, and from layer diets at £121.20 (fed 20–40 weeks) and £116.00 (fed 40–80 weeks) per tonne.

growth curve with significantly lighter birds compared with the control at six weeks of age, but similar growth rates thereafter. Treatment 7, on the other hand, gave birds of similar body weight to the controls at six weeks of age, but caused a marked depression in growth rate between six and 15 weeks and allowed compensatory growth in the final stages of rearing.

AN OVERALL ECONOMIC EVALUATION OF THE RESULTS

Economic evaluations of the results, based on prices prevailing during the respective experimental periods, are set out in *Tables 13.12* and *13.13*.

Table 13.13 MARGIN OF EGG INCOME OVER FOOD COST: EXPERIMENT 2

	1	*2*	*3*	*4*	*5*	*6*
		Treatment (see Table 13.3)				
Egg income†						
Average price per dozen eggs (p)	34.58	34.57	34.60	34.74	34.84	34.69
Egg income per bird housed (£)	8.51	8.41	8.37	8.46	8.52	8.46
Food cost per bird housed†† (£)						
Rearing (day-old–20 weeks)	0.94	0.90	0.89	0.80	0.79	0.83
Laying (20–80 weeks)	5.72	5.65	5.62	5.61	5.65	5.64
Total (day-old–80 weeks)	6.66	6.55	6.51	6.41	6.44	6.47
Margin per bird housed (£)						
20–80 weeks	2.79	2.76	2.75	2.85	2.87	2.82
Day-old–80 weeks	1.85	1.86	1.86	2.05	2.08	1.99

† Egg prices include packing station bonuses.
†† Calculated from starter and grower diets at £116.19 and £106.92 per tonne, respectively and from layer diets at £117.29 (fed 20–40 weeks) and £106.19 (fed 40–80 weeks) per tonne.

Financial margins for the entire period from day-old to 80 weeks of age showed substantial advantages over the full-fed control for treatments involving quantitative restriction of food in the middle of the growing period. These regimens provided the biggest savings in food cost during rearing, coupled with the highest margins of egg income over food cost in the laying period, which in the case of treatments 5 and 7 of Experiment 1 and treatments 4 and 5 of Experiment 2 amounted to a financial advantage, compared with the control, of at least 20 p/bird housed, while treatment 6 in Experiment 2 showed an advantage over the control of 14 p/bird housed.

With all the other restricted feeding programmes, the savings in food cost during rearing were either offset (treatments 2, 3 and 8 in Experiment 1 and treatments 2 and 3 in Experiment 2) or outweighed (treatments 4 and 6 in Experiment 1) by losses due to poorer economic performances in the subsequent laying period.

Conclusions

The way in which pullet growth is controlled in order to achieve a given target weight at point-of-lay appears to be important. Some of the programmes of restricted feeding applied in these experiments gave better overall biological and economic performance than conventional full-feeding, but other programmes resulted in poorer performance.

The greatest economic benefits arose from quantitative restriction of food intake which commenced at six weeks of age and markedly reduced growth rate in the middle of the rearing period, but allowed compensatory growth to take place in the later stages. Heavy mid-period rationing of this nature provided:

(1)   substantial reductions of food intake and corresponding savings in food cost during the rearing period with a comparatively small effect on body weight at point-of-lay;
(2)   small but consistent reductions in food intake during the laying stage with little or no adverse effect on egg output, which improved the biological efficiency of food conversion and increased the margin of egg income over food cost.

Acknowledgements

The contributions to this work of technical, plant and secretarial staff at the Harper Adams Poultry Husbandry Experimental Unit are gratefully acknowledged.

References

BALNAVE, D. (1973). A review of restricted feeding during growth of laying-type pullets. *Wld Poult. Sci. J.*, **29**, 354–362

BOLTON, W. and BLAIR, R. (1974). *Poultry Nutrition. MAFF Bull. 174.* London; HMSO

COGGINS, J.E. and WELLS, R.G. (1979). The effects of various patterns of growth between six and 18 weeks of age on the performance of layers. *Harper Adams Poult. Husbandry Exp. Unit, Report R/26*

EGGS AUTHORITY (1977). Accumulative basis conversion graph. *Supplement to Weekly Report No. 332*, pp. 3–4

LOW, E.M. (1970). The importance of grading in egg marketing. In *Factors Affecting Egg Grading*, pp. 3–16. Eds B.M. Freeman and R.F. Gordan. Edinburgh; British Poultry Science Ltd

OLDALE, P.M.D., COGGINS, J.E. and WELLS, R.G. (1977). Effect of variations in the pattern of growth of pullets on subsequent laying performance. *Harper Adams Poult. Husbandry Exp. Unit, Report R/24*

OVERFIELD, N.D. (1973). *Testing of Eggs for Quality. MAFF Bull. 28.* London; HMSO

14

RESPONSE OF LAYING HENS TO ENERGY AND AMINO ACIDS

R. M. GOUS and F. J. KLEYN
Department of Animal Science and Poultry Science, University of Natal, Pietermaritzburg, South Africa

Introduction

Twenty years ago Morris (1968) presented a review concerning the response of laying hens to nutrient density from which it is possible to optimize this characteristic of the feed for a flock of laying hens. Very little has been added to our knowledge of this subject since then. Significant advances have, however, been made during the past two decades in the characterization of responses among laying hens to the more important amino acids. In an elegant paper on this subject, Fisher, Morris and Jennings (1973) produced a model by which such responses could be used to determine the optimum intakes of amino acids in a population of laying hens. Responses are measured as changes in egg weight, rate of lay, body size and composition, and food intake. Information published on these responses is reviewed briefly here; some recent research is discussed pertaining to food intake regulation; and information is presented regarding the relative changes in egg weight and rate of lay that take place with changes in protein supply.

In spite of the financial and biological benefits that result from the use of the above research reports, there is a reluctance among many feed compounders to make use of such methods to optimize feeding strategies in laying flocks.

There are two particular reasons why all nutritionists do not make use of biological responses in feed formulation. Firstly, compounders tend to 'play safe' with feeds whose composition has been found to produce satisfactory results in the field, i.e. they do not wish to move outside the narrow bounds defined in their linear programming (LP) matrix. Secondly, the calculation of marginal costs of amino acids and the determination of optimum nutrient densities is both tedious and time-consuming, as least-cost solutions have to be obtained over a range of concentrations of amino acids and nutrient densities in order to obtain such values. It is easier to ignore changes in ingredient prices which may (but may not) make only small differences to profitability. More emphasis has been directed towards the development of linear programs to optimize bulk handling and usage of raw materials in the feed mill than in optimizing the feeding strategy of laying hens. Milling costs have certainly been reduced in this way, with a consequent increase in millers' margin, but further benefits should be possible on behalf of the egg producer. Such an approach would be especially beneficial in large integrated operations. An efficient and rapid computer program has been developed recently as a means of determining the optimum strategy

First published in *Recent Advances in Animal Nutrition – 1988*

for feeding flocks of laying hens, making use of the responses reviewed in this chapter, and some examples are given of the value of such a technique in making nutritional decisions.

Effect of amino acid and energy concentrations on egg output

AMINO ACIDS

It is common knowledge that dietary amino acid supply will affect egg production. The requirement for these amino acids should, however, not be seen as a fixed concentration in the feed but should take account of the relationship between amino acid intake and egg output, which is dependent on both feed intake and level of production. A series of response experiments, mainly at the University of Reading (Fisher, Morris and Jennings, 1973; Pilbrow and Morris, 1974; Morris and Wethli, 1978; Wethli and Morris, 1978; Morris and Blackburn, 1982) has improved our knowledge of efficiencies of utilization of amino acids for egg production, and culminated in a review by McDonald and Morris (1985) in which best estimates of the coefficients of response by laying hens to different amino acids were presented (*Table 14.1*).

With such information the Reading Model (Fisher, Morris and Jennings, 1973) can be used effectively to determine the optimum economic intake of each of the amino acids for existing stocks of laying hens as well as future stocks, which may differ in body weight, in potential egg output, or in both. Because this optimum is dependent on such variables as flock uniformity, marginal costs of amino acids and marginal revenue for eggs, the most economical intake of each amino acid will differ under the variable conditions experienced between farms, between flocks and between different countries. No single recommendation could be expected to be applicable under all circumstances.

The optimum intake of the first-limiting amino acid will influence the optimum intakes of all other amino acids. Because the marginal costs of essential amino acids are likely to vary one from the other, one, or possibly two, amino acids with the highest marginal costs will limit egg output of the flock. Under most circumstances it would, therefore, be unnecessary to feed the remaining amino acids at their optimum intakes, some economic advantage being derived by reducing the intakes of these amino acids to the level governed by the egg output sustainable by the first-limiting amino acid.

A major difficulty confronting feed formulators is the conversion of these optimum intakes of amino acids into dietary concentrations. This difficulty would be resolved if

Table 14.1 COEFFICIENTS OF RESPONSE (mg) FOR SOME ESSENTIAL AMINO ACIDS FOR INDIVIDUAL PULLETS

Lysine	10.0 E	+	73 W
Methionine	4.8 E	+	31 W
Tryptophan	2.6 E	+	11 W
Isoleucine	8.0 E	+	67 W
Valine	8.9 E	+	76 W

After McDonald and Morris (1985)
Where E = Egg output (g/bird day) and W = bodyweight (kg)

food intake could be predicted accurately. Further discussion on this subject is presented later.

Partitioning the response to amino acids between egg weight and egg number

Egg size can be manipulated only to a very limited extent by nutrition. A high energy diet containing supplementary fat will increase egg weight by a maximum of 1 g (Morris, 1985), provided that the protein:energy ratio is maintained. Dietary amino acid supply also has a small effect on egg weight, mainly on albumen content (Fisher, 1969) but because egg numbers are influenced also, there is little scope for manipulating egg size by adjusting amino acid supply.

For a detailed financial analysis of the consequences of varying protein supply it is necessary to predict separately the expected changes in rate of lay and egg size, since in most markets eggs of different sizes have different values per unit weight. Morris and Gous (1988) surveyed 42 sets of data in which responses in egg number and egg size to dietary amino acids were measured. Contrary to the common belief that only egg size is reduced when protein supply is marginally reduced (i.e. that slightly more protein is needed to maximize egg size than to maximize egg number) the evidence indicated that small increments in protein (or amino acids), close to the optimum, resulted in equal proportional responses in rate of lay and egg size. When protein supply is below the optimum (egg output below 0.95 of the maximum potential of the flock), the expected reduction in rate of lay is greater than the expected reduction in egg size. In the above analysis it was found that reduction in egg size did not fall below 0.9 of the maximum value, until amino acid supply was well below 0.5 of its optimum value, whereas rate of lay was reduced below 0.7 of its potential value when amino acid intake is half the optimum (*Figure 14.1*).

The proportion of eggs falling into different grades will differ with protein supply, and hence influence the marginal revenue for eggs. The equation relating egg size to protein supply (*Figure 14.1*) can therefore be used to determine marginal revenue. This is a prerequisite if the Reading Model is to be used accurately. However, the optimum intakes of amino acids appear to be relatively insensitive to changes in the marginal revenue for eggs (see later).

It can be concluded that, because under most commercial conditions the amino acid supply that maximizes profit in a layer operation will be that which allows the flock to produce at a near-maximal rate, any effort to use nutritional means to reduce egg size will result in a concomitant decrease in rate of lay, which would be likely to reduce profits. Egg size can be manipulated to a greater extent by adjusting lighting patterns (e.g. ahemeral cycles longer than 24 h, or 6 h repeating light–dark cycles (Sauveur and Mongin, 1983) than by altering nutrient levels (Morris, 1985) and the effects of such cycles on egg size can be switched on and off when necessary, a further advantage over the use of nutritional methods of altering egg size.

ENERGY

A certain amount of confusion exists regarding the terms 'energy concentration' and 'nutrient density'. It has been recognized for a long time that when the energy content of a poultry feed is increased, the concentration of most of the other nutrients in the

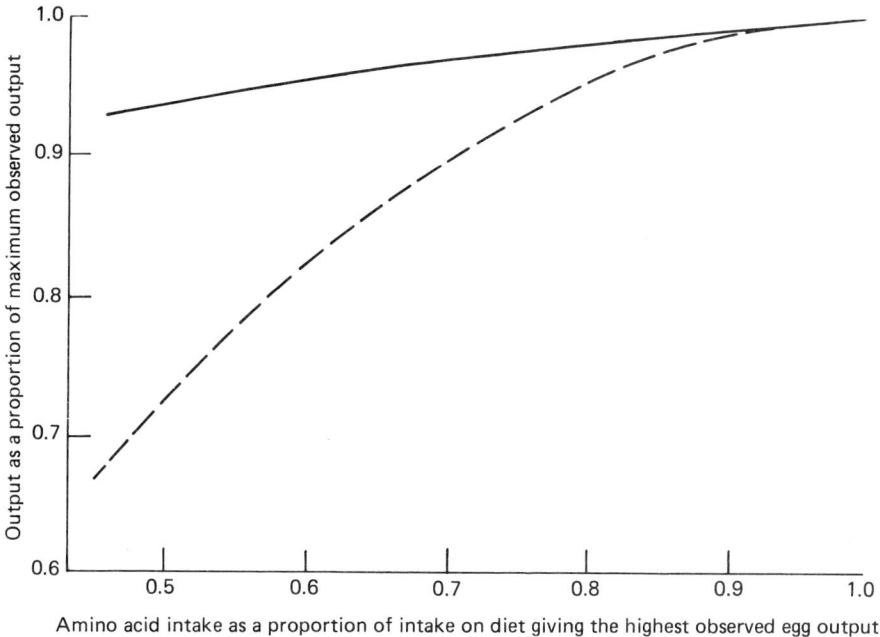

Figure 14.1 The relationship between intake of a limiting amino acid and rate of lay (–––) or egg weight (——). Equations for the two responses are: relative egg weight $= 1 - 0.07353x - 0.10424x^2$; relative rate of lay $= 1 - 0.03734x - 1.02927x^2$. (After Morris and Gous, 1988)

feed should be increased proportionately (Combs, 1962). Thus the more useful term 'nutrient density' should be used to describe the concentrations of dietary energy, assuming that nutrients have been adjusted accordingly. Some of the confusion has arisen as a result of the use of ME as the unit measure of nutrient density.

Although much of the following discussion concerns nutrient density, some interesting experiments involving changes in energy at constant amino acid concentrations will be mentioned also.

Effect of nutrient density on egg production

The production of eggs by laying hens is not influenced by nutrient density, except at the extremes (Morris, 1968). Evidence regarding the effect of nutrient density on egg weight is less well defined, recent reports suggesting that feeds high in nutrient density produce slightly larger eggs. McDonald (1984) found the regression of egg weight on nutrient density to be 0.2487 g/MJ kg. This effect is so small as to have virtually no influence on the optimum density of feeds.

The positive effect of nutrient density on egg weight could probably be ascribed to either a linoleic acid deficiency at very low nutrient densities, which is overcome with the use of high energy ingredients, or to the use of oil in the feeds of high nutrient density, this latter being known to increase egg size (Edwards and Morris, 1967).

Effect of dietary energy concentration on egg output

In three experiments, designed to measure the responses to lysine, methionine and isoleucine at different energy concentrations (Gous, Griessel and Morris, 1987) the response in egg output was such that a common curve for each amino acid adequately described the response at each of the dietary energy concentrations (*Figure 14.2*). Egg output was, therefore, not influenced by energy concentration, other than through its effects on food intake, with a consequent change in intake of the first-limiting amino acid. It was demonstrated that, although the main effect of energy concentration on egg weight and rate of lay in the methionine experiment was significant in both cases, these differences were brought about by the particularly low egg output associated with the low energy, low methionine treatment. When the response in egg weight and rate of lay were compared with the actual intakes of methionine it was evident that at

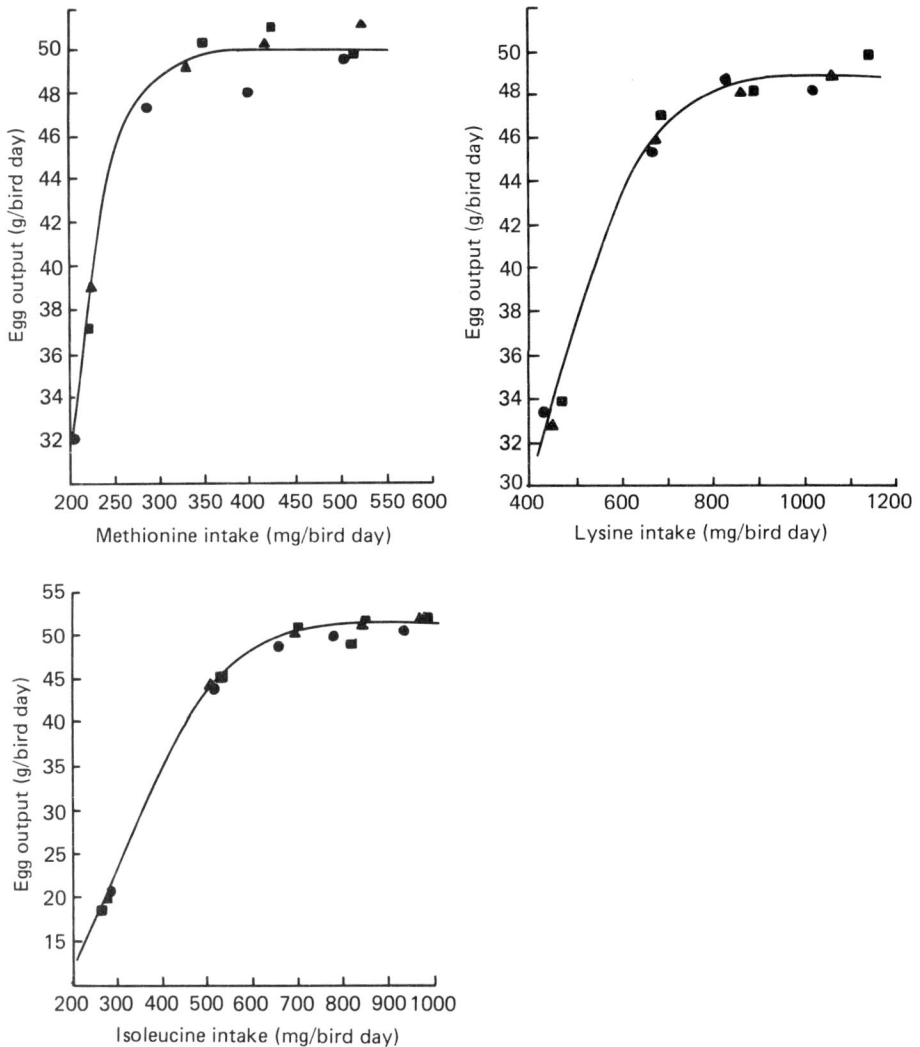

Figure 14.2 Response of laying hens to intakes of methionine, lysine and isoleucine at three energy concentrations (low ●, medium ◁ and high ■). (After Gous, Griessel and Morris, 1987)

the lowest methionine concentration the difference in productivity between energy levels had not been caused by the low energy concentrations *per se*, but rather by the low intake of methionine (*Figure 14.3*). The apparent effect of energy concentration on egg output can therefore be explained by expressing output as a function of intake of the limiting nutrient.

Effect of environmental temperature on egg output

The principle described above applies to environmental temperatures also. High temperatures will reduce intake of all nutrients, and output, which is a function of intake of the first-limiting nutrient, will be reduced by a corresponding amount. Bray and Gesell (1961) showed that output at high temperatures (30°C) could be sustained by increasing the concentration of nutrients in the feed. Subsequently it was suggested (Smith and Oliver, 1972) that at temperatures above 27°C it is likely that egg output is restricted by an inadequate intake of energy, although another factor which probably plays a role in reducing egg output is a reduction in blood supply to the ovary (Wolfenson *et al.*, 1979). At temperatures in excess of 30°C, if energy intake is the limiting factor in determining egg output, it would be unnecessary and counterproductive to increase amino acid supply still further as the excess protein, by the process of digestion and deamination, will increase heat production by the hen thereby further aggravating her heat load. Under such circumstances the use of dietary fat would serve the purpose of increasing energy intake and, because of its lower heat increment compared with other feed materials, would reduce the heat load on layers at extremely high temperatures (Marsden and Morris, 1981). Egg size, which is reduced at high temperatures, can be increased in this manner, although the effect would be expected to be very small (Marsden, Morris and Cromarty, 1987).

The obvious conclusion regarding the effect of constant high temperatures on egg production is that the hen becomes too hot and no nutritional manipulations will enable her to overcome this problem. In the short term, utilization of lipid reserves may enable the hen to produce eggs at a higher rate than might be expected considering her low energy intake, but in the long term such a buffering capacity could not persist and production would decline.

Effect of amino acid and energy concentrations on food intake

Egg output of laying hens is determined largely by the intake of the first-limiting nutrient in the feed (Almquist, 1952; Emmans, 1981). Some accurate method of predicting food intake under varying environmental and nutritional circumstances would therefore be a prerequisite in calculating optimum dietary concentrations of nutrients in feeds for laying hens. Despite this, very little effort has been made to predict food intake accurately, yet this is central to the issue of maximizing profit in a layer enterprise.

Because of the vast body of publications stating that 'birds eat to satisfy their energy requirement', a major effort has been made to determine the energy requirements of laying hens, usually by multiple regression analysis (Byerly, 1941; 1979) but scant attention has been paid to the effects on intake of, for example, low dietary amino acid or calcium concentrations. Some observations and suggestions regarding this important issue are reported below, with the effects of energy being discussed first.

ENERGY

When fed *ad libitum*, birds will consume different amounts of energy depending on the nature of the diet presented to them (Morris, 1968). Most experiments in which dietary energy has been partitioned between body maintenance, egg output and growth have involved one feed only and under such circumstances the 'goodness of fit' of the resultant empirical equation is usually statistically highly significant (e.g. Leeson, Lewis and Shrimpton, 1973; McDonald, 1977; Gous *et al.*, 1978). The outliers are regarded as 'experimental errors' and usually no attempt is made to explain these anomalies. Prediction equations of this nature, then, are valuable only as a means of roughly estimating the average food intake of a population. Some sophistication can be introduced by considering the effects of temperature and feather cover (Emmans, 1974), but there are basic inaccuracies in estimating energy intake in this manner.

The relationship between environmental temperature and energy intake is curvilinear, with food intake declining more steeply as ambient temperature approaches body temperature (Marsden and Morris, 1987). The partition equation of Emmans (1974) expresses energy required for maintenance under different environmental temperatures but does not predict how much energy will be consumed by layers under such circumstances. A quadratic term would have to be introduced into the equation if the effect of temperature on energy intake is to be predicted. Alternatively, by expressing energy intake as a function of metabolic body size it can be represented as a linear function of temperature within the range 15–30°C, with a slope of -141.8 kJ/ $kg^{0.75}$ day (Marsden and Morris, 1987).

A further source of error in partition equations is that the energy required for maintenance is expressed in terms of body weight. A bird with large lipid reserves probably does not need more energy than a bird of similar protein weight with no lipid reserves. Maintenance heat should thus be scaled according to protein weight (preferably feather-free) and possibly also to the degree of maturity (Emmans, personal communication). This would have a significant effect on the recommended energy requirement of broiler breeder hens and would reduce the discrepancy in maintenance energy requirements between brown and white strains of laying hens, although this will not be eliminated entirely because of behavioural (activity) and anatomical (comb and wattle size; rate of feather loss) differences between these strains. Bodyweight gain (or loss) could be a result of protein retention (or depletion) with the associated change in water, lipid retention (or loss) or both and the amount of energy required (or yielded) will differ in each case. A single coefficient for change in body weight is a best-estimate for a flock under certain circumstances, but the nutritionist should be aware that there are many individuals in the flock that will not conform to this estimate, nor would this coefficient be the same for all feeds or for all environmental conditions.

AMINO ACIDS

In the published reports of many trials designed to measure responses of laying hens to amino acid concentrations (e.g. Pilbrow and Morris, 1974; Wethli and Morris, 1978; Morris and Wethli, 1978; Gous, Griessel and Morris, 1987) food intake increased in almost all cases as the concentration of dietary amino acids decreased, the birds clearly attempting to eat more food to compensate for marginal deficiency of the first-limiting amino acid. As the deficiency became more severe food intake declined in virtually all cases. In the three experiments reported by Gous, Griessel and Morris (1987) in which reponses to lysine, methionine and isoleucine were measured at different energy concentrations, food intake was affected to a greater extent by the dietary amino acid concentrations than by the energy concentrations (*Figure 14.3*).

The reason for the decline in food intake at low amino acid concentrations is not known. Emmans (1981) lists three reasons for the failure of birds to consume sufficient feed to overcome the deficiency in the first-limiting nutrient—bulk, environmental heat demand and the presence of toxins.

The first and last of these seem unlikely to be involved here, as the feeds in the dilution series do not differ markedly in these respects. Environmental heat demand is likely to be responsible for the observed decline in food intake. A consequence of consuming the excessively large amounts of a food, limiting in an amino acid, necessary to maintain maximum egg output, would be that the bird would need to lose more heat than it is capable of losing, this maximum heat loss being some function of body size and enironmental temperature. The hen would thus have no alternative but to reduce food intake. Egg production could be sustained immediately after the switch to a feed of low amino acid content if the hen had protein reserves from which to draw. These would differ between birds, hence egg production would drop more rapidly in some hens than in others.

Differences in initial lipid stores and maximum storage capacity would also influence the extent to which food intake and egg output were affected immediately after a change in dietary amino acid concentration. A feed intake simulation model, based on these concepts, is in the process of being developed by Emmans and Gous (as yet unpublished).

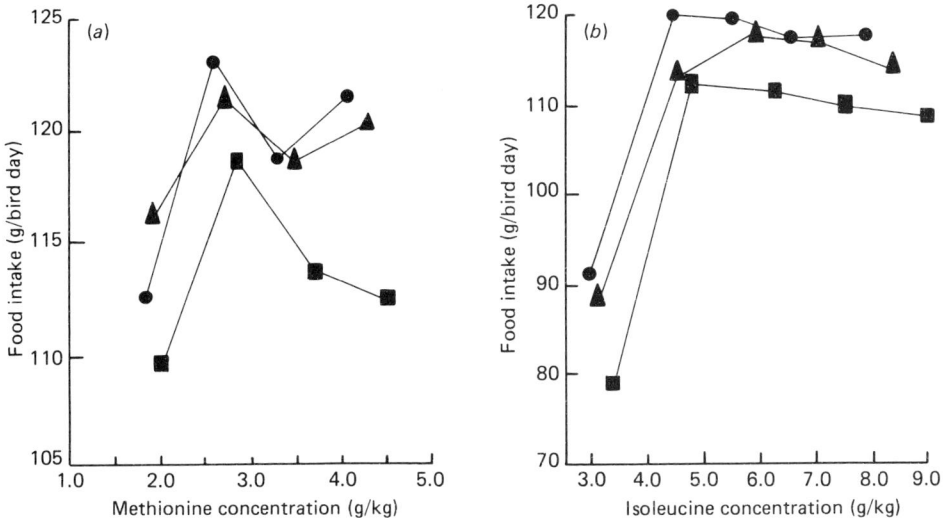

Figure 14.3 Food intakes of laying hens fed different concentrations of methionine (a) and isoleucine (b) at three energy concentrations (low ●, medium ▲ and high ■). (From Gous, Griessel and Morris, 1987)

Optimizing amino acid intakes and nutrient density

AMINO ACID INTAKES

The Reading model (Fisher, Morris and Jennings, 1973) is a population response model that optimizes the intake of amino acids for flocks varying in body weight and potential egg output, and for different relative values of amino acids and eggs.

The calculation of optimum intake of an amino acid is illustrated in *Tables 14.2* and *14.3*. At a dietary energy concentration the amino acid intake necessary to produce a specified mean egg output by the flock can be converted to a concentration in the feed, from a knowledge of the characteristic intake of food by the flock at that energy concentration. Feeds of increasing amino acid supply are formulated and the cost of feeding the bird is calculated as food cost, cents/g × food intake, g. Similarly the revenue derived can be calculated as value of egg, cents/g × mean egg output, g, the difference between these being the margin over food cost. The optimum intake is that which results in the highest margin.

A more rapid method of calculating intake of each amino acid in turn is described by Fisher, Morris and Jennings (1973), but the method described above, using basic

Table 14.2 AMINO ACID INTAKES REQUIRED TO SUSTAIN A RANGE OF EGG OUTPUTS IN A FLOCK OF LAYING HENS AND THE CONCENTRATIONS (g/kg) REQUIRED AT DIFFERENT DIETARY ENERGY CONCENTRATIONS

Mean egg output (g/bird day)	Amino acid intakes required[a] (mg/bird day)		Energy concentration (MJ/kg)						
			10.0	10.5	11.0	11.5	12.0	12.5	13.0
					Food intake (g/bird day)[b]				
			124	113	118	108	104	100	96
46	Lys	625	5.12	5.37	5.62	5.87	6.12	6.37	6.62
	Met	290	2.34	2.45	2.57	2.68	2.80	2.91	3.02
	Trp	145	1.17	1.23	1.28	1.34	1.40	1.46	1.51
48	Lys	652	5.26	5.52	5.77	6.03	6.29	6.54	6.80
	Met	303	2.44	2.56	2.68	2.80	2.92	2.04	3.16
	Trp	153	1.23	1.29	1.35	1.41	1.47	1.53	1.59
50	Lys	682	5.50	5.77	6.04	6.31	6.58	6.84	7.11
	Met	318	2.56	2.69	2.81	2.94	3.06	3.19	3.31
	Trp	161	1.29	1.36	1.42	1.48	1.55	1.61	1.67
52	Lys	720	5.81	6.09	6.38	6.66	6.94	7.22	7.51
	Met	336	2.71	2.84	2.98	3.11	3.24	3.37	3.50
	Trp	170	1.37	1.43	1.50	1.57	1.63	1.70	1.77
53	Lys	742	5.98	6.28	6.57	6.86	7.15	7.45	7.74
	Met	347	2.80	2.94	3.07	3.21	3.35	3.48	3.62
	Trp	175	1.41	1.48	1.55	1.62	1.69	1.76	1.82
54	Lys	772	6.23	6.53	6.84	7.14	7.44	7.75	8.05
	Met	360	2.90	3.05	3.19	3.33	3.47	3.61	3.75
	Trp	183	1.47	1.54	1.62	1.69	1.76	1.83	1.90
55	Lys	802	6.47	6.78	7.10	7.42	7.73	8.05	8.36
	Met	378	3.05	3.20	3.35	3.50	3.64	3.79	3.94
	Trp	192	1.54	1.62	1.70	1.77	1.85	1.92	2.00
56	Lys	875	7.06	7.40	7.75	8.09	8.44	8.78	9.12
	Met	413	3.33	3.49	3.65	3.81	3.98	4.14	4.30
	Trp	207	1.67	1.75	1.83	1.91	2.00	2.08	2.16

[a]Calculated from the coefficients published by McDonald and Morris (1985)
[b]Assuming a characteristic intake of 1243 kJ ME at a dietary energy concentration of 11.3 MJ/kg, using the equation of Morris (1968)

Table 14.3 MARGIN OVER FOOD COST[a] FOR A RANGE OF AMINO ACID
CONCENTRATIONS, REQUIRED TO SUSTAIN A GIVEN MEAN EGG OUTPUT IN A
LAYING FLOCK, AT DIFFERENT DIETARY ENERGY CONCENTRATIONS

Mean egg output[b] (g/bird day)	*Energy concentration* (MJ/kg)						
	10.0	*10.5*	*11.0*	*11.5*	*12.0*	*12.5*	*13.0*
46	91.4	91.7	91.9	92.1	92.0	91.9	91.7
48	96.3	96.6	96.9	96.9	96.8	96.7	96.2
50	98.5	98.8	99.0	99.1	98.9	98.7	98.1
52	99.2	99.6	99.8	99.7	99.5	99.3	98.5
53	99.4	99.7	100.0	99.8	99.6	99.2	98.5
54	99.3	99.6	99.8	99.6	99.4	98.9	98.2
55	99.1	99.5	99.5	99.3	99.1	98.4	97.6
56	98.3	98.6	98.6	98.3	97.9	97.1	96.3

[a]Calculated as (mean egg output, g × egg revenue, cents/g) − (food intake, g/bird × food cost cents/g) and expressed relative to the highest margin, which is given the value 100
[b]Mean egg output of a flock with E_{max} = 56 g/bird day, W = 2.0 kg and characteristic intake of energy at 11.3 MJ ME/kg of 1243 kJ/day. Amino acid concentrations required to sustain these egg outputs are given in *Table 8.2*

principles, will ensure a better understanding of the optimization procedure which follows.

The necessity of determining which amino acid is first-limiting and then to calculate all other amino acid intakes in terms of the limiting amino acid, is obviated if all amino acids are considered simultaneously in the above exercise. It is necessary to know the concentration of each amino acid required to sustain each level of egg output, and although the feeds formulated on this basis will not necessarily consist of an ideal balance of amino acids, nevertheless the feed resulting in the highest margin will be at the optimum economic amino acid balance, which will be dependent on the supply of protein-containing ingredients.

NUTRIENT DENSITY

From a knowledge of the characteristic intake of the strain or breed being considered, energy intakes, and hence food intakes, over a range of dietary energy concentrations can be calculated (Morris, 1968). If the amino acid supply is considered simultaneously, as described above, a matrix of energy and amino acid concentrations will result (*Table 14.2*). By calculating margin over food cost for each cell of the matrix (*Table 14.3*) the optimum nutrient density can be found which will maximize profitability under the prevailing economic conditions. The optimum feeding strategy under the conditions outlined in *Table 8.3* would be to supply sufficient amino acids to sustain a mean flock egg output of 53 g/bird day at an energy concentration of 11.0 MJ ME/kg. By adding a surcharge of 100 rand/ton to the cost of the feed the optimum strategy was shifted to an egg output of 52 g/bird day and an energy concentration of 12.0 MJ ME/kg, corresponding to a food intake for the flock of 105 g/bird.

It is interesting to note that profitability decreases only marginally in the immediate vicinity of the optimum combination of amino acid (egg output) and energy concentrations, with different combinations of these two dietary characteristics yielding similar profits.

USE OF MIXED INTEGER PROGRAMMING TO OPTIMIZE FEEDING STRATEGY

The method outlined above is time-consuming and repetitive. A rapid method of achieving the same goal has been described by Kleyn and Gous (1988) involving mixed integer programming, an extension of the usual linear programming method of feed formulation.

This technique 'offers' the computer a range of egg outputs, specified by the user, together with both the nutrient intakes required to sustain that level of output and the value of the output. This is referred to as an integer block of egg outputs, equivalent to the rows in *Table 14.2*. Because each level of output is assigned a maximum value of 1 and a minimum value of 0 the computer can be made to choose only one output under the conditions prevailing, the output chosen being that which maximizes profit. Similarly, in an integer block of energy concentrations (equivalent to the columns of the matrix in *Table 14.2*), the amount of each amino acid that would be consumed at each energy concentration is specified by the user. In the mixed integer program algorithm the computer to choose both the energy concentration and the egg output (amino acid supply) that will maximize profit, and will produce the same result as that in *Table 14.3*, which was the result of 56 least-cost feed formulations.

Because of the speed of such a method of optimizing the feeding strategy in a flock of laying hens, simulations are more readily carried out, and the remainder of this section will deal with examples of factors that may influence feeding strategy and the extent to which each factor is of importance. The standard situation from which changes were measured involved a flock of laying hens with a maximum egg output of 56 g/bird day and a mean body weight of 2.0 kg, with a characteristic intake of 110 g/bird day of a feed containing 11.3 MJ ME/kg. A surcharge of 100 rand/ton was added to food cost.

Changes in egg revenue

Changing the marginal revenue of eggs to values of 0.5 above and 0.5 below current revenue (*Table 14.4*) illustrate the relative inflexibility of optimum amino acid intake to this variable. Little financial benefit is derived by reducing feed quality when marginal revenue decreases, but when the ratio of feed cost to egg revenue is

Table 14.4 THE EFFECT OF CHANGES IN EGG PRICE ON THE OPTIMUM COMBINATION OF FOOD INTAKE AND EGG OUTPUT FOR A FLOCK OF LAYING HENS

Egg price	Feed intake (g/bird day)	Mean flock egg output (g)	Margin over food cost[a]
Less 50%	104	51.5	15
Standard price	105	52.0	100
Plus 50%	105	54.0	271

[a]Relative to standard conditions described in the text

substantially reduced, profit would be maximized by increasing the supply of amino acids thereby increasing egg output.

Changes in cost of availability of bulky ingredients

Ingredients such as wheat bran and middlings, lucerne and other bulky, low energy materials fluctuate in price and availability. By increasing the amount of these ingredients available in the daily food allocation from zero to 10 g/bird day the optimum nutrient density changed (*Table 14.5*) but the amino acid supply that maximized profit remained the same. Margin over food cost increased with wheat bran inclusion, implying that low density feeds are more cost-effective than feeds of high nutrient density. However, when surcharges are added to the cost of each ton of feed, a higher nutrient density is chosen as the surcharge increases (*Table 14.6*).

Change in characteristic food intake

Morris (1968) illustrated the differences in optimum food intake between a light, a medium and a heavy strain of laying hen which was the result of differences in their response to changes in dietary energy content. Such scenarios are modelled here to demonstrate how the optimum feeding strategy would differ for birds with characteristic intakes of 100, 110 and 120 g/day at an energy level of 11.3 MJ/kg (*Table 14.7*). The egg output sustainable at the optimum remained the same in all cases, with margin over food cost being highest for the bird with the lowest appetite. No

Table 14.5 EFFECT OF LOW DENSITY INGREDIENT (WHEAT BRAN) AVAILABILITY ON THE OPTIMUM COMBINATION OF FOOD INTAKE AND EGG OUTPUT FOR A FLOCK OF LAYING HENS

Availability of wheat bran (g/bird day)	Feed intake (g/bird day)	Mean flock egg output (g)	Margin over food cost[a]
0	100	52	99.4
2.5	102	52	99.7
5.0	103	52	99.8
7.5	104	52	99.9
10.0	105	52	100

[a]Relative to standard conditions described in the text

Table 14.6 EFFECT OF DIFFERENT SURCHARGES ON THE OPTIMUM COMBINATION OF FOOD INTAKE AND EGG OUTPUT FOR A FLOCK OF LAYING HENS

Surcharge (rand/ton)	Food intake (g/bird day)	Mean flock egg output	Margin over food cost[a]
0	113	53.0	116
+ 50	109	52.7	109
+ 100	105	52.0	100

[a]Relative to standard conditions described in the text

Table 14.7 EFFECT OF STRAIN OF BIRD WITH DIFFERENT CHARACTERISTIC FOOD INTAKES ON THE OPTIMUM COMBINATION OF FOOD INTAKE AND EGG OUTPUT FOR A FLOCK OF LAYING HENS

Characteristic intake of energy[a] (kJ/bird day)	Food intake (g/bird day)	Mean flock egg output (g)	Margin over food cost[b]
1130	100	52	103
1243	105	52	100
1356	117	52	97

[a]When given a feed containing 11.3 MJ ME/kg
[b]Relative to standard conditions described in the text, and assuming the same egg output

allowance was made for differences in egg size or potential output in the three strains, although this is possible, which would have influenced the relative margins.

Summary

Responses among laying hens to dietary energy and amino acids have been well characterized by quantitative nutritionists, who have simultaneously developed sophisticated techniques for optimizing the feeding strategy in flocks of laying hens. Although feed compounders are aware of these techniques there has been a reluctance on their part to make full use of them. Most feed formulators have however been made aware of certain feeding principles emanating from this research, such as the need to feed higher concentrations of amino acids to birds in hot climates or whose characteristic intake is low. It is possible, too, to ascertain the optimum feeding strategy of future laying flocks that might differ from those of today in potential egg output, in body weight, or both.

One of the most controversial aspects discussed in this chapter relates to the partitioning of the response to amino acids between egg weight and rate of lay, in which it is shown that egg weight cannot be manipulated by altering the dietary amino acid concentrations without simultaneously affecting rate of lay, the effect on rate of lay being more severe than that on egg weight.

A rapid technique for determining the feeding strategy that will maximize the profitability of a laying flock has been developed recently and this method is outlined, together with some practical examples of factors that may influence this strategy. This method provides a rapid and simple means of making use of the developments that have taken place in this area of quantitative nutrition.

References

ALMQUIST, H.J. (1952). *Archives of Biochemistry and Biophysics*, **59**, 197–202
BRAY, D.J. and GESELL, J.A. (1961). *Poultry Science*, **40**, 1328–1335
BYERLY, T.C. (1941). *Bulletin Maryland Agricultural Experimental Station*, A–1, 1–29
BYERLY, T.C. (1979). In *Food Intake Regulation in Poultry*, pp. 327–363. Ed. Boorman, K.N. and Freeman, B.M. British Poultry Science Ltd, Edinburgh
COMBS, G.F. (1962) In *Nutrition of Pigs and Poultry*, pp. 127–147. Ed. Morgan, J.T. and Lewis, D. Butterworths, London

EDWARDS, D.G. and MORRIS, T.R. (1967). *British Poultry Science*, **8,** 163–168

EMMANS, G.C. (1974). In *Energy Requirements of Poultry*, pp. 79–90. Ed. Morris, T.R. and Freeman, B.M. British Poultry Science Ltd, Edinburgh

EMMANS, G.C. (1981) In *Computers in Animal Production*, pp. 103–110. Ed. Hillyer, G.M., Wittemore, C.T. and Gunn, R.G. British Society of Animal Production

FISHER, C. (1969). *British Poultry Science*, **10,** 149–154

FISHER, C., MORRIS, T.R. and JENNINGS, R.C. (1973). *British Poultry Science*, **14,** 469–484

GOUS, R.M., BYERLY, T.C., THOMAS, O.P. and KESSLER, J.W. (1978). *16th Worlds Poultry Congress, Rio de Janeiro*, **2,** 1–8

GOUS, R.M., GRIESSEL, M.J. and MORRIS, T.R. (1987). *British Poultry Science*, **28,** 427–436

KLEYN, F.J. and GOUS, R.M. (1988). *Agricultural Systems*, **26,** (in press)

LEESON, S., LEWIS, D. and SHRIMPTON, D.H. (1973). *British Poultry Science*, **14,** 595–602

MARSDEN, A. and MORRIS, T.R. (1981). In *Intensive Animal Production in Developing Countries*, pp. 299–309, Ed. Smith, A.J. and Gunn, R.G. British Society of Animal Production

MARSDEN, A. and MORRIS, T.R. (1987). *British Poultry Science*, **28,** 693–699

MARSDEN, A., MORRIS, T.R. and CROMARTY, A.S. (1987). *British Poultry Science*, **28,** 361–380

MCDONALD, M.W. (1977). *Research Bulletin* 1/77, School of Environmental Studies, Queensland Agricultural College, Lawes, Australia

MCDONALD, M.W. (1984). *British Poultry Science*, **25,** 139–144

MCDONALD, M.W. and MORRIS, T.R. (1985). *British Poultry Science*, **26,** 253–264

MORRIS, T.R. (1968). *British Poultry Science*, **9,** 285–295

MORRIS, T.R. (1985). *South African Journal of Animal Science*, **15,** 120–122

MORRIS, T.R. and BLACKBURN, H.A. (1982). *British Poultry Science*, **23,** 405–424

MORRIS. T.R. and GOUS, R.M. (1988). *British Poultry Science*, (in press)

MORRIS, T.R. and WETHLI, E. (1978). *British Poultry Science*, **19,** 455–464

PILBROW, P.J. and MORRIS, T.R. (1974). *British Poultry Science*, **15,** 51–73

SAUVEUR, B. and MONGIN, P. (1982). *British Poultry Science*, **24,** 405–416

SMITH, A.J. and OLIVER, J. (1972). *Rhodesian Journal of Agricultural Research*, **10,** 43–60

WETHLI, E. and MORRIS, T.R. (1978). *British Poultry Science*, **19,** 559–565

WOLFENSON, D., FREI, Y.F., SNAPIR, N. and BERMAN, A. (1979). *British Poultry Science*, **20,** 167–174

15

CLIMATIC ENVIRONMENT AND POULTRY FEEDING IN PRACTICE

G.C. EMMANS
East of Scotland College of Agriculture, Pennicuik, Midlothian

and

D.R. CHARLES
ADAS, East Midland Region, Derby

Introduction

Feed cost now accounts for about 70% of the cost of production of eggs. Payne (1967) and Smith and Oliver (1972 a and b) found that the feed intake of *ad lib.* fed birds is substantially reduced at high environmental temperatures. Yet Payne (1967) and Mowbray and Sykes (1971) found that up to about 30°C this reduction of feed intake was not associated with a depression of egg production if the intake of the nutrients other than those needed for energy was maintained. Therefore, there has rightly been considerable commercial and experimental interest in the control of laying house temperature in recent years (Marsden *et al.*, 1973; Marsden and Morris, 1975; Spencer, 1975; Clark *et al.*, 1975).

In the UK poultry industry laying house temperature control is always achieved by means of thermostatic variation of the ventilation rate. A minimum air supply intended to be adequate for gas exchange is provided in cold weather, and as house temperature rises the thermostatic control equipment calls for more air, up to a maximum which is considered to be adequate for the removal of metabolic heat in hot weather. This maximum rate is readily calculated from engineering heat balance equations (e.g. Longhouse, Ota and Ashby, 1960; Payne, 1961) and has long been undisputed and successfully applied in practice. However, the minimum rate has been the object of much more contention. Its clarification is extremely important because small changes in winter ventilation rate result in large differences in house temperature and therefore feeding cost.

The series of experiments reported below was intended to provide practical recommendations on the more important aspects of the temperature/nutrition/ventilation interactions. Large groups of birds were used in experiments designed to simulate closely commercial conditions whilst also exploring the underlying biological generalisations.

First published in *Nutrition and the Climatic Environment* (1976)

The first experiment of the series involved a detailed examination of the interaction between diet and climate, and also compared two minimum ventilation rates (one above and one below the commercial rate prevalent at the time of designing the experiment). The second experiment was intended to define the biological and economic optimum temperatures. The third experiment investigated the effects of temperature on birds restricted to less than *ad lib.* feed intake, following the work of Sykes (1972) and considerable commercial interest in restricted feeding. Since both temperature and restricted feeding affect energy balance, it was considered worth exploring the possibility of interaction between them.

For reasons of space the report below deals with only some of the parameters examined. However, the conclusions at the end of this paper are intended to provide some fairly definitive recommendations about house temperature, its control through ventilation, and its implications for feed formulation.

Methods

Eight rooms each containing 1026 birds in cages were separately temperature- and ventilation-controlled.. Ventilation rate was thermostatically varied from minimum to maximum as in commercial houses. In all rooms the maximum ventilation rate was 3 m^3 s^{-1} per 1000 birds (6 ft^3 min^{-1} per bird) and in experiments 2 and 3 the minimum was 0.3 m^3 s^{-1} per 1000 birds (0.75 ft^3 min^{-1} per bird). In Experiment 1, minima of 0.25 and 0.50 m^3 s^{-1} per 1000 birds (0.5 and 1.0 ft^3 min^{-1} per bird) were compared. Experimental temperature treatments were defined as minimum temperatures since heating was provided thermostatically when necassary, but cooling was not provided because it is never used commercially in the UK. Room temperatures were permitted to rise above the normal temperatures when outside temperature exceeded the nominal. Thus the temperature at which the treatments were run closely simulated the temperature patterns which occur in commercial houses at a selection of thermostat settings. At nominal temperatures of 21°C and below the heaters were in practice hardly ever used.

White birds were stocked at 5 per 20 cm cage and brown birds at 4 per 20 cm cage in all three experiments.

EXPERIMENT 1 (1971/72)

This was intended to test Payne's hypothesis (that production is not depressed at high temperature provided that nutrient intake is maintained) for two stocks, and also to find the minimum ventilation requirements for layers. Treatments were as follows:

Stock   Shaver 288, Warren SSL
Minimum temperatures (°C)   16, 24
*Mean temperatures (°C)   19, 24
Minimum ventilation rates
(m^3 s^{-1} per 1000 birds)   0.25, 0.5
Dietary protein levels (%)   14, 17 (both at
                                                   2.75 kcal per g
                                                   determined metabo-
                                                   lisable energy level,
                                                   2.7 N-corrected)

Diets used are given in *Table 15.1*

EXPERIMENT 2 (1972/73)

This was designed to test the response of *ad lib.* fed birds to temperature
more completely by using four temperatures, and including an economic
appraisal.

Stock   Babcock B300, Warren SSL
Minimum temperatures (°C)   15, 18, 21, 24
Mean temperatures (°C)   18, 20, 22, 24
Dietary protein levels (%)   15, 18 (both at 2.75 kcal per g
                               classical ME 2.7 N-corrected)

Diets used are given in *Table 15.2*

EXPERIMENT 3 (1973/74)

This experiment investigated the effects of temperature on restricted
feeding as applied by the method of limiting time of access to the
trough (the method used by Patel and McGinnis (1970) Bougon (1972)
and Bougon and Mevel (1972)).

Stock	Shaver 288, Warren SSL	
Minimum temperatures (°C)	12, 18, 21, 24	
Mean temperatures (°C)	16, 20, 22, 25	
Diets	Metabolisable energy level, classical (kcal per g)	Crude Protein (%)
	2.8	15
	2.8	17
	3.0	15.5
	3.0	18

Diets are given in *Table 15.3*

*Temperature was recorded every two hours by data logger at 2 points per room
outside the cages, and at 12 points in one room in which horizontally the
temperature range was within 0.5°C. There was 0.9°C difference between tiers.

Feeding regimes were *ad lib.*, 8 hours per day access to the trough, and 6 hours per day access to the trough. The access periods were from 08.00 hours to 16.00 and 10.00 hours to 16.00. Lights went off at 17.30. The feeding troughs were covered with wooden lids except for the periods when access was allowed.

All birds in all three experiments were reared on a 10 hour day length and from 20 weeks of age day length was increased by 20 minutes per week up to 17 hours. Light intensity was not less than 5 lux at any point in the room for each experiment.

Energy partitions were calculated in order to find total body heat loss, which was the balancing term in the relationship:

$$\text{Metabolisable energy intake} = \text{egg energy} + \text{bodyweight change energy} + \text{total heat loss} \quad (15.1)$$

Table 15.1 Diets used in Experiment 1

Ingredients	%	%
Maize	13.7	11.2
Wheat	50.0	50.0
Barley	10.0	20.6
Ext. soya bean meal	6.2	3.1
White fish meal	6.2	5.0
Meat and bone meal	6.2	2.5
Limestone granules	6.2	6.2
Mineral and vitamin supplement	1.2	1.2
Calculated Analysis		
Metabolisable energy (kcal per g)	2.75	2.75
Crude protein	17.0	14.0
Calcium	3.4	3.1
Phosphorus	0.8	0.6
Fibre	2.0	2.2
Methionine	0.31	0.25
Cystine	0.29	0.26
Lysine	0.84	0.63
Determined analysis		
Crude protein	17.3	14.4
Calcium	3.8	3.4
Phosphorus	0.79	0.6
Crude fibre	3.4	3.3

% Crude protein	Determined % crude protein		Metabolisable energy (kcal per g, N-corrected)
	28–47 weeks	20–60 weeks	
14	15.0 (6)*	14.8 (14)*	2.72 ± 0.014 (7)*
17	17.4 (6)*	17.9 (14)*	2.69 ± 0.019 (7)*

*Number of observations

Table 15.2 Diets used in Experiment 2

Ingredients		%	%
Maize		55.0	55.0
Wheat		15.6	7.1
Hycal		4.5	5.0
Ext. soya bean meal		6.1	9.1
White fish meal		5.0	8.0
Meat and bone meal		5.0	8.0
Limestone granules		7.5	5.9
Minerals and vitamin supplement		1.2	1.8
Calculated Analysis			
Crude protein		15.0	19.0
Calcium		4.0	4.0
Phosphorus		0.6	0.8
Fibre		1.8	1.8
Methionine		0.33	0.44
Cystine		0.21	0.25
Lysine		0.71	0.98
Metabolisable energy (kcal per g)		2.9	2.9
Determined analysis			
Crude protein		14.8	18.2
Calcium		3.95	3.89
Methionine		0.32	0.39
Cystine		0.22	0.28
Lysine		0.78	1.02

Age (weeks)	% Crude protein	Determined % crude protein	Metabolisable energy (kcal per g, N-corrected)
28–43	15	14.6	2.623 (7)*
	19	18.1	2.664 (7)*
44–59	15	14.5	2.683 (8)*
	19	18.3	2.679 (7)*

*Number of observations

Using the energy contents of eggs (1.6 kcal per g) and of bodyweight gain (4 kcal per g) given by Emmans (1974), then:

Heat loss (kcal per bird per day) = ME intake (N-corrected determined value) − [egg ouput (g per bird per day) × 1.6] − [bodyweight gain (g per bird per day) × 4]   (15.2)

Maintenance requirement was calculated assuming an efficiency of 80% for converting dietary metabolisable energy to egg and carcase energy (Emmans, 1974). Thus:

Maintenance requirement (kcal per bird per day) = ME intake − [egg output × 2] − [bodyweight change (g) × 5]   (15.3)

Table 15.3 Diets used in Experiment 3

Ingredients	%	%	%	%
Maize	11.2	11.0	55.0	55.0
Wheat	17.1	12.6	17.3	11.8
Barley	47.75	47.9	–	–
Soya bean oil	1.0	1.0	2.0	2.0
Ext. soya bean meal	6.2	7.25	7.0	8.2
White fish meal	5.0	6.2	5.0	7.3
Meat and bone meal	2.5	5.8	5.0	8.0
Limestone granules	8.0	6.75	7.5	5.9
Min. and vit. mix	1.25	1.5	1.25	1.8
600 g Methionine per ton				
Calculated analysis				
ME kcal per g	2.6	2.6	3.0	3.0
CP	14.5	17.0	15.5	18.5
Calcium	3.97	3.91	3.95	3.9
Phosphorus	0.64	0.82	0.70	0.91
Fibre	3.1	3.0	1.9	1.8
Methionine	0.30	0.36	0.33	0.42
Cystine	0.24	0.27	0.22	0.25
Lysine	0.70	0.86	0.74	0.94
Tryptophane	0.16	0.18	0.15	0.17
Determined analysis		*20–36 weeks of age*		
Crude protein	14.9	17.0	15.1	17.8
Calcium	3.78	3.76	3.99	3.89
		36–56 weeks of age		
Crude protein	15.4	17.4	15.8	18.3
Calcium	3.73	3.63	4.03	3.81
		56–72 weeks of age		
Crude protein	15.3	17.3	15.5	17.9
Calcium	3.75	3.89	3.92	3.94
		20–72 weeks of age		
Crude protein	15.2	17.2	15.5	18.0
Calcium	3.75	3.77	3.98	3.89
Dry matter	89.3	89.0	89.0	88.9
Ether extract	3.16	3.30	4.84	5.3
Total ash	12.0	11.3	11.2	11.7
Available carbohydrates	44.8	43.5	45.6	43.6
Metabolisable energy (kcal per g, N-corrected)	2.5	2.6	3.0	2.8

An adjustment was made to the estimate of N-corrected ME intake used in the above relationships to allow for the fact that it was determined on sample cages not the whole flock. Values were corrected back to the N-balance of the flock since the ME determination was for the N-balance of the sample birds.

Table 15.4

Stock		Shaver 288				Warren SSL				Standard error
% Crude protein		14		17		14		17		
Minimum ventilation rate (m³ s⁻¹ per 1000 birds)		0.25	0.5	0.25	0.5	0.25	0.5	0.25	0.5	
Eggs per hen	19°C	271	264	270	273	276	272	283	280	
housed	24°C	253	256	262	265	213	235	216	223	±6.57
Eggs per hen	19°C	0.738	0.731	0.745	0.762	0.720	0.711	0.732	0.732	
per day	24°C	0.716	0.713	0.725	0.738	0.582	0.627	0.579	0.603	±0.0107
Feed intake	19°C	108	110	109	111	116	116	117	117	
(g per bird per day)	24°C	96	96	96	96	98	98	99	100	±1.27
Egg weight	19°C	57.8	58.1	58.9	58.9	58.5	58.5	58.6	59.1	
(g per egg)	24°C	53.6	74.7	54.3	55.0	52.4	53.5	52.8	53.1	±0.45
Mortality	19°C	13.3	15.6	13.3	14.0	5.3	5.0	5.3	6.1	
(%)	24°C	19.5	13.2	14.7	14.7	14.0	10.1	10.7	13.6	±2.75

Heat loss (kcal per bird)		(means for both ventilation rates)				
28–48 weeks	19°C	223.1	224.5	242.3	246.9	
of age	24°C	197.5	192.6	211.2	217.8	±6.91
56–76 weeks	19°C	243.9	241.6	251.5	253.3	
	24°C	209.6	206.0	221.0	215.6	±7.33
Heat loss (kcal per kg)						
28–48 weeks	19°C	134.5	129.1	107.0	107.8	
	24°C	126.9	116.9	108.7	107.9	±5.85
56–76 weeks	19°C	134.3	127.0	101.8	103.0	
	24°C	122.3	113.2	101.4	95.7	±8.10
Maintenance requirement (kcal per bird)						
28–48 weeks	19°C	205.9	205.8	224.0	227.8	
	24°C	182.4	175.5	197.0	203.3	±7.24
56–76 weeks	19°C	226.8	224.3	237.2	239.7	
	24°C	194.1	191.2	211.4	206.0	±7.67
Maintenance requirement (kcal per kg)						
28–48 weeks	19°C	124.2	118.3	98.9	99.5	
	24°C	117.2	106.6	101.4	100.7	±5.76
56–76 weeks	19°C	124.9	117.9	96.0	97.5	
	24°C	113.3	105.1	97.0	91.5	±7.95

At the end of the trials samples of birds were weighed and, after killing, the feathers, combs and wattles removed for separate weighing. Data can be supplied by the authors.

Results and Discussion

Each experiment is dealt with separately in this section but in the Conclusions some general principles have been derived from all three experiments.

EXPERIMENT 1 (20–76 WEEKS OF AGE)

At the lower temperature there was no effect of ventilation rate on any performance trait (*Table 15.4*). In some sub-treatments production was depressed by the low ventilation rate at the high temperature, notably with Shavers on the high protein diet. This may have been due to ammonia levels in the cages, since birds on the high protein diet would have excreted more nitrogen. Manure removal was two or three times per week.

For the Shaver 288, Payne's hypothesis concerning the interaction between temperature and nutrition was confirmed since the effect of temperature on egg production appeared to be through its effect on nutrient intake. This is demonstrated by *Figure 15.1*. Egg production on the high protein diet at high temperature was similar to that on the low protein diet at low temperature, these two treatments having similar protein intakes.

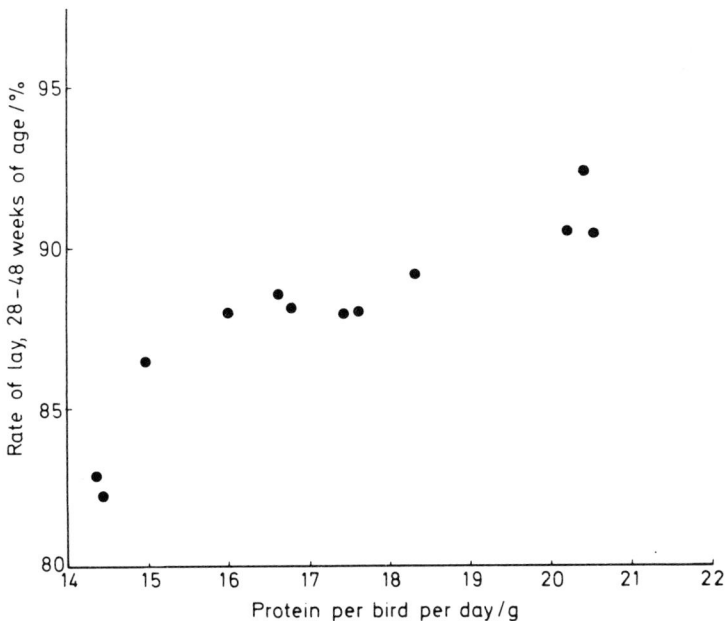

Figure 15.1 Effect of protein intake on rate of lay

Table 15.5

Stock % Crude protein	Babcock B300		Warren SSL		Standard error
	18	15	18	15	
Mean temperature (°C)	*Eggs per hen housed*				
18	251	245	252	244	
20	253	251	249	247	
22	256	247	259	247	
24	250	249	251	243	± 6.95
	Eggs per hen per day				
18	0.748	0.726	0.725	0.705	
20	0.750	0.752	0.713	0.704	
22	0.760	0.736	0.726	0.709	
24	0.742	0.732	0.706	0.692	± 0.0127
	Feed intake per bird per day (g)				
18	111	111	120	122	
20	110	111	117	119	
22	105	107	114	116	
24	99	101	107	109	± 1.85
	Mean egg weight (g)				
18	59.7	59.0	62.6	61.9	
20	59.7	58.4	62.0	61.6	
22	59.0	58.5	61.9	60.7	
24	57.9	57.6	60.1	59.5	± 0.51
	Mortality (%)				
18	19.6	19.6	10.1	11.2	
20	18.4	19.8	8.8	8.8	
22	17.5	19.3	5.0	10.1	
24	14.2	14.9	5.0	7.5	± 3.78
	Egg output (g per bird per day)				
18	44.7	42.8	45.4	43.6	
20	44.8	43.9	44.2	43.0	
22	44.8	43.1	44.9	43.0	
24	43.0	42.2	42.4	41.2	
	*Energetic efficiency** (cal egg output per cal ME intake)				
	A/B	*A/B*	*A/B*	*A/B*	
18	0.242	0.231	0.227	0.216	
20	0.245	0.238	0.227	0.220	
22	0.257	0.242	0.238	0.223	
24	0.261	0.252	0.238	0.227	

*Calculation of energetic efficiency
   grams egg output x 1.6 kcal per egg per bird per day = A
   grams food intake x determined ME (N-corrected) (kcal per g) = B

In the Warren, production was depressed at 24°C, whatever diet was fed. The difference between strains in response to temperature may reflect differences in insulation. The Shavers had less weight of feathers and more weight of comb and wattles per unit of bodyweight than the Warrens (Emmans and Dun, 1973).

At the higher temperature heat loss and maintenance requirement per bird were lower, but in some cases they were similar per kg, as if body-weight change had been thermoregulatory. The birds on the high protein diet had less heat loss per kg presumably because they were better feathered. (Feathers were weighed at the end of the trial and data can be supplied). Heat losses per bird were generally higher at the end of lay but this was not true per kg of bodyweight.

EXPERIMENT 2

Hen housed egg production was maximal in both stocks at 22°C on the high protein diet and at 20°C on the low protein diet (*Table 15.5*). Hen day rate of lay in Warrens was not much affected by temperature except for a depression at 24°C compared with the other temperatures. In Babcocks, rate of lay was highest at 22°C on the high protein diet and 20°C on the low (*Table 15.5*). Presumably these optimum temperatures were specific to the intakes of the first limiting nutrients with these particular diets.

Feed intake decreased with increasing temperature but not linearly. The rate of decrease was faster the higher the temperature. Over the temperature range studied the decrease in feed intake averaged approximately 1.5% per °C temperature rise (*Table 15.5*).

There was a progressive decrease in egg weight with increase in temperature. At least part of this decrease could probably be attributed to decreased nutrient intake reflected in the reduced feed intake.

Mortality was progressively reduced as temperature was increased. In both stocks birds fed the higher protein diet grew at a faster rate.

There were some cases of depression of shell density due to temperature, some of, but not all of, which probably reflected calcium intake. The percentage of cracked eggs as expressed by packing station seconds was not affected by temperature. In this experiment all treatments received at least 4 g calcium per bird per day.

Four egg price and four feed price markets were arbitrarily chosen and the margin of egg value less feed cost was calculated for each treatment. The egg and feed prices used which were typical at the time of completion of the experiment (early 1974) are given in *Tables 15.6* and *15.7* with their associated margins. Note that this economic appraisal does not take into account seconds. Egg weights were converted to egg grades from the experimental data using the standard deviation method of Wills (1973).

Table 15.6 Babcock B300, Margin of egg income over food cost from 20 to 72 weeks of age (p per bird)

Mean temperature (°C)			18		20		22		24	
% Crude protein	18	15	18	15	18	15	18	15	18	15
Egg market, pence per dozen	Food market, cost per ton (£)		Margin of egg income over food cost (p)							
22	92	80	165	194	170	207	191	207	192	221
	86	80	187	194	192	207	212	207	212	221
	82	70	202	231	207	243	226	242	225	255
	76	70	224	231	229	243	246	242	245	255
28	92	80	302	327	308	343	330	340	327	355
	86	80	324	327	330	343	350	340	347	355
	82	70	339	364	345	379	364	376	360	389
	76	70	361	364	367	379	385	376	380	389

Table 15.7 Warren Sex-Sal-Link, margin of egg income over food cost from 20 to 72 weeks of age (p per bird)

Mean temperature (°C)			18		20		22		24	
% Crude protein	18	15	18	15	18	15	18	15	18	15
Egg market, pence per dozen	Food market, cost per ton (£)		Margin of egg income over food cost (p)							
24	92	80	196	224	198	233	222	241	219	244
	86	80	221	224	222	233	246	241	241	244
	82	70	237	266	238	274	262	280	256	282
	76	70	262	266	262	274	286	280	279	282
30	92	80	340	363	340	373	369	380	360	380
	86	80	364	363	364	373	393	380	382	380
	82	70	381	404	380	414	409	420	397	418
	76	70	405	404	404	414	433	420	420	418

EXPERIMENT 3 (20–72 WEEKS OF AGE)

The third experiment had a complex incomplete factorial design containing so many treatments that it is impossible to report the results fully within the scope of this presentation. Therefore only the important effects and interactions have been selected for detailed reporting. (Complete tables of data can be obtained from the authors).

There was generally no significant depression of hen housed production due to restriction (*Table 15.8*) although at the temperatures associated with the highest levels of production on *ad lib.* feeding there was a non-significant depression of production due to restriction. This was probably not an effect of protein intake as judged by comparison of protein ibtakes with those of *ad lib.* treatments on the low protein diets. The interaction was not significant, suggesting that temperature did not modify the effects of restriction of feed.

Table 15.8

	Shaver 288								Standard error
Energy level (kcal per g)		2.8				3.0			
Feeding regime	Ad lib.	Ad lib.	8h	6h	Ad lib.	Ad lib.	8h	6h	
% protein	15	17	17	17	15.5	18	18	18	
°C				Eggs per hen housed					
16	265.5	275.3	275.2	269.9	249.1	261.0	263.2	265.1	
20	266.6	278.6	265.4	259.5	243.4	275.4	257.6	255.7	
22	251.2	268.4	271.1	267.5	241.3	274.9	257.4	257.2	
25	253.2	268.6	261.4	267.4	230.2	249.6	265.9	250.6	±10.91

	Warren SSL								
Energy level (kcal per g)		2.8				3.0			
Feeding regime	Ad lib.	Ad lib.	8h	6h	Ad lib.	Ad lib.	8h	6h	
% protein	15	17	17	17	17.5	18	18	18	
°C				Eggs per hen housed					
16	266.3	266.8	270.1	269.5	258.4	263.3	265.4	257.3	
20	274.6	265.2	269.9	271.8	265.9	262.9	260.7	261.8	
22	254.7	273.1	270.1	268.7	252.4	270.7	259.8	262.6	
25	264.2	265.7	262.8	262.8	253.2	265.4	268.3	267.6	±10.19

In the Shaver 288, production averaged over all feeding treatments fell as temperature was increased (*Table 15.9*). Presumably this was because feed intake was so low that some nutrient became limiting (*Table 15.11*). In the Warren, temperature had no effect on production averaged over all feeding treatments (*Table 15.9*).

Table 15.9

	16°C	20°C	22°C	25°C	Standard error
		Eggs per hen housed			
Shaver 228	265.5	262.8	261.1	255.9	± 3.38
Warren SSL	264.6	266.6	264.1	263.7	± 4.75
		Eggs per hen per day			
Shaver 288	0.753	0.747	0.749	0.783	± 0.0079
Warren SSL	0.741	0.744	0.743	0.736	± 0.0079

The restriction progressively depressed production averaged over all temperatures but not significantly (*Table 15.10*).

Table 15.10

Feeding regime		Ad lib.	Ad lib.	8h	6h	Standard error
Protein level		Low	High	High	High	
	Energy level (kcal per g)		*Eggs per hen housed*			
Shaver	2.8	259.1	272.7	268.3	266.1	
	3.0	241.5	265.2	261.0	257.1	± 10.91
Warren	2.8	264.9	267.7	268.2	268.5	
	3.0	257.5	265.6	263.6	262.3	± 10.19
			Eggs per hen per day			
Shaver	2.8	0.739	0.773	0.763	0.742	
	3.0	0.713	0.759	0.748	0.738	± 0.0215
Warren	2.8	0.736	0.753	0.742	0.747	
	3.0	0.736	0.746	0.739	0.731	± 0.0219

Feed intake fell by 1.3 and 1.0% per $^{\circ}$C for Shavers and Warrens respectively in both cases with no interaction between temperature and feed restriction (*Tables 15.11* and *15.12*).

Table 15.11 Feed intake (g per bird per day)

	16°C	20°C	22°C	25°C	Standard error
Shaver	99.8	96.8	93.2	88.5	± 0.47
Warren	123.3	118.4	115.9	111.7	± 0.62

Restriction depressed feed intake in the Shaver significantly by 3% on 8h access to the feed and by 5% on 6h. For the Warren corresponding figures were 4% and 6% (*Table 15.12*). Note that these are based on the cumulative feed intakes from 20 to 72 weeks of age but the restriction was only applied from 40 weeks; thus the percentage reductions in intake were slightly higher during the period when the restriction was applied.

Table 15.12 Feed intake (g per bird per day)

Feeding regime		Ad lib.	Ad lib.	8h	6h	Standard error
Protein level		Low	High	High	High	
	Energy level (kcal per g)		*Feed intake (g per bird per day)*			
Shaver	2.8	96.6	99.7	96.3	94.4	
	3.0	92.5	94.9	92.3	90.0	± 2.62
Warren	2.8	122.5	122.3	117.2	115.8	
	3.0	117.5	118.7	113.5	111.1	± 2.26

Energy intake of the shaver increased by 6.4 kcal per bird per day for a 100 kcal per kg increase in dietary energy level. For the Warren the corresponding value was 8.7. Both are higher than the increases expected from literature (e.g. Morris, 1968).

Restriction generally had no effect on egg weight and there was no interaction between temperature and restriction. Egg weight was depressed as temperature was increased but it was increased with increasing dietary energy level (*Table 15.13*).

Table 15.13

		16°C	*20°C*	*22°C*	*25°C*	*Standard error*
		Egg weight (g)				
Shaver		57.4	56.8	56.3	54.9	± 0.32
Warren		61.1	61.6	60.7	59.9	± 0.60
Feeding regime		*Ad lib.*	*Ad lib.*	8h	6h	
Protein level		Low	High	High	High	
	Energy level (kcal per g)					
Shaver	2.8	55.9	56.5	53.8	55.9	
	3.0	56.0	56.8	57.0	56.9	±0.63
Warren	2.8	60.5	60.2	60.9	60.5	
	3.0	60.6	60.9	61.8	61.1	±0.84
		Mortality (%)				
Shaver		7.4	9.3	9.2	8.7	± 1.05
Warren		4.7	5.0	5.7	4.0	± 2.07
Feeding regime		*Ad lib.*	*Ad lib.*	8h	6h	
Protein level		Low	High	High	High	
	Energy level (kcal per g)					
Shaver	2.8	8.3	7.2	7.5	3.6	
	3.0	13.9	8.9	10.3	9.4	±4.01
Warren	2.8	4.2	5.2	3.1	4.5	
	3.0	7.3	5.6	5.2	3.8	±3.37

The general conclusion from the production results of Experiment 3 is that the time of access to the trough method resulted in a mild level of intake restriction which could probably be usefully commercially exploited. Production and egg size were not generally affected by it so that margin of egg value less feed cost could be slightly improved by imposing the restriction. In answer to the question the experiment was intended to clarify it would appear that there is little effect of temperature over the range tested on moderate restriction of feed intake by regulating the time of access to the trough.

Supplementary Field Survey — Predicting Feed Intake in Commercial Flocks

Emmans (1974) gave a prediction equation for ME intake, which includes a term for the effect of temperature based partly on the data reported here. Prediction of intake is essential in feed formulation and this equation emphasises the need to take account of temperature. The equation was validated by comparing predicted and measured intake for 49 8-week periods for 20 commercial laying flocks.

The prediction equation was:

$$ME = W(a + bT) + 2E + 5\Delta W \qquad (15.4)$$

where: ME = ME intake kcal per bird per day
W = mean period bodyweight, kg
T = mean environmental temperature, $°C$
E = egg output, g per bird per day
ΔW = body weight gain g per bird per day
a and b have values of 170, 155 and 140 and −2.2, −2.1 and −2.0 respectively for white, tinted and brown egg laying stocks respectively.

For each 8 week period, measurements were made of W, ΔW, T, E and feed intake. Feed intake was predicted from the equations by assuming that all feeds contained 2.7 kcal ME per g.

Actual feed intake averaged 122.9 g per bird per day with a co-efficient of variation (cv) of 9.13%. The ratio of actual to predicted feed averaged 1.030 with a cv of 7.05%.

All flocks were scored for feather loss, L, on a scale from 1 to 6. It was expected that the value of $(a + bT)$, i.e. maintenance per kg per day, in the equation given above might depend on feathering. The ratio of actual maintenance to that predicted was regressed on $\log_e (6 - L)$ and the resulting regression equation was:

$$\frac{\text{Actual maintenance}}{\text{Predicted maintenance}} = 1.405 - 0.28959 \log_e (6 - L) \qquad (15.5)$$

This equation had r = 0.641 ($P<0.01$).

Including feather score in the prediction equations reduced the cv of the ratio of actual to predicted feed intake to 5.46%. Sixty per cent of the flocks had actual feed intakes within ±4% of those predicted.

As a rule of thumb maintenance increased by about 9% for each unit increase in feather loss score. The original production equations fitted best to birds with a feather loss score of 2.0. To derive feed intakes for other feather loss scores the adjustment factors given in *Table 15.14* are used.

Table 15.14

Feather loss score	Multiply estimated maintenance by:
1	0.94
2	1.00
3	1.08
4	1.20
5*	1.40

*Beyond range of data

Supplementary Experiment – Physiology of Temperature Effects

It was not the function of these experiments to explore the physiology in detail, but it was observed that rectal temperature increased as air temperature was increased (*Table 15.15*). It is possible that this is relevant to the depression of egg production frequently found at the highest temperatures used.

Table 15.15

Air temperature (°C)	Rectal temperature (°C)	Standard error
16	41.0	
20	41.1	± 0.065
22	41.3	
25	41.5	

Conclusions and Application

Taking into account all three experiments, it is now possible to make some general conclusions and specific recommendations about laying house temperature.

1. Feed intake of *ad lib.* fed birds falls by about 1.5% per °C temperature rise, but not linearly. The effect is greater the higher the temperature.

2. Payne's hypothesis is tenable, namely that the effect of temperature depends upon the diet offered to the birds. The higher the dietary concentration of the first limiting non-energy nutrient, the higher the temperature associated with maximum egg production. This is a reflection of the effect of temperature on feed intake and emphasises the need to describe diets in terms of nutrient intake rather than just percentage composition.

3. If no nutrient is limiting, there is no depression of production up to at least 22° and in some experiments in this series up to 25°. The precise optimum probably depends on factors such as feather cover, comb size and number of birds per cage, as well as diet. Above the critical level of temperature, production is depressed whatever diet is fed and this may reflect a rise in body temperature.

4. On a given diet egg weight falls as temperature rises, but this is at least partly simply a reflection of nutrient intake.

5. Under most UK market conditions the margin of egg value less feed cost is maximised at a minimum house temperature of 21° (mean 22°), but where the difference in price between egg grades is not important, then higher temperatures would yield more money.

6. Energetic efficiency (defined as yield of egg energy per unit of feed metabolisable energy) improves roughly linearly as temperature rises. Thus it is interesting that maximum biological efficiency and maximum economic efficiency do not necessarily occur at the same temperature.

7. In order to exploit the best temperature, low levels of minimum ventilation rate can safely be used in winter, and 0.25 m³ s⁻¹ per 1000 birds appears to be enough. Practical ventilation systems capable of sufficient wind exclusion to permit this are now available (further information can be obtained from the authors). Note that small errors in the precision of ventilation control cause large depressions of house temperature below the economic optimum (*Figure 15.2*).

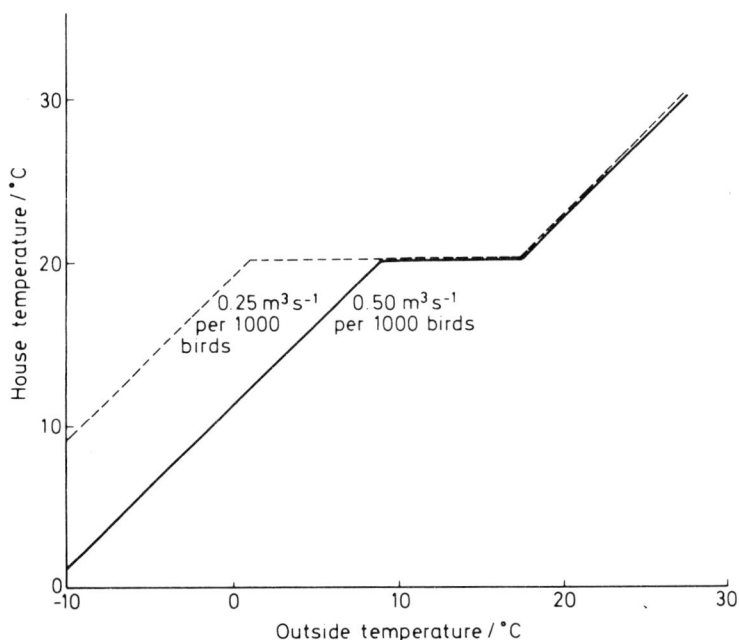

Figure 15.2 Effect of ventilation rate on house temperature

8. There has been widespread uptake of this work by the poultry industry and presumably better and more consistent control of house temperature should permit feed compounders to feed more precisely since feed intake will probably be less variable and more predictable than in the past.

9. Feed intake can be predicted fairly precisely if the standard of feather cover is taken into account as well as the house temperature. Maintenance requirement increases about 9% for each unit increase in feather loss score.

10. To some extent the benefits of temperature and feed restriction can be combined. A mild (less than 5%) restriction can be achieved by limiting time of access to the trough to 6 or 8 hours per day. This feed saving can possibly be exploited without depressing egg production or egg size, but our experiment was not definitive in this respect. There were some non-significant depressions of egg production at the temperatures associated with the highest production for *ad lib.* fed birds.

References

Bougon, M. (1972). 'Influence du rationnement des aliments distribués pendant la periode de production sur les performances des pondeuses.' *Bull. Inf., Ploufragon,* **12** (4), 118

Bougon, M. and Mevel, M. (1972). 'Influence du rationnement en periode de production sur les performances de trois variétés commerciales des pondeuses.' *Bull. Inf., Ploufragon,* **12**, (4), 154

Clark, J.A., Charles, D.R., Wathes, C.M. and Arrow, J. (1975). 'Heat transfer from housed poultry, its implications for environmental control.' *Wld's Poult. Sci. J.,* **31**, 312

Emmans, G.C. (1974). 'The effects of temperature on the performance of laying hens.' In *Energy Requirements of Poultry,* pp. 79–90. Ed. by T.R. Morris and B.M. Freeman. Edinburgh; British Poultry Science Ltd.

Emmans, G.C. and Dun, P. (1973). Temperature and ventilation rate for laying fowls; supplementary report on feathering. Unpublished ADAS data.

Longhouse, A.D., Ota, H. and Ashby, W. (1960). 'Heat and moisture data for poultry housing.' *Agric. Engng St Joseph, Mich.,* **41**, 567

Marsden, A. and Morris, T.R. (1975). 'Comparisons between constant and cycling environmental temperatures applied to laying pullets.' *Wld's Poult. Sci. J.,* **31**, 311

Marsden, A., Wethli, E., Kinread, N. and Morris, T.R. (1973). 'The effect of environmental temperature on feed intake of laying hens.' *Wld's Poult. Sci. J.,* **29**, 286

Morris, T.R. (1968). 'The effect of dietary energy level on the voluntary calorie intake of laying hens.' *Br. Poult. Sci.,* **9**, 285

Mowbray, R.M. and Sykes, A.H. (1971). 'Egg production in warm environmental temperatures.' *Br. Poult. Sci.,* **12**, 25

Patel, P.R. and McGinnis, J. (1970). 'Effect of restricting feeding time on feed consumption, egg production and body weight gain on Leghorn pullets.' *Poult. Sci.,* **49**, 1425

Payne, C.G. (1961). 'Studies on the climate of broiler houses. 2. Comparison of ventilation systems.' *Br. Vet. J.,* **117**, 106

Payne, C.G. (1967). 'Environmental temperature and egg production.' In *The Physiology of the Domestic Fowl,* pp. 235–241. Ed. by C. Horton-Smith and E.C. Amoroso. Edinburgh; Oliver and Boyd

Smith, A.J. and Oliver, J. (1972a). 'Some nutritional problems associated with egg production at high environmental temperatures. 1. The effect of environmental temperature and rationing treatments on the productivity of pullets fed on diets of different energy content.' *Rhod. J. Agric. Res.*, **10**, 3

Smith, A.J. and Oliver, J. (1972b). 'Some nutritional problems associated with egg production at high environmental temperatures. 4. The effect of prolonged exposure to high environmental temperatures on the productivity of pullets fed on high energy diets.' *Rhod. J. Agric. Res.*, **10**, 43

Spencer, P.G. (1975). 'A cost benefit analysis of temperature controls in housing laying hens.' *Wld's Poult. Sci. J.*, **31**, 309

Sykes, A.H. (1972). 'The energy cost of egg production.' In *Egg Formation and Production*, pp. 187–196. Ed. by B.M. Freeman and P.E. Lake. Edinburgh; British Poultry Science Ltd

Wills, J.R. (1973). Personal communication

16

DIETARY PHOSPHORUS FOR LAYING HENS

J.R. HOPKINS
ADAS, Leeds

A.J. BALLANTYNE
Gleadthorpe EHF, Mansfield
and
J.L.O. JONES
ADAS, Cambridge, UK

The UK egg industry involves about 41.5 million layers producing 13 610 million eggs. Over 40% of the birds are kept in units of over 50 000 and about 96% are housed in cages: all birds are now brown egg layers and consume about 1.5 million tonnes of feed per annum. The UK is essentially self-sufficient in egg production and this has led the purchasing public to become much more discerning about egg quality and value for money. It is estimated that 90% of all egg downgrading is due to cracking. Since most cracked eggs receive 15–20p less per dozen than those of first quality eggs, this is equivalent to a reduction in annual revenue of about 4.5–5.0p per bird per 1% 'seconds'. The current loss to the industry is estimated to be worth M£7 per annum. Several factors have been shown to influence downgrading and the causes most frequently cited are disease, strain differences, environmental/ system factors, physiology and nutrition. Egg size and age of bird are also known to be influential.

Over the last 20 years nutrition has been shown to have important consequences in both the quantity and quality of eggshell produced by laying fowls. Since the eggshell contains about 2.2 g calcium, in the form of 5.5 g calcium carbonate, and 20 mg phosphorus, the calcium and phosphorus levels in the diets of laying fowls have received special attention. The pattern of changing recommendations for

Table 16.1 NRC RECOMMENDATIONS OF DIETARY CALCIUM AND PHOSPHORUS FOR LAYING HENS

Year	Calcium (g/hen/day)	(% diet)	Total phosphorus (g/hen/day)	(% diet)	Available phophorus (g/hen/day)	(% diet)
1944	2.27	2.25	0.75	0.75	—	—
1950	—	2.25	—	0.75	—	—
1954	2.46	2.25	0.66	0.60	—	—
1960	2.46	2.25	0.66	0.60	0.43	0.39
1966	3.00	2.75	0.66	0.60	0.43	0.39
1971	3.00	2.75	0.66	0.60	0.43	0.39
1977	3.60	3.25	0.66	0.50	0.32	0.29
1984	3.75	3.40	—	—	0.35	0.32

After Roland (1986)

First published in *Recent Advances in Animal Nutrition – 1987*

these minerals is well illustrated in *Table 16.1* which gives the NRC recommendations for dietary calcium and phosphorus for laying fowls over the last 40 years, as
quoted by Roland (1986). These figures indicate a recommendation for increasing
levels of dietary calcium and decreasing levels of dietary phosphorus over these
years.

Dietary phosphorus is normally present as both inorganic and organic phosphorus. Whilst most of the common forms of inorganic phosphorus can be considered to
be almost 100% available, the organic phosphorus which is linked to phytin can
have a very low availability to the birds. Since about two-thirds of the phosphorus
in plants is present with a phytin linkage, it is important that the chemical form of
the phosphorus is considered when assessing minimal dietary phosphorus levels.
Traditionally the poultry nutritionist has achieved this by calculating the 'available
phosphorus' content of a feed by summing the phosphorus from inorganic supplements, the phosphorus in animal feeds and 30% of the phosphorus in plant feeds.
However, for experimental and quality control purposes it is desirable to use
specific chemical methods of assay. To this end work on suitable routine methods of
chemical analysis for both inorganic and organic phosphorus has been carried out
by ADAS analytical chemists.

Analytical determination of phosphorus

Total phosphorus content of feedingstuffs is routinely determined in ADAS spectrophotometrically (MAFF, 1973) as the yellow phospho–vando–molybdate complex in a hydrochloric acid extract of the feed residue remaining after the
destruction of the organic matter of plant material. This technique measures both
inorganic and organic forms of phosphorus. Inorganic phosphorus can be determined by the method of Pons and Guthrie (1946) in which the inorganic phosphorus is first extracted from the feedingstuff with 0.75 M trichloroacetic acid, the
extract is then decolorized with charcoal and the concentration of phosphorus in
the extract is determined spectrophotometrically as the yellow phospho–vando–
molybdate complex. Recent work has concentrated on examining procedures to
produce a selective, rapid and accurate method for the determination of phytic
phosphorus.

The first published method for the determination of phytic phosphorus was by
Heubner and Stadler (1914). This depended upon the extraction of phytic acid by
means of hydrochloric acid and titration of the extract with an acid solution of ferric
chloride in the presence of ammonium thiocyanate. The phytic acid was precipitated as its insoluble iron salt and the end point determined by the appearance of a
red colour. Various workers experienced problems in determining the end point
and minor modifications were made in the years between 1917 and 1932. In 1935,
McCance and Widdowson published a method in which the phytic acid was
extracted with hydrochloric acid, precipitated as the ferric salt and the phosphorus
in the precipitate determined, after a sulphuric–perchloric digest, by spectrophotometric measurement as the yellow phospho–vando–molybdate complex. This
method which was generally accepted as a reference method, presented some
analytical problems and was very time consuming due to the many intricate stages.
In 1983, Haug and Lantzsch published a method for the rapid determination of
phytin phosphorus. This method consisted of an acid extraction of the phytin and
its precipitation with an acidic ferric iron solution of known iron content. The

decrease in iron in the supernatant was used to measure the phytin content. In 1982, Graf and Dintzis published a method for the determination of phytin phosphorus in feeds by high pressure liquid chromatography (HPLC). In this method the dried defatted sample was extracted with hydrochloric acid, the resulting slurry centrifuged and the supernatant further clarified by passing through a millipore filter. The filtrate was diluted with water and passed through an anion exchange resin column. The phytin was then eluted from the column with hydrochloric acid and after further preparation a small aliquot was analysed by reverse phase liquid chromatography on a C18 column with 5 mM sodium acetate as the mobile phase. The phytin was measured with a refractometer. The phytin phosphorus content of 15 feedstuffs determined by the methods of McCance and Widdowson (1935), Haug and Lantzsch (1983) and Graf and Dintzis (1982) are given in *Table 16.2*. The results indicate that the three methods gave comparable results and hence the rapid method of Haug and Lantzsch (1983) has been adopted in ADAS laboratories for routine purposes.

Table 16.2 PHYTIN PHOSPHORUS CONTENT OF COMMON FEEDINGSTUFFS BY THE METHOD OF MCCANCE AND WIDDOWSON (1935), HAUG AND LANTZSCH (1983) AND GRAF AND DINTZIS (1982)

Feedingstuffs	*Phytin phosphorus* (P/kg)		
	McCance and Widdowson	*Haug and Lantzsch*	*Graf and Dintzis*
Wheat	2.0	2.1	1.9
Maize	2.6	2.3	2.5
Barley	2.1	2.1	2.0
Maize gluten feed	2.4	2.3	2.3
Rice bran meal	15.3	15.8	14.6
Rapeseed meal	2.3	2.1	2.1
Ex soyabean meal	3.5	3.5	3.4
Full fat soyabean	2.8	2.9	2.5
Peas	1.8	2.0	1.7
Poultry feed 1	2.5	2.4	2.4
2	2.5	2.3	2.3
3	2.7	2.7	2.4

Phosphorus content of feedingstuffs

Thirteen common feedingstuffs used in poultry rations have been analysed for total phosphorus (MAFF, 1973), inorganic phosphorus (Pons and Guthrie, 1946) and phytin phosphorus (Haug and Lantzsch, 1983) and the results expressed on a dry matter basis are given in *Table 16.3* together with an 'available phosphorus' value calculated as 30% of total plant phosphorus. These results demonstrate that the non-phytin phosphorus in the feedstuffs examined provided between 33 and 83% of the total phorphorus value and would indicate that the arbitrary assumption that only 30% of the phosphorus in plant feeds is not combined with phytin and hence is 'available' to the bird could be misleading. The difference between the non-phytin phosphorus and inorganic phosphorus values suggest a variable level of organic phosphorus which could be expected to have a high availability to the bird. To assess the phosphorus requirements of birds it is essential that the form of the

Table 16.3 PHOSPHORUS CONTENT OF COMMON FEEDINGSTUFFS

Feedstuffs	Total P (g/kg)	Phytin P (g/kg)	Non-phytin P (g/kg)	Inorganic P (g/kg)	Available P[a] (g/kg)
Wheat	3.9	2.2	1.7	1.1	1.2
Barley	4.3	2.4	1.9	1.2	1.3
Oats	3.6	1.9	1.7	1.2	1.1
Triticale	4.4	3.0	1.4	1.5	1.3
Maize	3.0	2.0	1.0	1.2	0.9
Peas	4.3	2.1	2.2	1.3	1.3
Beans	5.8	1.5	4.3	1.5	1.7
Maize gluten feed	7.2	2.9	4.3	2.6	2.2
Prairie meal	2.6	1.1	1.5	1.2	0.8
Ex soyabean meal	7.0	3.0	4.0	1.4	2.1
Full fat soyabean	5.3	2.3	3.0	1.2	1.6
Ex sunflower meal	9.4	3.0	6.4	1.4	2.8
Ex rapeseed meal	12.4	2.1	10.3	1.7	3.7

[a] Available P for plant materials = 0.30 total P

phosphorus is taken into account; the arbitrary assumption that only 30% of plant phosphorus is available is not satisfactory and diets should be analysed for their total and phytin phosphorus contents.

Dietary phosphorus for layers

In hens producing large numbers of eggs, Antillon (1976) showed that when dietary available phosphorus was reduced from 5.5 to 2.5 g/kg, egg production was as high as that with the control diet and the breaking strength of the eggs of the hens receiving 2.6 g/kg available phosphorus was significantly superior to that of those receiving 5.5 g/kg. A reduction in eggshell quality from the inclusion of dietary phosphorus in excess of that required for maximum egg production has also been reported by Damron, Eldred and Harms (1974), Bletner and McChee (1975), Summers, Grandhi and Leeson (1976), Harms and Miles (1980) and Guenter (1980). Detailed studies over a 12-month laying period were carried out at the ADAS Poultry Unit at Gleadthorpe with white birds in 1979–80 and with brown birds in 1982–83 with the following results.

In the first study 540 Shaver 288 birds were fed, from 32–80 weeks of age, one or other of nine dietary treatments, differing only in their calcium and phosphorus levels. The base diet was formulated to contain 20 g/kg calcium, 4.4 g/kg total phosphorus and 2.0 g/kg non-phytin phosphorus. The dietary treatments were obtained by adding graded levels of limestone and dicalcium phosphate to the base diet. The diets comprised three calcium levels (25, 35 and 45 g/kg) and three levels of total phosphorus (4.6, 6.0 and 7.4 g/kg) combined in a full factorial design. The mineral composition of the diets, daily intakes and eggshell densities are given in *Table 16.4* whilst the effects upon laying performance are given in *Table 16.5*.

Mean daily intakes of non-phytin phosphorus ranged from about 0.23 to 0.70 g/day whilst the calcium intakes ranged from about 3.0 to 5.3 g/day. Eggshell density was maintained at the highest levels with diets providing 0.23 g/day of non-phytin phosphorus. A daily intake of about 4.1 g calcium and no more than 0.55 g total

Table 16.4 DIETARY MINERAL COMPOSITION, MEAN DAILY INTAKES AND MEAN SHELL DENSITY FOR SHAVER 288 BIRDS RECEIVING DIETS OF DIFFERENT CALCIUM AND PHOSPHORUS CONTENT

Diet	*Composition and mean intake*						*Mean eggshell density (mg/cm²)*
	Calcium (g/kg)	*(g/day)*	*Total phosphorus (g/kg)*	*(g/day)*	*Non-phytin phosphorus (g/kg)*	*(g/day)*	
1	26	3.0	4.6	0.52	2.0	0.23	71.0
2	35	4.1	4.7	0.55	2.0	0.23	74.2
3	47	5.5	4.7	0.56	2.0	0.24	74.3
4	25	3.0	6.0	0.71	4.0	0.47	70.9
5	36	4.3	5.8	0.68	4.0	0.47	72.9
6	45	5.2	6.0	0.68	4.0	0.46	74.2
7	26	3.0	7.4	0.87	6.0	0.71	70.9
8	36	4.2	7.4	0.87	6.0	0.70	70.4
9	45	5.2	7.3	0.84	6.0	0.69	72.9

Table 16.5 EGG PRODUCTION AND MORTALITY FOR SHAVER 288 BIRDS FROM 33 TO 80 WEEKS OF AGE RECEIVING DIETS OF DIFFERENT CALCIUM AND PHOSPHORUS CONTENT

Diet	*Eggs per hen housed*	*Egg weight (g)*	*Mean egg mass (g/day)*	*Mortality (%)*
1	234	60.8	45.3	10.1
2	247	61.6	49.2	5.2
3	244	63.2	50.5	5.0
4	249	62.1	49.4	8.3
5	258	62.1	49.8	3.3
6	249	62.1	48.7	3.3
7	242	62.2	48.8	9.0
8	229	61.9	46.0	10.5
9	254	61.7	49.1	1.7
Mean	245	62.0	48.5	6.3
SED±	12.0	0.89	1.49	

phosphorus containing 0.23 g non-phytin phosphorus was found to produce optimum eggshell quality. At higher phosphorus intakes there was an indication that more calcium was needed to maintain shell quality. There were no significant differences between the treatments for feed intake, egg numbers or mortality. Non-systematic significant differences were observed for both rate of lay and mean egg output. These differences were in keeping with those already observed prior to the start of the investigation.

In the second study 720 ISA brown birds were fed, from 20 to 80 weeks of age, one or other of 12 dietary treatments. The base feed on this occasion was formulated to contain 21 g/kg calcium, 3.3 g/kg total phosphorus and 1.6 g/kg non-phytin phosphorus. All of the dietary phosphorus was supplied from plant sources. The dietary treatments were generated by adding graded amounts of limestone and dicalcium phosphate to the base mix. The dietary treatments comprised three calcium levels (30, 40 and 50 g/kg) and four levels of total phosphorus (3.4, 4.2, 5.1

Table 16.6 DIET MINERAL COMPOSITION, MEAN DAILY INTAKES AND MEAN SHELL DENSITY FOR ISA BROWN BIRDS RECEIVING DIETS OF DIFFERENT CALCIUM AND PHOSPHORUS CONTENT

Diet	Composition and mean intake						Mean eggshell density (mg/cm²)
	Calcium		Total phosphorus		Non-phytin phosphorus		
	(g/kg)	(g/day)	(g/kg)	(g/day)	(g/kg)	(g/day)	
1	31	3.9	3.5	0.44	1.6	0.20	77.3
2	41	5.3	3.3	0.43	1.6	0.20	81.1
3	49	6.3	3.3	0.42	1.6	0.20	81.2
4	30	3.9	4.2	0.54	2.6	0.33	78.3
5	40	5.1	4.2	0.54	2.6	0.33	79.2
6	49	6.4	4.2	0.55	2.6	0.33	80.4
7	29	3.7	5.2	0.66	3.6	0.45	78.6
8	40	5.3	5.0	0.65	3.6	0.46	79.6
9	49	6.4	5.0	0.66	3.6	0.47	80.4
10	31	4.1	6.0	0.79	4.6	0.60	77.6
11	38	4.8	5.9	0.75	4.6	0.58	78.9
12	51	6.6	6.1	0.79	4.6	0.59	80.5

Table 16.7 EGG PRODUCTION AND MORTALITY FOR ISA BROWN BIRDS FROM 33 TO 80 WEEKS OF AGE RECEIVING DIETS OF DIFFERENT CALCIUM AND PHOSPHORUS CONTENT

Diet	Eggs per hen housed	Egg weight (g)	Mean egg mass (g/day)	Mortality (%)
1	283	63.3	49.5	1.7
2	271	64.0	49.7	11.7
3	270	64.1	49.6	8.3
4	279	64.3	50.9	6.7
5	285	63.8	50.0	1.7
6	273	64.2	49.3	3.3
7	280	64.7	49.8	0
8	282	63.6	51.2	6.7
9	286	63.9	50.7	3.3
10	284	63.9	50.6	1.7
11	284	63.3	50.3	3.3
12	286	63.4	50.0	3.3
Mean	280	63.9	50.1	4.3
SED±	8.7	0.61	1.25	3.48

and 6.0 g/kg) combined in a full factorial design. The mineral composition of the diets, daily intakes and eggshell densities are given in *Table 16.6* whilst the effects on laying performance are given in *Table 16.7*

Mean daily intake of non-phytin phosphorus ranged from about 0.20 to 0.59 g/day whilst the calcium intake ranged from about 3.9 to 6.4 g/day. Eggshell density was maintained at the highest levels with diets providing 0.20 g non-phytin phosphorus. A daily intake of about 5.3 g calcium and no more than 0.43 g total phosphorus containing 0.20 g non-phytin phosphorus was found to produce optimum eggshell quality. Eggshell density was reduced at higher dietary phosphorus intakes but this did not appear to be improved by higher calcium intakes. There were no significant differences between the treatments for feed intake nor egg production although the highest levels of mortality occurred on diets with the lowest phosphorus content, 50% of these deaths occurred past 57 weeks of age. Postmortem examination and analyses of 'cause of death' failed to demonstrate any relationship with mineral disorders. No significant differences were found in tibial bone ash nor the calcium and phosphorus content of bone ash of the birds at 80 weeks of age. It was therefore concluded that the deaths were of a random nature with egg peritonitis and prolapse being the most frequently cited cause of death.

Dietary phosphorus requirements for layers

Daghir, Farran and Kaysi (1985) fed maize-soya based diets containing 1.5, 2.5, 3.5 and 4.5 g/kg available phosphorus continuously from 26 to 74 weeks of age. Egg production was significantly reduced on the lowest phosphorus diet providing 0.16 g/day available phosphorus and shell thickness was depressed on diets providing more than 0.39 g/day available phosphorus. Rodriguez, Owings and Sell (1984) also reported reduced egg production and increased eggshell thickness on diets providing 0.15 g/day of available phosphorus. In both of these studies it was assumed that 30–33% of the phosphorus in the maize and soyabean meals was available to the bird. Using the analysis data given in *Table 16.3* for the phosphorus content of maize, soya and wheat, the non-phytin phosphorus content of a maize-soya diet could be expected to be about 39% whilst that for a wheat-soya diet would be 47%. This difference in non-phytin phosphorus of base feeds would provide a real difference in dietary available phosphorus between maize-soya and wheat-soya diets which at the margin of dietary phosphorus adequacy might explain the depression in egg production found in the studies compared with those at Gleadthorpe. This clearly illustrates the need to determine both total and phytin phosphorus when assessing the adequacy of low phosphorus diets for poultry.

Table 16.8 CURRENT RECOMMENDATIONS FOR DIETARY CALCIUM AND PHOSPHORUS FOR THE LAYING HEN COMPARED WITH GLEADTHORPE FINDINGS

	Calcium		Total phosphorus		Available phosphorus	
	(g/day)	(g/kg)[a]	(g/day)	(g/kg)[a]	(g/day)	(g/kg)[a]
ARC 1975	4.0	3.3	—	—	0.35	2.9
NRC 1977	3.6	30.0	0.66	5.5	0.32	2.7
NRC 1984	3.75	31.3	—	—	0.35	2.9
WPSA 1984	4.6	38	—	—	0.38	3.2
Gleadthorpe	5.3	44	0.43	3.6	0.20	1.7

[a] Assumes an average daily feed intake of 120 g/day

Comparison of current recommendations for dietary calcium and phosphorus with the Gleadthorpe findings for a brown layer eating 120 g feed per day are given in *Table 16.8*. These data suggest that the recommendations for phosphorus are probably still generous but work to assess the adequacy of lower dietary phosphorus should be accompanied by analytical assessments of both total and non-phytin phosphorus levels in the diets used.

References

ANTILLON, A. (1976). PhD Thesis, Cornell University, New York

AGRICULTURAL RESEARCH COUNCIL (1975). *The Nutrient Requirements of Farm Livestock No. 1 Poultry*. Agriculture Research Council, London

BLETNER, J.K. and MCCHEE, G.C. (1975). *Poultry Science*, **54**, 1736

DAGHIR, N.J., FARRAN, M.T. and KAYSI, S.A. (1985). *Poultry Science*, **64**, 1382–1384

DAMRON, B.L., ELDRED, A.R. and HARMS, R.H. (1974). *Poultry Science*, **53**, 1916

GRAF, E. and DINTZIS, F.R. (1982). *Journal of Agricultural and Food Chemistry*, **30** (6)

GUENTER, W. (1980). *Poultry Science*, **59**, 1615

HARMS, R.H. and MILES, R.D. (1980). *Poultry Science*, **59**, 1618–1619

HAUG, W. and LANTZSCH, H. (1983). *Journal Science, Food and Agriculture*, **34**, 1423–1426

HEUBNER, W. and STRADLER, H. (1914). *Biochemical Journal*, **64**, 422–437.

MAFF (1973). *The Analysis of Agricultural Materials. Technical Bulletin 27*. HMSO, London

MCCANCE, R.A. and WIDDOWSSON, E.M. (1935). *Biochemical Journal*, **29**, 2694–2699

PONS, W.A. and GUTHRIE, J.D. (1946). *Industrial and Engineering Chemistry*, **18** (3), 184–186

RODGRIGUEZ, M., OWINGS, W.J. and SELL, J.C. (1984). *Poultry Science*, **63**, 1553–1562

ROLAND, D.A. (1986). *World Poultry Science Journal*, **42** (2), 154–165

SUMMERS, J.D., GRANDHI, R. and LEESON, S. (1976). *Poultry Science*, **55**, 402–412

World Poultry Science Journal (1984). **40** (2), 183–187

17

NATURAL PRODUCTS FOR EGG YOLK PIGMENTATION

C.G. BELYAVIN
Harper Adams Poultry Research Unit, Newport, Shropshire, UK
and
A.G. MARANGOS
Peterhand (GB) Ltd, Stanmore, Middlesex, UK

'Natural – of or according to or provided by nature'
The Pocket Oxford Dictionary

Egg yolk colour has always been regarded as an important egg quality characteristic and recently has had an even more important role in the marketing of eggs. Traditionally, consumers have associated strong yolk colour with good quality because eggs from free range, farmyard hens generally showed a rich yolk colour derived from the carotenoid content of grass and weeds. Marketing companies requested that their producers with intensively kept hens also produced eggs with a similar rich yolk colour in order that the eggs could be promoted with a good country image. Consumers generally prefer yolk colours ranging from golden yellow to orange.

The colour of egg yolk is a dietary response. The development and maintenance of a suitable degree of pigmentation is usually under the control of the poultry nutritionist who should ensure that the feed contains adequate levels of pigmenting materials (Adams, 1985). The relevant pigmenters are the xanthophylls which are consumed by the chicken and are readily transferred to the blood and deposited in subcutaneous fatty tissue and egg yolks. There are effectively two ways in which yolk pigmenters can be included in poultry feed. Firstly, via feed ingredients, i.e. raw materials, or secondly, in a commercially available concentrated form composed of either a synthetic pigment or a so-called 'natural' one, or a combination of both. An attempt has been made to classify the pigmenting materials in *Figure 17.1*.

In the UK the legal control of colourants in feed is contained within the Feeding Stuffs Regulations. Regulation 15 of Schedule 4 states that 'no material intended for use as a feeding stuff shall contain any colourant other than a colourant named or described in Column 1 of Part II'. For poultry and laying hens, the permitted colourants are shown in *Table 17.1*. The Feeding Stuffs (no. 2) Regulations (1986) which came into operation on 3 December 1986 also changed the way that feeding stuffs are labelled. It is now no longer sufficient to state that the feed contains 'permitted colourant'. It is necessary to declare the name of the colourant.

Over recent years there has been a change in the attitudes of consumers towards what they eat. There is now concern expressed about foods which contain artificial additives or foods which come from animals which have been given feeds containing additives. It is in this second category that eggs can be included. There has

First published in *Recent Advances in Animal Nutrition – 1987*

Pigmenting materials

'Natural'	'Nature identical'	'Nature related'
Capsanthin (paprika)		Ethyl ester of
Beta-apo-8'-carotenol (oranges)		beta-apo-8'-
Lutein (grass, lucerne, marigold)		carotenoic acid (3)
Cryptoxanthin (maize, alfalfa, rosehips)		
Violaxanthin (pumpkins)		
Canthaxanthin (flamingo, mushroom)	Canthaxanthin (1)	
Zeaxanthin (maize)		
Citranaxanthin (citrus fruits)	Citranaxanthin (2)	

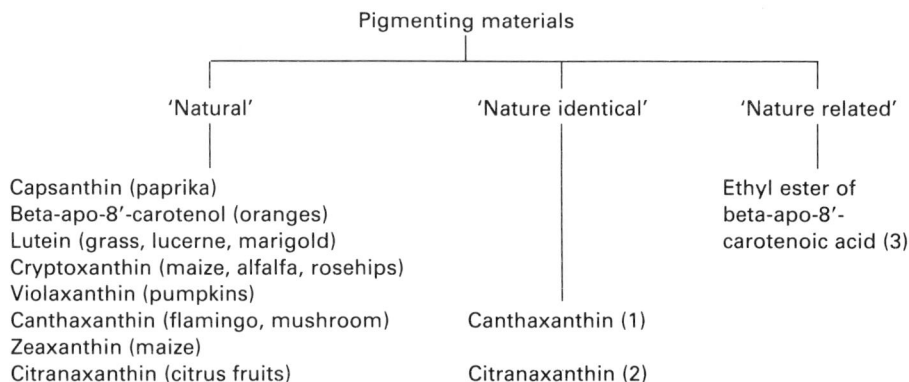

Registered trade names: (1) Carophyll red (Hoffmann La Roche)
(2) Lucantin CX (BASF)
(3) Carophyll yellow (Hoffman La Roche)

Figure 17.1 A classification of pigmenting materials

Table 17.1 PERMITTED COLOURANTS FOR USE IN POULTRY FEEDSTUFFS (FEEDING STUFFS (NO. 2) REGULATIONS (1986))

EEC no.	Name or description	Chemical formula	Kind of animal	Maximum content (mg/kg in complete feeding stuff)	Conditions
E160c	Capsanthin	$C_{40}H_{56}O_3$			
E160c	Beta-apo-8'-carotenal	$C_{30}H_{40}O$			
E160f	Ethyl ester of beta-apo-8'-carotenoic acid	$C_{32}H_{44}O_2$	Poultry	80: alone or together	None
E161b	Lutein	$C_{40}H_{56}O_2$			
E161c	Cryptoxanthin	$C_{40}H_{56}O$			
E161c	Violaxanthin	$C_{40}H_{56}O_4$			
E161g	Canthaxanthin	$C_{40}H_{52}O_2$			
E161h	Zeaxanthin	$C_{40}H_{56}O_2$			
E161i	Citranaxanthin	$C_{33}H_{44}O$	Laying hens		

recently been a trend towards the production of feeds for laying hens which the manufacturers claim contain no artificial additives; preservatives and synthetic pigmenters being singled out. In such feed the yolk colourants would therefore have to be of 'natural' origin.

It is necessary at this stage to consider in a little more detail the basis of egg yolk pigmentation.

The mechanics of egg yolk pigmentation

It seems that only carotenoids which possess functional groups in the molecule, which contain oxygen, such as hydroxyl, keto or ester groups, will pigment egg yolks (*Figure 17.2*). These are commonly known as the xanthophylls, the chemistry of which will not be discussed further.

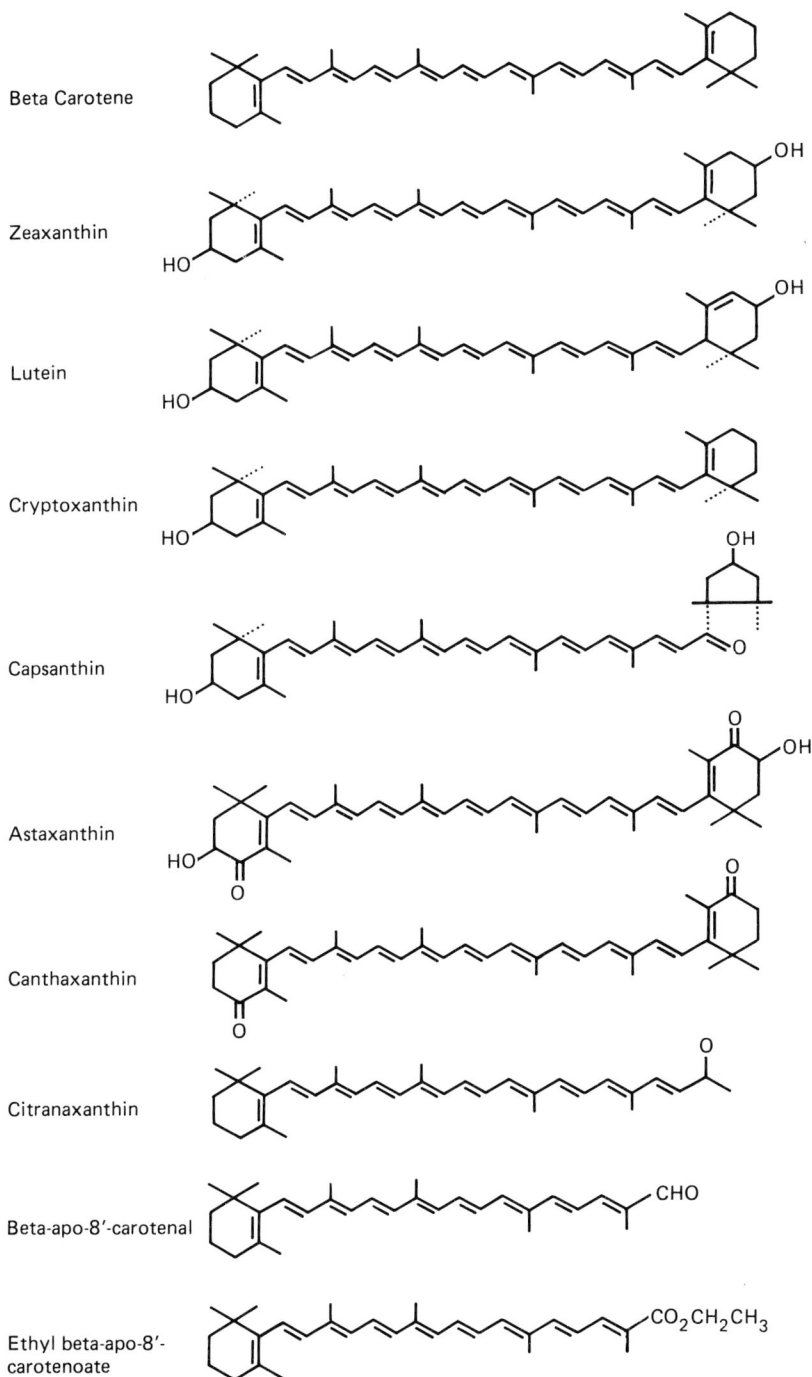

Figure 17.2 Carbon-skeleton formulae of carotenoids, showing functional substituents for yolk pigmentation (after Karunajeewa *et al.*, 1984)

Deposition of xanthophylls into egg yolks is quite a rapid process, responses to ration changes occurring quickly. However, it can take up to three weeks for a stable and final response to pigments in the diet to be achieved (Adams, 1985). The colour perceived by the consumer is a function of both the quantity and quality of the xanthophylls in the diet. A substantial base of yellow xanthophylls is required for good pigmentation, but Bougon, Protais and Launay in work undertaken at Ploufragon, France in 1982 and cited by Adams (1985) showed how the efficiency of transfer of xanthophylls by the hen from the feed to the egg yolk decreases as the dietary yellow xanthophyll content increases (*Figure 17.3*). It is reasonable to suggest that it would not be possible to achieve the yolk colour desired by the consumer from yellow pigments alone. Some judicious use of red pigment is also necessary and this has been demonstrated clearly by De Groote (1970) (*Figure 17.4*). However, too much red pigment can lead to 'blotchy yolks' if it is not properly balanced with yellow pigment. In practice, the necessary blend of yellow and red has been achieved by deriving the yellow base from feed ingredients and/or the ethyl ester of beta-apo 8'-carotenoic acid and the red base from the 'nature identical' canthaxanthin or citranaxanthin, the manufactured synthetic products (*see Figure 17.1*).

There are a number of other factors which will influence egg yolk pigmentation (Karunajeewa *et al.*, 1984). Since yolk colour is derived from the feed, feed intake will influence the quantity of material available for yolk pigmentation. Also, yolk colour is influenced by the genotype and the rate of egg production in hens. There is some variation between breeds, strains and individual hens in their ability to absorb and deposit oxycarotenoids in egg yolk. Age, health and dietary factors such as fats, calcium and vitamins also have an effect (Karunajeewa *et al.*, 1984).

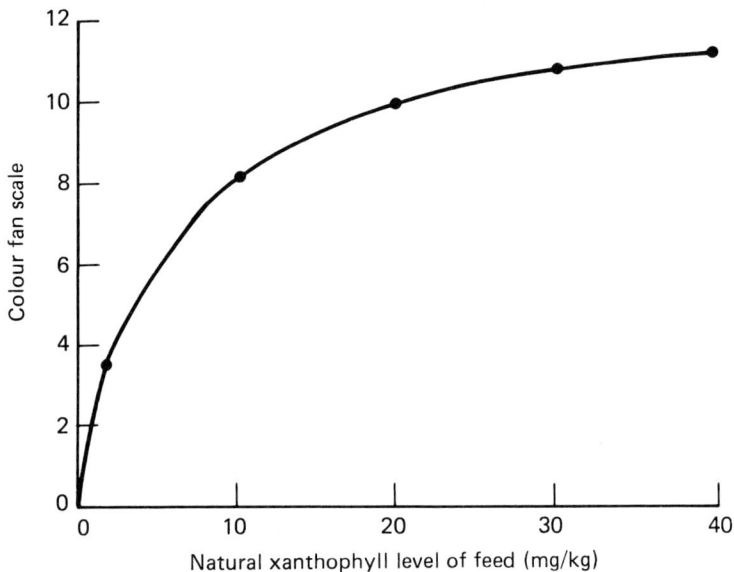

Figure 17.3 Relationship between yellow xanthophyll level of feed and yolk pigmentation as measured by a colour fan (after Adams, 1985)

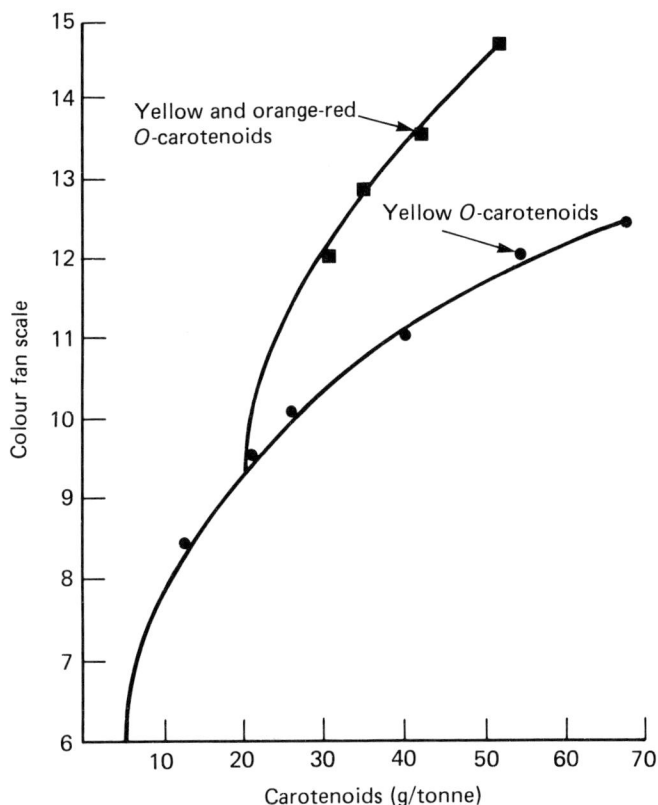

Figure 17.4 Influence of yellow and orange-red carotenoids on egg yolk colour (after De Groote, 1970)

The absorption and deposition of oxycarotenoids in egg yolks is enhanced by dietary lipids, the saturated lipids being more effective than the unsaturated lipids. The oxidation of dietary lipids resulting in a high level of peroxides has been found to reduce yolk colour. This can be prevented by the inclusion of antioxidants and vitamin E in the diet of the laying hen.

The use of natural products for egg yolk pigmentation

In natural plant materials the major xanthophylls of the type suitable for yolk pigmentation are lutein and zeaxanthin. This means that suitable feed ingredients are maize, maize products, grass meal and lucerne, the xanthophyll content of which are given in *Table 17.2*. Other feed ingredients, such as wheat and barley, are very low in xanthophylls. As shown in *Table 17.2* considerable variation can occur even within a particular raw material and this may relate to the source of a particular delivery as well as to the variety, harvest and storage conditions. The xanthophylls are chemically related to fats — they are lipids, and consequently are subject to the same degradative processes of autoxidation as fats and oils. The

Table 17.2 XANTHOPHYLL CONTENTS OF THE PRINCIPAL PIGMENT CARRIERS IN MIXED FEEDS

	Xanthophylls (lutein and zeaxanthin fraction)	
	Mean content (mg/kg)	Range of variation (mg/kg)
Lucerne meal (15–17% protein)	140	40 to 620
Grass meal	320	140 to 500
Yellow maize	17	8 to 40
Plate maize	28	10 to 50
Maize gluten meal (42% protein)	110	60 to 340

After Anon (1974)
(The total xanthophyll contents of feeding stuffs reported in the literature tend to be some 10 to 30% higher than the figures in this table since they refer to unseparated xanthophyll fractions which include other carotenoids of unknown pigmenting powers)

process of the oxidation of lutein and zeaxanthin is illustrated in *Figure 17.5*. It is beyond the scope of this chapter to discuss this in more detail but suffice to say that it is an important aspect of their use. An indication as to the rate of loss of xanthophyll from stored poultry feed was provided by Vuilleumier (1963) cited by Anon (1974) and is shown in *Figure 17.6*. Once the natural materials are ground and incorporated into the feed, the opportunity for degradation to occur is greater.

Putnam (1985) discussed in some detail the practical problems of deriving yolk pigmenters from feed ingredients when formulating rations. The xanthophyll content of the feed is generally a minor consideration when formulating rations. The content of energy and other nutrients such as protein, calcium and phosphorus are important considerations as well as cost. Putnam (1985) pointed out that in order to match the yolk colour of free range eggs a layers' ration has to be formulated with a minimum of 40% yellow maize and 5% dried grass or lucerne meal. He illustrated how this could be modified with the introduction of prairie meal into the formulation but pointed out that this approach was unlikely to meet the other nutrient and economic requirements of the ration. Because practical ration formulation hinges on economic considerations once the ration specification has been established, it may often be uneconomic to include such ingredients in a poultry ration. Their rate of inclusion will also fluctuate with price and their contribution of yolk pigmenters.

A further consideration is the availability of the carotenoids for yolk pigmentation from these raw ingredients. This has been studied by a number of workers and the findings are summarized in *Table 17.3* and *Figure 17.7* (Anon, 1974). The variation between workers probably reflects differences due to variety and storage conditions already discussed and supports concern about relying solely on these ingredients for yolk pigmentation, apart from the obvious lack of red pigment.

This, together with consumer attitudes, has led to the development of 'natural' pigmenters that are extracts from natural raw materials such as marigold (yellow) and paprika (red). Theoretically, these products can be used in the same way as the more traditional manufactured pigmenters and can therefore be added to the feed via a commercially manufactured premix. An advantage of using the raw materials is that they are generally added to the feed in fairly large quantities making mixing easier, whereas the manufactured pigmenters are added in micro-quantities making even dispersion in finished feed more difficult.

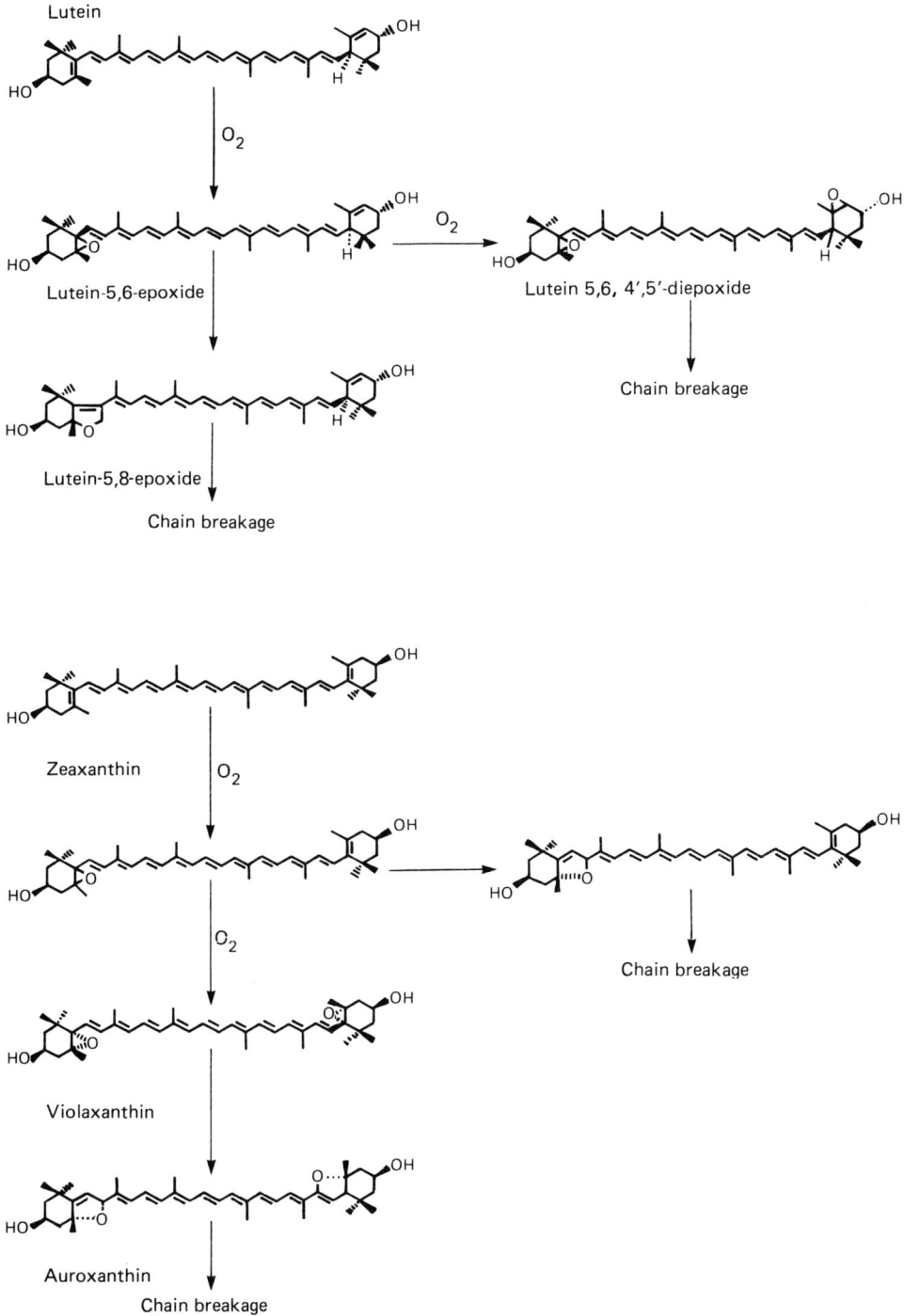

Figure 17.5 Oxidation products of lutein and zeaxanthin (after Nelson, 1985)

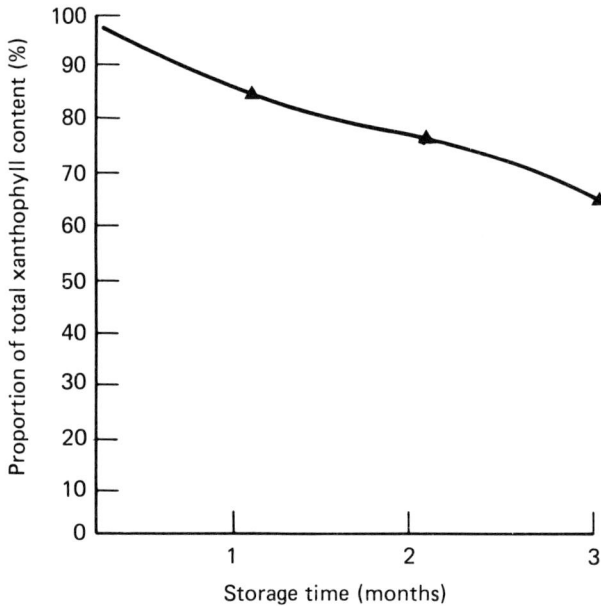

Figure 17.6 Percentage losses of total xanthophylls from the feed of laying poultry during storage (after Anon, 1974)

Table 17.3 RELATIVE UTILIZATION OF CAROTENOIDS FROM MAIZE, AND MAIZE–GLUTEN AND LUCERNE MEAL, FOR YOLK PIGMENTATION

Maize	Maize–gluten meal	Lucerne meal	Workers and year	
100	111	78	Morehouse and Hanson	1961
100	31	47	Thumim	1962
100	—	69	Bartov and Bornstein	1966
100	—	31	Prohászka and Toth	1967
100	80	60	De Groote	1968
100	96	80	Marusich and Wilgus	1968
100	80	37	Härtel and Ostendorff	1971

After Anon (1974)

THE MANUFACTURE OF A 'NATURAL' PIGMENTER

It is not intended to give a detailed description of the manufacturing process for all the 'natural' pigments available, but rather to give an outline of the various processes involved, the potential losses in xanthophyll content which can take place and the possible causes of the losses at each stage. The information is summarized in *Figure 17.8* and relates to a marigold extract (Nelson, 1985).

There are seven stages in the process from start to finish and total xanthophyll losses can range from 10 to 70%. The first step is the harvest of the raw material at which stage losses can amount to as much as 5% due to oxidative enzymes and light exposure. Manufacturers claim to be very selective in their choice of plant material

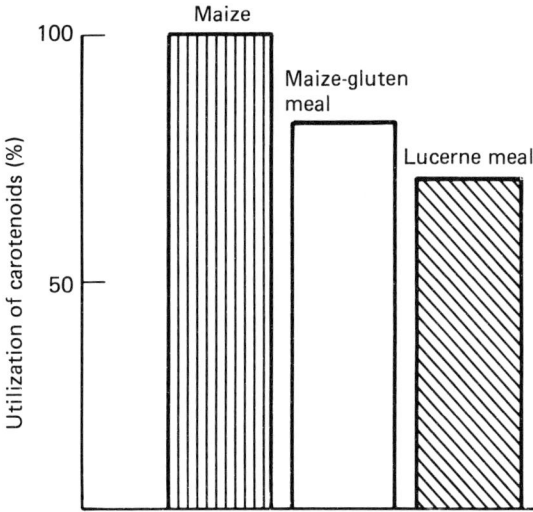

Figure 17.7 Comparison of the utilization of carotenoids from various feed ingredients (after Anon, 1974)

	Possible total xanthophyll losses	Reason
Harvest	1–5%	Oxidative enzymes, light exposure
↓		
Drying	5–10%	Heat damage, oxygen exposure
↓		
Extraction	1–10%	Heat damage during distillation/extraction
↓		
Saponification	1–15%	Caustic catalysed oxidation, metal ions present
↓		
Product production	1–5%	Exposure to oxygen. Metal ion catalysed autoxidation
↓		
Storage of finished product	1–25%	Air exposure

Figure 17.8 The manufacture of concentrated marigold pigments and losses of xanthophyll during processing under moderately stringent oxidation prevention strategies

in an attempt to maximize the quality of the raw material. Stage two involves the drying of the harvested flowers which can lead to further losses of 5 to 10% of the xanthophyll content resulting from heat damage and oxygen exposure. The extraction process is based on distillation and here further losses of up to 10% can take place as a result of heat damage. The next two stages in the process are described as saponification and product production and losses can range from 2 to 20%, primarily as a result of oxidation. The greatest potential for loss occurs during the storage of the finished product where losses due to exposure to air can amount to 25%. It is normal for these products to be stored in light-proof containers.

Because these products are manufactured it is fair to assume that they should be less variable than the raw materials, because of the opportunities for quality control during the manufacturing processes. In fact, evidence suggests that this does not appear to be the case. Data are summarized in *Tables 17.4* and *17.5* which show that these 'natural' products have varied considerably in content of pigments both between samples of the same product and also between extracts of the same raw material. *Table 17.4* relates to products derived from marigold and *Table 17.5* to a product derived from paprika.

Table 17.4 CAROTENOID CONTENTS (mg/kg) OF PIGMENTERS EXTRACTED FROM MARIGOLD

| Sample no. | Product A | | Product B |
	1	2	1
Monohydroxycarotenoids	11300	600	54
Dihydroxycarotenoids	6500	20200	15910
Total	17800	20800	15964

Table 17.5 CAROTENOID CONTENTS (mg/kg) OF A PRODUCT EXTRACTED FROM PAPRIKA

Sample	1	2
Monohydroxycarotenoids	120	1382
Dihydroxycarotenoids	2650	4436
Total	2770	5818

STABILITY OF THE EXTRACTED 'NATURAL' PIGMENTERS

There is a distinct shortage of information on the stability of these products after manufacture except that available from the manufacturers. It is common practice therefore for manufacturers to include antioxidants to stabilize the finished product. There are three stages at which degradation can take place, namely, during storage as the raw product, when included in the vitamin–mineral supplement and when included in the finished feed. (Adams, C.A., personal communication) The information is summarized in *Tables 17.6, 17.7* and *17.8*. Because the data are limited it would be wrong to attempt to draw too many conclusions from them. However, the indications are that losses are less than those described by Vuilleumier (1963) (cited by Anon, 1974) for poultry feed containing natural raw ingredients. This seems logical considering the fact that one set of 'products' is 'natural' whereas the other is manufactured.

Table 17.6 TOTAL XANTHOPHYLL CONTENT OF A NATURAL PIGMENTER DURING STORAGE DETERMINED BY STANDARD AOAC METHOD

Storage time (days at room temperature)	Xanthophyll content (g/kg)	Xanthophyll loss (%)
0	22.4	0
16	23.0	0
36	21.2	5
87	21.3	5
121	20.4	9

Table 17.7 STABILITY OF A NATURAL PIGMENTER IN A VITAMIN–MINERAL PREMIX (XANTHOPHYLL mg/kg)

Storage time (days)	Pigmenter inclusion (%)		
	0.5	1.0	5.0
0	105	210	1050
5	109	236	1255
21	142	271	1305
53	106	162	1122

Table 17.8 STABILITY OF A NATURAL PIGMENTER IN A DIET FOR LAYING HENS (XANTHOPHYLL mg/kg)

Storage time (days)	Pigmenter inclusion (g/tonne feed)			
	0	200	500	800
0	2	7	15	23
5	2	9	17	23
20	2	9	17	24
53	2	6	13	19

Trial work investigating the use of extracted, 'natural' yolk pigmenters in the feed of laying hens

As a consequence of the change in consumer attitudes there has been a marked increase in interest in the feasibility of deriving yolk pigmentation from exclusively 'natural' sources in finished poultry feeds. As, in the UK particularly, it is usually uneconomical to add adequate pigmenting raw materials, it is necessary also to include concentrated forms of pigmenting substances. The marigold and paprika extracts have been of particular interest along with the 'nature identical' citranaxanthin and canthaxanthin products. As long ago as 1981 Fletcher and Halloran showed that a paprika extract (red pigment) when used in combination with a marigold extract (yellow pigment) could increase visual yolk colour score in greater proportion than could corresponding increases in a marigold extract alone. Much of the more recent work has been undertaken at the Harper Adams Poultry Research

Unit in the UK as well as at other centres. Data are presented not in an attempt to compare products, but more to assess the progress that has been made and to try to identify possible problem areas. Therefore, products are not named in any of the experiments described but are referred to by a number preceded by the letter 'S' in the case of the traditional pigmenting additives and the letter 'N' in the case of the natural ones. In the studies described there is also the opportunity to examine one or two more fundamental aspects of yolk pigmentation, namely the effect of time on the accumulation of pigment and also the subjectiveness of measuring yolk colour in the traditional way using the colour fan.

It has to be pointed out that the work undertaken at Harper Adams Poultry Research Unit comprised a number of commissioned trials for commercial companies and not a series planned at the Unit. Consequently, there is not necessarily a logical relationship between the various pieces of work.

TRIAL 1

This work was undertaken in Australia (Hall, G.R., personal communication) with white birds and studied the addition of a product based on both marigold and paprika extracts to a wheat based diet containing no maize or maize products or other raw materials contributing pigments. The rate of inclusion of this blend was 333 g/tonne of finished feed.

The first eggs were laid 38 days after the introduction of the trial diet when the birds were 22 weeks old and a daily, random sample of 10 eggs was broken out and measured with the traditional colour fan. Results are presented in *Table 17.9* and *Figure 17.9*.

It can be seen that reasonable yolk colour was achieved by the use of this particular blend of pigment and that within any particular day the variation in yolk colour between eggs was generally small with standard deviation values of one or less and showing no trend. What is more interesting is the relationship between yolk colour and time. Even though eggs were not measured until 38 days after the trial diet was introduced, the trend for yolk colour score was upward and markedly linear over the 36-day recording period (*Figure 17.9*). It is likely that this reflects the trend for feed intake over the period of measurement and the purpose of illustrating it here is to emphasize the caution required when undertaking yolk colour studies because of the influence of the other performance characteristics.

TRIAL 2

This larger scale study involving many treatments was undertaken in the UK with brown birds (Farley, R., personal communication). Combinations of the traditional pigmenters as well as the 'natural' ones were used in order to achieve two target yolk colours (9 and 12) as measured with the traditional colour fan. Details of some of the treatments are summarized in *Table 17.10*.

An interesting feature of this study was that more than one scorer assessed the same ten eggs at breaking out time. The results are summarized in the table.

At both target yolk colour scores there were significant differences ($P < 0.01$) between treatments. At the lower target score (9) all pairs of differences were significant whereas at the higher target yolk colour (12), treatments 1 and 3 (which

Table 17.9 YOLK COLOUR SCORES FOR EGGS FROM HENS FED A WHEAT BASED DIET WITH AN ADDED PIGMENTER BLEND EXTRACTED FROM MARIGOLD AND PAPRIKA (N1)

	No. of eggs in yolk colour categories								Mean yolk colour score for 6 day bands
Day	7	8	9	10	11	12	Mean	SD	
1	—	2	3	5	—	—	9.3	0.823	
2	—	3	5	2	—	—	8.9	0.738	
3	—	2	6	2	—	—	9.0	0.667	9.0
4	1	4	3	2	—	—	8.6	0.966	
5	1	2	3	2	2	—	9.2	1.317	
6	—	2	6	2	—	—	9.0	0.667	
7	—	—	6	2	2	—	9.6	0.843	
8	—	—	3	3	3	1	10.2	1.033	
9	1	—	3	3	2	1	9.8	1.398	9.73
10	1	1	3	3	2	—	9.4	1.265	
11	—	1	3	4	2	—	9.7	0.949	
12	—	1	3	4	2	—	9.7	0.949	
13	—	—	2	6	1	1	10.1	0.876	
14	—	1	3	4	2	—	9.7	0.949	
15	—	—	5	2	3	—	9.8	0.919	9.85
16	—	1	5	2	2	—	9.5	0.972	
17	—	—	2	3	3	2	10.5	1.080	
18	1	1	3	2	3	—	9.5	1.354	
19	—	—	5	4	1	—	9.6	0.699	
20	—	—	2	3	4	1	10.4	0.966	
21	—	—	—	4	6	—	10.6	0.516	10.18
22	—	1	1	2	3	3	1C.6	1.350	
23	—	—	2	4	4	—	10.2	0.789	
24	—	1	1	8	—	—	9.7	0.675	
25	—	1	2	4	3	—	9.9	0.994	
26	—	—	2	2	3	3	10.7	1.160	
27	—	—	—	3	5	2	10.9	0.738	10.5
28	—	—	—	1	9	—	10.9	0.316	
29	—	—	1	3	5	1	10.6	0.843	
30	—	1	2	4	2	1	10.0	1.155	
31	—	—	1	2	7	—	10.6	0.699	
32	—	—	2	3	5	—	10.3	0.823	
33	—	—	—	3	6	1	10.8	0.632	10.63
34	—	—	—	3	5	2	10.9	0.738	
35	—	—	1	1	8	—	10.7	0.675	
36	—	—	1	3	6	—	10.5	0.707	

both contained pigment S1) produced higher values. The treatments which did not contain pigmenter S1 (treatments 2 and 4) showed a lower yolk colour score than those which did contain it. Treatment 4, which did not contain either pigment S1 or S2 did not show the lowest yolk colour scores. However the treatment still contained a synthetic red pigmenter.

The target yolk colour appeared to be the stronger influence over the differences between scorers rather than the material used to achieve the target yolk colour. At

$$y = 10.0081 + 0.0497 (x - 19)$$

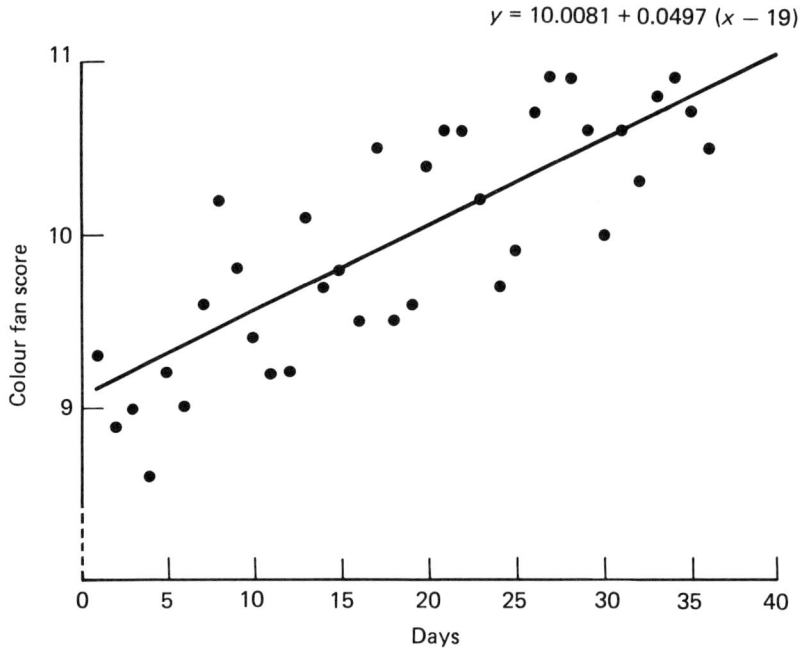

Figure 17.9 Regression analysis on yolk colour scores for eggs from hens fed a wheat based diet with an added pigmenter blend extracted from marigold and paprika (*see also Table 17.9*)

Table 17.10 YOLK COLOUR VALUES AS RECORDED BY DIFFERENT SCORERS FOR RATIONS BASED ON A NUMBER OF PIGMENTERS AIMED AT PRODUCING TWO YOLK COLOUR SCORES (9 AND 12)

Treat-ment no.	Pigment inclusion (g/tonne)				Yolk colour fan score				
	S1	S2	S3	N3	Target	ScA	ScB	ScC	OM
1a	9.0	22.0	0	0	9	10.15	10.90	10.15	10.40
2a	0	17.0	16.0	0	9	8.10	7.30	8.00	7.80
3a	9.0	0	0	110	9	13.10	12.80	11.45	12.45
4a	0	0	18.0	110	9	9.45	8.55	8.00	8.67
1b	28.0	42.0	0	0	12	13.10	13.65	—	13.38
2b	0	24.0	44.0	0	12	11.05	11.15	—	11.10
3b	28.0	0	0	210	12	13.50	13.35	—	13.43
4b	0	0	56.0	210	12	11.80	11.30	—	11.55

ScA = Scorer A etc., OM = Overall mean

the higher target score (12) there were no significant differences between scorers in any treatment. It should be noted that only two scorers were used in this case as opposed to three at the lower target score (9). At this lower level there were significant differences between scorers in three out of the four treatments (treatment 2 was the exception) but there was no clear trend in that in treatment 1 scorer 2 was significantly different (higher), whereas in treatment 3 it was scorer 3 (lower),

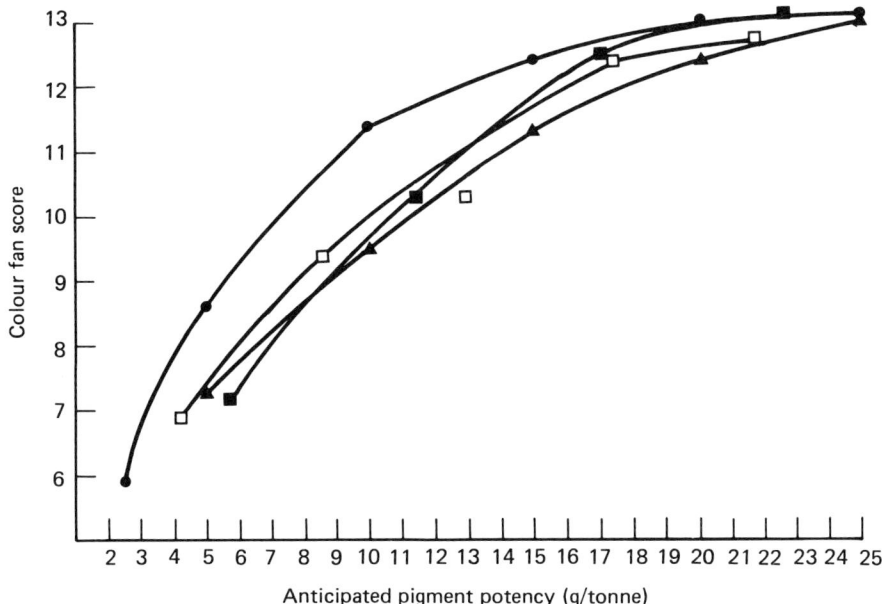

Figure 17.10 Response curves for four treatments based on different ratios of three synthetic pigmenters (S2:S1, 2:1 (●); S2:S3, 2:1.8 (▲); S2:S3, 2:1.4 (□); S2:S3, 2:2.2 (■))

and in treatment 4 it was scorer 1 (higher). Therefore, no individual scorer tended towards a high or low measurement of yolk colour.

The two trials reported so far clearly indicate the need to standardize, or at least note, the circumstances under which yolk colour studies are undertaken.

Although yolk pigments can be extracted and analysed spectrophotometrically against colour standards such as carotene (AOAC method), yolk colour traditionally is measured visually (subjectively) against a colour fan. This in itself can lead to scorer differences simply because of the difficulties in sometimes interpreting a match between the colour fan and the yolk. There are those who argue that there is clearly a need for the development of a more precise objective measurement of yolk colour and the use of reflectometers has been examined (Wells, 1966). It should be remembered that the consumer for whose benefit a specific yolk colour is achieved assesses the egg visually and under a variety of circumstances.

The traditional fan was developed by Hoffmann La Roche for use with their products which tend to produce an orange/red yolk colour. It has been suggested that consumers prefer more golden-yellow coloured yolks to orange/red (Anon, 1985) and these are probably more readily obtained using the 'natural' pigmenters rather than the traditional ones. It also raises the question as to the suitability of the traditional fan for measuring yolk colour of eggs derived from the 'natural' pigmenters.

An influence of the scorer on yolk colour assessment has already been demonstrated. Further influences (Harper Adams Poultry Research Unit, unpublished data) are the source of light under which the measurement was done (fluorescent or natural) and the background colour onto which the eggs were broken (glass plate or white dish for example).

Table 17.11 EXPERIMENTAL TREATMENTS TO TEST THE SUITABILITY AND POTENCY OF A SYNTHETIC COMPOUND FOR COLOURING EGG YOLKS

Treat-ment	Anticipated Colour fan score	Pigment potency (g/tonne)	Pigment ratios S2:S1	S2:S3	Pigment inclusion (g/tonne) S2:S1	S2:S3	Yolk fan colour scores Days 19	20	21	Mean	26	27	28	Mean	OM
1	5–7	2.5	2:1	—	6.7:3.4	—	5.85	6.00	6.67	6.17	5.8	5.43	5.87	5.70	5.9
2		—		2:1.4	—	13.3:9.3	8.00	6.17	6.43	6.87	7.83	6.33	6.50	6.89	6.9
3	8–9	5		2:1.8	—	13.3:12.0	7.00	7.37	8.00	7.46	7.00	6.67	7.71	7.13	7.3
4		—		2:2.2	—	13.3:14.6	6.90	6.25	7.40	6.85	7.30	7.50	7.78	7.53	7.2
5	8–9	5	2:1	—	13.3:6.7	—	7.71	8.29	8.67	8.22	7.80	8.63	10.50	8.98	8.6
6		—		2:1.4	—	26.6:18.8	9.14	9.40	10.67	9.74	9.22	8.33	9.67	9.07	9.4
7	10–11	10		2:1.8	—	26.6:24.2	9.60	8.89	10.29	9.59	8.87	8.87	10.17	9.30	9.5
8		—		2:2.2	—	26.6:29.7	10.71	9.83	11.00	10.51	11.00	9.40	9.83	10.08	10.3
9	10–11	10	2:1	—	26.6:13.5	—	10.80	11.58	11.50	11.29	11.17	11.90	11.50	11.52	11.4
10		—		2:1.4	—	39.9:28.1	10.71	11.00	9.87	10.53	9.63	11.33	9.25	10.07	10.3
11	12	15		2:1.8	—	39.9:36.2	10.17	11.33	11.14	10.88	11.17	12.22	11.67	11.69	11.3
12		—		2:2.2	—	39.9:44.2	12.25	12.29	12.71	12.42	13.14	12.50	11.87	12.50	12.5
13	12	15	2:1	—	39.9:20.1	—	12.44	12.33	12.00	12.26	12.44	12.29	13.14	12.62	12.4
14		—		2:1.4	—	53.2:37.5	12.25	12.43	12.43	12.37	12.89	12.73	11.67	12.43	12.4
15	12–13	20		2:1.8	—	53.2:48.2	12.00	12.00	12.91	12.30	12.22	12.29	12.89	12.47	12.4
16		—		2:2.2	—	53.2:59.0	12.20	12.42	12.33	12.32	13.00	13.00	13.33	13.11	13.1
17	12–13	20	2:1	—	53.2:26.8	—	13.00	13.14	13.00	13.05	13.00	13.00	13.00	13.00	13.0
18		—		2:1.4	—	66.5:46.9	12.30	12.00	13.22	12.51	12.78	13.14	12.50	12.81	12.7
19	13–14	25		2:1.8	—	66.5:60.3	13.33	12.70	13.00	13.01	13.30	12.89	13.00	13.06	13.0
20		—		2:2.2	—	66.5:73.7	13.90	13.33	13.50	13.58	13.75	13.75	13.14	13.55	13.7
21	13–14	25	2:1	—	66.5:33.5	—	13.00	13.90	13.13	13.34	13.00	13.37	13.29	13.22	13.1
22	0–2	Basal diet with no added pigments					3	3.33	2.50	3.01	2.63	2.67	3.17	2.82	3

OM = Overall mean

TRIAL 3

This trial did not include any of the 'natural' pigmenters but the data are presented to illustrate how relatively predictable yolk colour scores can be when the synthetic red and yellow pigments are used as the basis of achieving egg yolk colour. The trial was undertaken by MAFF/ADAS using brown laying strains (Farley, R. personal communication). The treatments and results are summarized in *Table 17.11* and the response curves are illustrated in *Figure 17.10*.

Included in Table 17.11 are the anticipated colour fan scores along with the inclusion levels of pigment per tonne of finished feed and the actual yolk colour scores achieved. The results have not been adjusted for feed intake but it can be seen that the values achieved were close to those anticipated and the response curves shown in *Figure 17.10* follow an expected curvilinear pattern with the rate of response slowing down as the expected pigment potency (g/tonne) increased beyond a certain value (15).

With the change in consumer attitudes developing and the increasing availability of the 'natural' yolk pigmenters, there has recently been a number of trials undertaken evaluating these 'natural' pigments.

TRIAL 4

Trial 4 was undertaken with brown egg layers and involved eight treatments (*Table 17.12*). Treatments 1 to 4 were based on different inclusions of pigmenters S1 and S2 and in treatment 5 pigmenter S1 was replaced with a similar inclusion of S3 with a reduction in the inclusion of S2. The remaining three treatments were based on increasing inclusions of 'natural' pigmenter N2, a marigold extract (Farley, R., personal communication). A further feature of this study was an economic evaluation of the various treatment combinations. Information on feed intake and egg production were not available.

Taking treatments 1 to 4 and 6 to 8 separately, within the two sets of data there were the expected responses to the increasing amounts of pigment inclusion. The scores were measured on mirrored glass under fluorescent lighting which tends to reduce the scores by about one point on the colour fan (Harper Adams Poultry

Table 17.12 DETAILS OF A STUDY EVALUATING FOUR YOLK PIGMENTERS (SYNTHETIC AND NATURAL) IN GRADED INCLUSIONS IN A RATION

	Pigmenter inclusion			*Yolk colour score (colour fan)*				
				Experimental period				
Treatment	*Product*	*(g/tonne)*	*Cost (£)*	*Day 13*	*Day 14*	*Day 15*	*Mean*	*Range*
1	S1 + S2	5 + 17	0.96	6.8	5.7	6.1	6.2	4 to 8
2	S1 + S2	8 + 20	1.25	8.8	6.8	7.4	7.7	5 to 11
3	S1 + S2	12 + 24	1.64	9.6	9.9	10.2	9.9	7 to 12
4	S1 + S2	18 + 32	2.31	10.8	11.1	11.8	11.2	9 to 13
5	S3 + S2	9 + 15	0.88	6.1	4.9	6.2	5.7	3 to 8
6	N2	600	1.89	6.9	6.2	7.2	6.7	5 to 9
7	N2	900	2.83	8.6	8.2	8.9	8.6	7 to 10
8	N2	1500	4.72	9.5	10.0	10.2	9.9	9 to 11

Research Unit, unpublished data). Taking the UK requirements as the standard (Roche Fan 11), the only combination which achieved this was treatment 4. Treatments 3 and 8 came close to this and it is significant that one was based on the traditional synthetic pigments while the other was a 'natural' extract. The data suggested less variation within the treatments based on 'natural' pigments than in the other but the cost of achieving the desired colour fan score based on the 'natural' pigmenters was over double that of achieving it with the traditional synthetic pigmenters. It must be noted here that pigmenter N2 only supplies yellow-type pigment and it is known that much more pigment is required to achieve adequate yolk colour scores when only yellow is used as opposed to a blend of yellow and red.

TRIAL 5

A further study completed at Harper Adams Poultry Research Unit (Clarke, A.M., personal communication) is summarized in *Table 17.13*. A pigment-free diet was fed to the moulted brown birds (100 weeks old) for 14 days to achieve a base yolk colour. The six treatments were then introduced for a further 21 days and eggs were broken out on the last three days of the experimental period for assessment under fluorescent lighting using the conventional yolk colour fan. The results show that a very low yolk colour score was achieved by feeding the pigment-free diet and there were no significant differences between the treatment groups at this stage.

Over the experimental period there were no significant differences between groups for egg production or feed intake. However, there were significant differences ($P < 0.01$) between treatments for yolk colour by the end of the experimental period. On the three days that eggs were broken out, eggs from the treatments based on 'natural' pigmenters were paler than those from the treatments based on synthetic ones. Within each of the last three days there were also significant differences but these did not follow any logical pattern and no explanation for this can be offered. On day 33 eggs from treatments 3 and 5 had paler yolks than eggs from treatment 4. On day 34 there were significant differences between all pairs of treatments except between treatment 1 and 2 and between treatment 3 and 6. Eggs from treatment 5 had paler yolks than eggs from treatment 4 on day 35.

Before drawing conclusions from this study it is important to look at the results for days 33 to 35, particularly for treatments 1 and 2. There is a downward trend for yolk colour score over this period for these two treatments and this could reflect a lack of stability of the product in the finished feed for reasons which have already been discussed. This emphasizes the importance of using fresh material in poultry feeds. It is unlikely that the age of the birds influenced the findings, unless older birds are for some reason less able to utilize the 'natural' pigments. If this is the case it has implications for the use of these products.

TRIAL 6

This study was completed at Harper Adams Poultry Research Unit using commercial brown egg layers (103 weeks of age). A pigment-free diet was fed for the first 14 days and then various pigment combinations were added to the pigment-free diet as shown in *Table 17.14*. Essentially, the objective of this study was to test the

Table 17.13 RESULTS OF A STUDY BASED ON TWO NATURAL (ONE MARIGOLD AND ONE PAPRIKA EXTRACT) AND THREE SYNTHETIC YOLK PIGMENTERS

Treatment	Pigment addition (g/tonne)					Yolk colour fan score by day							Feed intake (g/bird/day)	Eggs/bird (experimental period)
	N3	N4	S1	S2	S3	12 (pigment free)	13	14	33	34	35	Mean (days 33–35)		
1	100	25				1.6	2.0	2.4	5.5	3.7	3.2	4.1	133	13.1
2	150	75				1.5	2.4	1.5	5.7	4.1	2.8	4.2	131	13.5
3			9	22		1.6	2.5	2.1	9.3	9.4	9.5	9.4	122	13.7
4			19	34		1.9	2.5	2.1	11.5	12.5	12.1	12.0	126	15.0
5				22	16.2	2.7	2.7	2.7	8.6	7.3	8.5	8.1	133	15.1
6				34	34.2	2.0	3.2	2.1	9.9	10.0	10.5	10.1	131	13.9

Table 17.14 A STUDY OF THE REPLACEMENT OF A SYNTHETIC YELLOW PIGMENTER WITH A MARIGOLD EXTRACT NATURAL YELLOW PIGMENTER

Treatment	Pigment addition (g/tonne)			Yolk colour fan scores				SD	Feed intake (g/bird/day)	Egg numbers (eggs/bird/experimental period)
	S2	S1	N3	Days (a) 12+13+14	Days (b) 26+27+28	Days (c) 40+41+42	$\frac{(b+c)}{2}$			
Group I	20	10	—	2.3	9.7	9.5	9.6	±0.765	108	16.8
	—	10	100	2.4	9.5	9.6	9.6	±0.991	119	18.9
Group II	30	15	—	2.3	11.5	11.5	11.5	±1.310	116	19.4
	—	15	150	2.0	11.0	10.5	10.8	±1.446	111	15.4
Group III	17	17	—	2.2	11.3	11.2	11.3	±1.139	116	18.1
	—	17	85	2.4	11.5	10.2	10.9	±1.011	116	17.9
Group IV	15	25	—	2.1	12.7	11.6	12.2	±1.225	113	16.4
	—	25	75	1.9	12.8	12.4	12.6	±1.520	112	15.9

replacement of pigmenter S2 with pigmenter N3 at a rate of 5:1, N3:S2 at increasing inclusions of pigmenter S1 to give darker yolk colour scores progressing from Group I to Group IV. The objective of the study was achieved and within each group there was no significant difference between the yolk score for each pair. The tendency was for the standard deviation and hence the variation to be greater for the treatments containing the 'natural' pigmenter than those containing the synthetics. The age of the birds did not appear to have any adverse effect on these results (*see above*).

TRIAL 7

The final study to be reported, also completed at Harper Adams Poultry Research Unit, is summarized in *Table 17.15*. The procedures followed for this study were identical to those described above for Trial 6. The birds were 24 weeks of age. In this case combinations of pigmenters N3 and S3 were examined, treatments 1 and 2 and 3 and 4 showing graded levels of S3 at two inclusions of N3. Treatment 5 was based on a higher inclusion of N3 plus the lower inclusion of S3.

In all cases adequate yolk colour scores were achieved but there were no significant differences between treatments, a difference of 0.824 being necessary to achieve significance. There were no significant treatment differences for feed intake or egg numbers. Standard deviations appeared higher in this study than the previous one which may be a reflection of the higher rate of lay recorded in this study or the fact that pigment S3 was used as opposed to pigment S1.

By looking at treatments 1, 3 and 5 it is possible to get an indication of the effect of increasing inclusions of pigment N3 in the diet. The results suggest little or no response which probably means that the upper limit of the effectiveness of the yellow pigment (N3) has been reached at an inclusion of 100 g/tonne of feed. When considering the red pigment (S3) the results suggested that increasing the dietary content from 30 to 40 g/tonne did lead to a response but only at the lower inclusion of the yellow pigment, the increase in yolk colour score nearly achieving significance.

Conclusions

The information available at present suggests that it is not possible to achieve acceptable yolk colour (Fan no. 11) from natural pigmenters, either from feed ingredients or marigold extracts, alone. The addition of a red pigment is necessary and to date a suitable natural red pigmenter is not available. However, results from the latest study completed at Harper Adams Poultry Research Unit, but not reported here, suggest that a good natural red pigmenter may soon be available.

The Feeding Stuffs Regulations, consumer attitudes and economics must be taken into account. It is unlikely that consumer attitudes will revert to not being concerned about what they eat and therefore the industry must be prepared to manufacture feedingstuffs which can genuinely be described as 'natural'. Within reason the economics of achieving this may have to be ignored in that a small rise in feed prices may have to be accepted in order to achieve truly 'natural' feeds. There is no evidence at the moment that a premium is available for eggs produced from 'natural' feeds unless they are sold through specialist outlets such as health food shops.

Table 17.15 RESULTS OF A TRIAL EXAMINING THREE INCLUSIONS OF A NATURAL MARIGOLD EXTRACT YELLOW PIGMENT AND TWO INCLUSIONS OF A SYNTHETIC RED PIGMENTER

Treatment	Pigment addition (g/tonne)			Yolk colour fan scores					Feed intake (g/bird/day)	Egg numbers (eggs/bird/experimental period)
	N3	+	S3	Days (a) 12+13+14	Days (b) 26+27+28	Days (c) 40+41+42	$\frac{(b+c)}{2}$	SD		
1	100	+	30	2.5	10.3	11.9	11.1	±1.770	114.6	27.0
2	100	+	40	3.1	11.3	12.6	11.9	±1.228	116.5	25.7
3	125	+	30	3.2	10.9	12.0	11.5	±1.614	116.9	26.3
4	125	+	40	2.5	10.9	12.1	11.5	±1.431	116.4	25.7
5	150	+	30	2.5	10.5	11.8	11.2	±1.674	116.4	25.3

It is the implications of the Feeding Stuffs Regulations which are probably most important. It is possible that some extracts of marigold contain not only lutein and zeaxanthin but also lutein dipalmitate and lutein dimyristate, both of which are not permitted colourants (*see Table 17.1*). Other elements which are considered harmful such as anthacins, pesticides and solvent residues have also been found.

The law clearly defines what is and is not permitted. Therefore, if 'natural' products are used that also contribute non-permitted substances then these are definitely not legal. The supplier must, therefore, clearly declare all the active ingredients of their product and such products should not contain substances which can be deemed illegal.

If the requirement for 'natural' pigmenters increases then the manufacturers must be able to guarantee the stability and the effectiveness of the products available. There is a clear lacking of dose response curves for the 'natural' products compared with the synthetic ones. Therefore, a research priority must be to establish such curves in order to make the information readily available to feed compounders and supplement manufacturers.

Acknowledgements

The authors would like to thank the following for their assistance with the preparation of this chapter: BP Nutrition (UK) Limited, Kemin (UK) Limited, Roche Products Limited, Four F Nutrition Limited.

References

ADAMS, C.A. (1985). *Pigmenters and poultry feeding.* Kemin Europa NV, Herentals, Belgium

ANON (1974). *Egg yolk pigmentation with carophyll* (2nd Edition). F. Hoffmann-La Roche and Co. Ltd, Basle, Switzerland

ANON (1985). *Animal Nutrition News*, **30**, (January), 1

DE GROOTE, G. (1970). *World's Poultry Science Journal*, **26**, 435

FEEDING STUFFS (NO. 2) REGULATIONS (1986). HMSO, London

FLETCHER, D.L. and HALLORAN, H.R. (1981). *Poultry Science*, **60**, 1846–1853

KARUNAJEEWA, H., HUGHES, R.J., MCDONALD, M.W. and SHENSTONE, F.S. (1984). *World's Poultry Science Journal*, **40**, 52

NELSON, C.E. (1985). *Stability of concentrated marigold preparations.* Pigmenter General Information, Kemin Industries, Inc., USA

PUTNAM, M. (1985). *Farm Mixer*, February/March, 13

WELLS, R.G. (1966). In *Egg Quality – A Study of the Hen's Egg*, pp. 207–250. Ed. Carter, T.C. Oliver and Boyd, Edinburgh

18

EGGSHELL FORMATION AND QUALITY

K.N. BOORMAN, J.G. VOLYNCHOOK
University of Nottingham School of Agriculture, Sutton Bonington, UK
and
C.G. BELYAVIN
Harper Adams Poultry Husbandry Experimental Unit, Edgmond, Newport, England, UK

Introduction

The problems associated with eggshells have been much reviewed (see for example Beuving, Scheele and Simons, 1981; Washburn, 1984) and much is known about the shell and its formation in fundamental terms (see for example Simkiss and Taylor, 1971; Taylor and Dacke, 1984). Much of the work aimed at redressing the problems has however been essentially empirical, i.e. applying a likely treatment and observing the outcome. While this approach has been successful in identifying the gross nutritional and management needs for producing eggs, a high proportion of which are well enough shelled to be marketable, it is less likely to be successful in identifying remaining problems and ameliorating new problems which may arise. In the case of nutritional treatments, for example, response will be related to the degree to which the nutrient is involved in the problem, the extent of the problem and the initial dietary concentration of the nutrient. Thus nutrient treatments which produce promising results in one circumstance may not do so elsewhere and equivocal results are common in studies on shell quality (see for example Belyavin and Boorman, 1981). Although there has been application of fundamental knowledge to speculation about limiting processes in shell formation (see for example Mongin, 1968; Hurwitz, Bar and Cohen, 1973; Buss and Guyer, 1981), relationships between such observations and shell faults as they occur in practice are not clear.

The studies described in this chapter started from the idea of trying to observe shell quality in essentially commercially-kept flocks and trying to relate the quality of the shell to the individual bird which produced it. This has developed into deeper studies of individuals and causes of variation in shell quality within an individual and between individuals. Some of these observations are described in the context of relevant observations of others, together with some comments on their implications in a broadly nutritional context.

'Shell quality' is used as a convenient term for the attributes measured by specific gravity, deformation, breaking strength, shell mass and density and in this sense relates mostly to the mineral components. It is recognized that quality in its fullest sense means more than this. In addition, it is not intended to imply that current problems in shell quality all stem from this one cause.

First published in *Recent Advances in Animal Nutrition – 1985*

Variation in shell quality

The variation in shell quality as the laying year progresses is well enough known, especially the accelerated decline which usually occurs in the last quarter. This curve for a flock is an integration of the trends for individuals. Belyavin (1979) isolated 48 individuals *in situ* in a flock of about 1000 Warren SSL hens and studied their characteristics over a laying year. Eggs were sampled, as far as possible, on a regular weekly basis. The mean trend in egg specific gravity for the flock and the patterns for three individuals are shown in *Figure 18.1a*. Although some individuals (bird 41) showed a fair degree of similarity with the mean trend, others showed sudden large differences (bird 12) or consistent smaller differences (bird 33). It is evident that the flock curve is not a simple integration of similar curves for individuals or of sudden thresholds in shell quality at different times in individuals. It is also evident that shell quality in an individual may be largely unrelated to the mean of the flock and that an individual's shell quality at one time may (bird 33) or may not (bird 12) be generally characteristic of that bird's relation to the mean. It is of interest that a sudden decline, at least in the middle period, is not necessarily irredeemable (bird 12). The individuals illustrated are not typical in that the final decline in shell quality is not incontrovertibly evident in any. It was usually evident and became clear as soon as trends were combined for several individuals. Its apparent inevitability is illustrated by the fact that, if the five best individuals with respect to production characteristics and shell quality are selected, their average trend shows this decline despite their overall superiority when compared with the flock mean (*Figure 18.1b*).

These data can also be used to show that there are no simple rules for predicting which individuals are likely to produce poor shells. *Table 18.1* shows mean characteristics of production, together with various attributes of shell quality, of the five best birds and of five which were among the worst. It cannot be concluded that it is the highest producers which produce the worst shells and, although there are more birds producing heavier eggs among the group producing poor shells, there is no simple relationship between these characteristics. Belyavin (1979) investigated correlations between the main production characteristics and shell quality and demonstrated using data for all individuals that there were no simple relationships.

Since eggs are laid in sequences, separated by pauses, and eggs are laid later in the day as the sequence progresses, it became evident that if there were structured changes in shell characteristics and/or egg mass during a sequence, some of the variation seen in individuals would arise from this cause. The monitoring described above was made without any knowledge of positions of sampled eggs in their respective sequences. However, there are indications of such structured changes. Early studies showed that the last egg of the sequence has a thicker shell than eggs laid earlier in the sequence (Wilhelm, 1940; Berg, 1945) and that the first egg of a sequence is heavier than subsequent eggs (Atwood, 1929). More recently Roland, Sloan and Harms (1973) found that eggs laid in the afternoon were of higher specific gravity than those laid in the mornings and the several subsequent observations from that laboratory culminated in the report of Choi *et al.*(1981) which identified and substantiated several important trends. This report is contrasted with our own findings below.

Belyavin (1979) carried out a limited examination of sequence effects in 12 Warren SSL hens isolated among a larger flock. He was not able to establish any trends with statistical significance but found good indications that the first egg of the

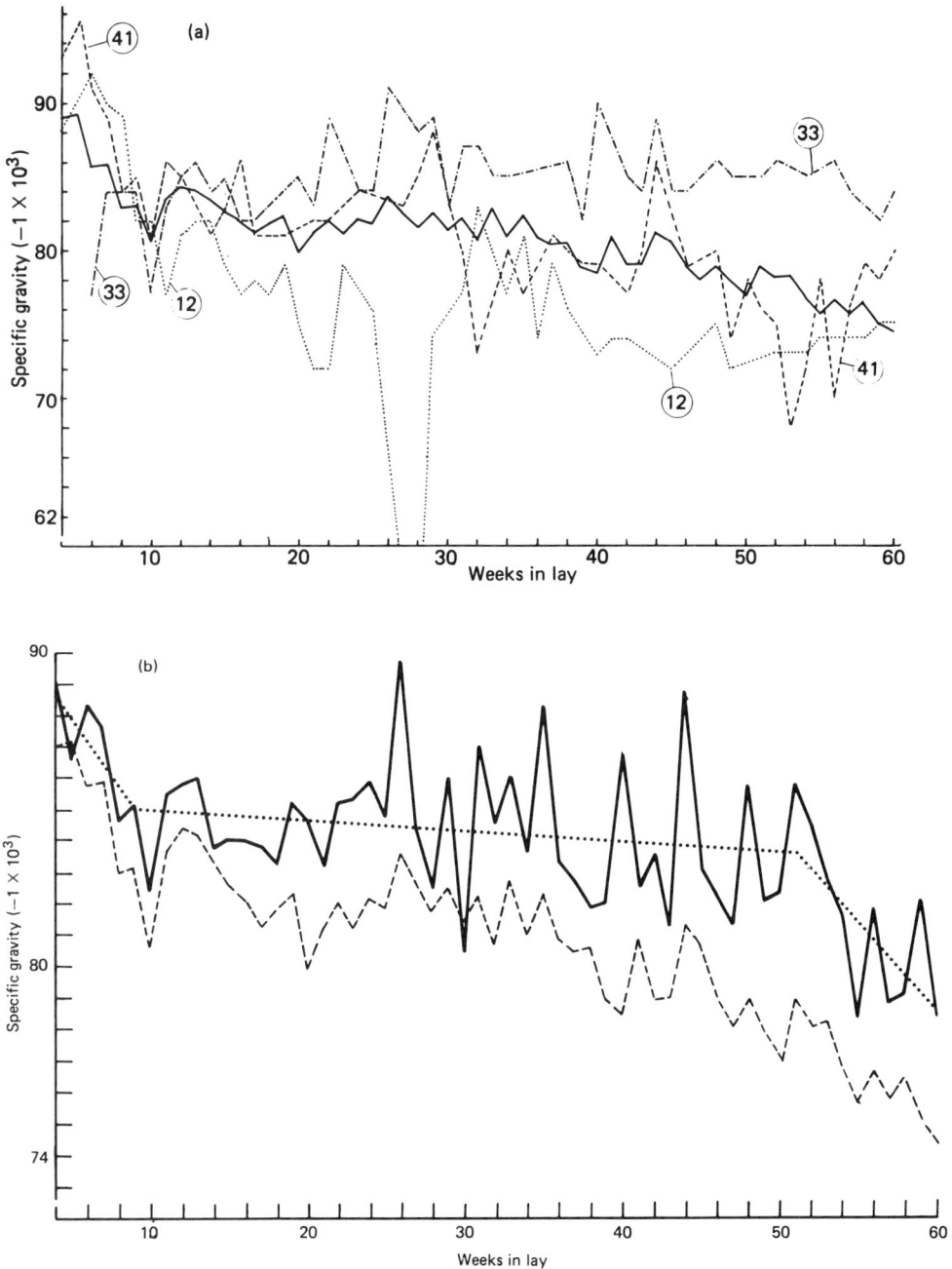

Figure 18.1 Variation in egg specific gravity among individual Warren hens distributed within a commercially-kept flock. (a) For three individuals (as numbered), together with the mean (solid line) of 48 birds. (b) Mean of the five best birds (solid line), together with the best-fit line (dotted) allowing two nodes, compared with the overall mean of 48 birds (broken line). From Belyavin (1979)

Table 18.1 SOME PRODUCTION CHARACTERISTICS OF TEN WARREN LAYING HENS HAVING ABOVE OR BELOW AVERAGE SHELL QUALITY (MEANS FOR 20 TO 80 WEEKS OF AGE)

	Above average birds						Below average birds					
	1	2	3	4	5	Mean	6	7	8	9	10	Mean
Mean food intake (g/bird daily)	111	110	123	125	137	121	122	119	141	115	124	124
Eggs produced	341	306	342	314	335	328	294	269	366	314	329	314
Mean egg mass	52.7	52.4	59.3	64.1	69.3	59.6	57.9	63.5	65.6	62.6	65.8	63.1
Total egg mass produced (kg)	18.0	16.0	20.3	20.1	23.2	19.5	17.0	17.1	24.0	19.7	21.6	19.9
Mean egg specific gravity (-1×10^3)	91	89	88	87	86	88	57	73	73	77	77	71
Mean dry shell mass per egg (g)	5.00	5.02	5.51	5.86	6.21	5.52	2.20	4.49	4.41	4.98	5.06	4.23
Total shell produced (kg)	1.71	1.54	1.88	1.84	2.08	1.81	0.65	1.21	1.61	1.56	1.66	1.34
Shell (mg) as proportion of egg mass (g)	97	98	95	94	92	95	40	73	69	82	79	69

Examples of individuals distributed randomly in a commercially-housed flock under conventional feeding (38 g Ca/kg) and lighting (17 h light/24 h) conditions.

Table 18.2 EFFECT OF POSITION OF THE EGG IN THE SEQUENCE ON ITS MASS AND SHELL QUALITY IN 24 SHAVER 585 HENS IN LATE LAY

	Slope[a]	Difference[b]	
		First egg	Last egg
For individuals:			
mass (g)	-4.59×10^{-3} (2.58×10^{-3})*	1.452 (0.522)**	-1.241 (0.406)***
specific gravity	2.32×10^{-5} (1.61×10^{-5})	-4.1×10^{-3} (1.0×10^{-3})****	5.2×10^{-3} (1.2×10^{-3})****
shell density[c] (mg/cm^3)	2.29×10^{-2} (1.71×10^{-3})	-3.66 (0.920)****	4.77 (1.083)****
For all sequences:			
mass (g)	-3.13×10^{-3} (3.06×10^{-3})	1.553 (0.205)****	-1.089 (0.174)****
specific gravity	3.20×10^{-5} (1.07×10^{-5})***	-4.0×10^{-3} (3.2×10^{-4})****	4.9×10^{-3} (2.0×10^{-4})****
shell density (mg/cm^3)	3.24×10^{-2} (1.22×10^{-2})***	-3.49 (0.262)****	4.43 (0.252)****

*$P<0.1$, **$P<0.05$, ***$P<0.01$, ****$P<0.001$.

Only data for sequences of three or more eggs were analysed and values (with standard errors in parentheses) refer to means of averages for individuals or means of all sequences (867 analysed).

[a]Slope refers to rate of change in egg characteristic with respect to number of the egg in the sequence.

[b]Difference refers to difference in egg characteristic between first or last egg and the mean for that characteristic of the rest of the sequence.

[c]Shell density (DN) from a relationship developed with these birds: DN = $-1098.76 + 1062.58SG + 0.4062M$, where SG is specific gravity and M is egg mass (g).

sequence was more poorly shelled than subsequent eggs. Volynchook, Boorman and Belyavin (1982) studied a much larger number of sequences (867) in 24 individually-caged Shaver 585 hens from 57 to 81 weeks of age in an experimental facility. When data for all sequences were combined there were highly significant positive changes in specific gravity and a derived measure, shell density, with position of the egg in the sequence (*Table 18.2*). It should be stressed that for one sequence the change from egg to egg will be very small and may not be consistent. For their identification these trends require measurement of large numbers of sequences. Much more consistently observed in any sequence however was that the first egg was poorly shelled, relative to the mean of the rest of the sequence, while the last egg was relatively well shelled (*Table 18.2*). There was a tendency towards a decrease in weight of egg with sequence, which could not be identified with statistical significance, although differences between first eggs (heavier) and last eggs (lighter) and the means of the sequences were clearly evident.

Choi *et al.* (1981) measured laying times, egg weight and shell characteristics in 552 eggs from 184, 50-week-old Babcock B-300 hens which laid eggs on three consecutive days. Thus the first day's collection included a proportion of first eggs of sequences, while the subsequent days' collections included none. Trends in egg weight with time of lay showed the heavier nature of first eggs, although there was no clear indication of a decrease in egg weight subsequently in the sequence. Trends in shell characteristics showed that shell mass and proportion of shell (100 × shell mass/egg mass) are not at their minimum values at the start of the day, but decline during the morning before increasing to much higher values in the afternoon. There were steady linear increases on the second and third days. These trends indicate that first eggs of the sequence do not have the poorest shells, as our observations indicate, but that shell quality declines from that of the first egg before improving later. The difference between the two studies cannot be resolved yet; it may be that our simple linear analysis ignored a small curvilinear effect, there may be differences in intensities of such trends due to strain or age (although both flocks were post-peak) and the methods of approach and measurement were different in the two studies. However, both studies show clearly that there are structured changes in shell deposition and egg mass.

Shell formation and calcium supply

Hens have the ability to store extra calcium (and phosphorus) in the skeleton to augment dietary supply during shell formation. This extra calcium, in the form of medullary ('spongy') bone, is deposited especially within the hollow interiors of the limb bones in response to the hormonal changes arising from stimulatory lighting patterns and sexual maturation. It is maintained, more or less, during laying (see Taylor and Dacke, 1984 for a recent review). During laying each egg is formed over a period which, for the average hen in a traditional lighting pattern, will be an hour or two longer than 24 h. Somewhat less than one-quarter of the formation time is spent in the upper tract (see Melek, Morris and Jennings, 1973), while the remainder is spent in the shell gland, although active shelling probably doesn't start until about 3 h after the egg enters the gland (Simkiss and Taylor, 1971). The first egg of a sequence is usually laid about the time of the dark–light change ('lights on'), and thus virtually all of the later shelling of this egg occurs in the following dark period. As the sequence progresses for each egg a greater proportion of shelling will occur after the lights come on. Roland, Sloan and Harms (1973) invoked this gradual change during the sequence as an explanation of the better shell quality later in the day, on the basis that calcium can be derived directly from the diet in the light whereas it must be mobilized from the skeleton in the dark when the bird does not feed. This explanation, its assumptions and consequences are explored further below.

Whether the large proportion of late shelling in the dark is sufficient to explain the special nature of the shell of the first egg of the sequence, or whether there is a more fundamental explanation is not known. It is clear that simple progression through the sequence does not entirely explain the curvilinear response described by Choi *et al.* (1981). The better shell of the last egg is consistent with the large proportion of its shelling which occurs in the light. It is also known that the last egg of the sequence is retained longer in the body—a feature first documented by Fraps (1955) and since incorporated into a model of the ovulatory cycle by Etches and Schoch (1984)—so that there is a longer interval (for example, about 2 h) between penultimate and final ovipositions than between earlier ovipositions. It is tempting to speculate that the extra time is spent in the shell gland and that this is reflected in shell quality. Longer residence in the shell gland is thought to account for the improved shell quality induced by ahemeral patterns of more than 24 h (Melek, Morris and Jennings, 1973). However, there is an inconsistency here; such ahemeral patterns also induce greater egg size, whereas in the case of the last egg of the sequence the size is usually less than that of the rest of the sequence. Indeed, the question of change in size during the sequence has not been much considered: why should the first egg be heavier, why should there be a change as the sequence progresses and what is the nature of the change in mass in terms of egg components?

The idea that the shell is better if calcium can be supplied from the diet directly *via* the gut, whereas it is poorer if calcium is provided from the skeleton, is based on an assumption; there is no direct proof that supplying calcium from the skeleton is less 'efficient' in some way. As described above, however, if the assumption is accepted it explains the gradual change in shell quality within the sequence and would also underlie possible mechanisms for differences between individuals. Scott, Hull and Mullenhorf (1971) suggested that mobilization of calcium from the skeleton might be a limiting factor in shell formation in the dark, to explain their observation that forms of calcium which persist in the gut ('grit') tend to improve shell quality. It is also possible to imagine that some individuals would be more affected by this constraint than others, leading to differences in shell quality.

Volynchook and Boorman (unpublished), have measured plasma total calcium at the end of the dark period, when skeletal mobilization should be most intense, in birds during complete sequences and have not found any correlation with shell mass. There is, however, a large flux of calcium through the plasma during shell formation and plasma concentration may be a poor indicator of rate of delivery to the shell gland. We have also attempted to measure plasma ionic ('free') calcium, but the variation in the measurement is very large, which defeats the object of trying to detect small differences between individuals. A difference between individuals in respect of the ability to mobilize calcium from the skeleton or to utilize such calcium once mobilized cannot therefore be excluded and provision of a form of calcium which persists in the gut should assist shelling in such individuals. It may also be that relationships are more complex and that the phosphorus released during skeletal mobilization is important.

Phosphorus and shell formation

Phosphorus is an important element for shell formation, not because eggshell contains much phosphorus (there is about 100 times as much calcium as phosphorus in eggshell), but because of the special relationship between calcium and phosphorus in bone formation. Calcium is stored in the skeleton probably almost entirely as calcium phosphate and therefore synthesis of medullary bone requires dietary phosphorus. This phosphorus is however involved in an essentially 'futile' process, because if the calcium is used for shell formation the phosphorus must be excreted.

The pattern of change in plasma inorganic phosphate during egg formation has been described by Miller, Harms and Wilson (1977a,b) and Mongin and Sauveur (1979). There are differences of detail between the findings, but they are essentially similar and the pattern found by the latter authors is shown in *Figure 18.2*. In hens receiving a conventional diet (ground limestone included in mash) phosphate starts to increase in the plasma prior to the dark period, increases to a maximum during the dark period and is declining to its normal concentration by the end of the dark period (*Figure 18.2a*). Since skeletal mobilization is necessary to provide calcium during the dark period and release of calcium must be accompanied by release of phosphate, this rising tide of phosphate is to be expected. However, as Mongin and Sauveur rightly emphasized, the start of this increase before onset of the dark period is not explained simply by skeletal mobilization during the dark period. They favoured increased intestinal absorption of phosphate as the mechanism of this initial increase. Mongin and Sauveur also showed that if a persistent source of

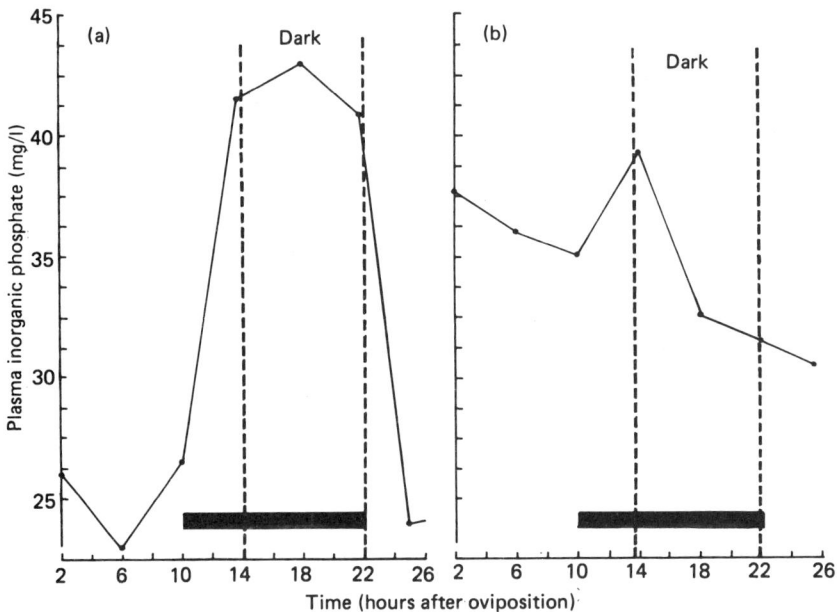

Figure 18.2   Plasma phosphate in relation to shelling (represented by solid bar) where birds were fed on a conventional diet (a) or a calcium-low diet with access to calcareous grit (b). Redrawn after Mongin and Sauveur (1979), and reproduced with permission

calcium ('sea shell' grit) was made available to hens, the response in plasma phosphorus was modified (*Figure 18.2b*). In this condition the initial increase in plasma phosphorus occurred, demonstrating its inevitability, but enhancement during the dark period did not occur and the concentration declined rapidly. This the authors explained on the basis that the continuing supply of calcium from the gut would obviate the need for skeletal mobilization.

One other feature commented upon by these authors and reflected in our own findings was the marked variation there was in the plasma phosphate response among individuals.

Nutritional interest in phosphorus has been stimulated by the several observations that dietary excesses of this element have a detrimental effect on shell quality (Arscott *et al.*, 1962; Taylor, 1965) and Harms and colleagues have studied this phenomenon intensively recently (*see* Harms, 1982a and b). It is not clear whether this phosphorus excess, by accumulating in the blood, interferes with mobilization of skeletal reserves of calcium phosphate during shelling in the dark or whether there is a direct antagonistic effect of the blood phosphorus on the shelling process. Whatever the mechanism however, there is no doubt that dietary treatments which lead to increases in plasma phosphate cause a decline in egg specific gravity. Miles and Harms (1982) showed a clear linear negative correlation between specific gravity and plasma phosphate over a range of treatments.

The involvement of phosphorus in shell formation is now thought to underlie the sometimes beneficial effect of dietary sodium bicarbonate on shell formation (Howes, 1966; Harms, 1982b; Washburn, 1984). Earlier interpretation of this effect concentrated more on possible effects of the bicarbonate ion in relation to the carbonate need for shell, but more recently the effect of the addition of the sodium ion has been seen as important. The data of Miles and Harms (1982) show clearly that dietary treatments without added sodium bicarbonate produce higher plasma phosphate concentrations (with consequent effects on specific gravity) than comparable treatments with added bicarbonate. It is assumed that the sodium exerts its effect by facilitating the renal loss of phosphate, as a balancing cation.

Unfortunately, involvement of sodium in calcium–phosphate relationships cannot be studied in isolation from the whole subject of ion ('acid–base') balance, especially with respect to dietary Na^+, K^+ and Cl^-. There are limits to the amounts of one of these ions which can be included in the diet without adjustment to the intakes of the other ions. Thus dietary additions of $NaHCO_3$ represent additions of Na^+ without balancing additions of Cl^-. A high $Na^+ : Cl^-$ ratio, especially at very low chloride intakes can cause severe overall effects on laying hens and their production (Junqueira *et al.*, 1984), while high dietary chloride, without concomitant increases in sodium, can have a deleterious effect on shell quality (see Harms, 1982b). The proposed effect of sodium, as $NaHCO_3$, in decreasing blood phosphorus cannot necessarily be reproduced by additions of sodium with other anions (Harms, 1982b; Washburn, 1984).

It is evident that the nature and extent of any effect of dietary sodium bicarbonate on shell quality will depend on the extent to which phosphorus is in excess, and whether the sodium bicarbonate can be added without disturbances to the overall ion balance ($Na^+ + K^+ - Cl^-$; see Saveur and Mongin, 1978). Ionic balance in laying hens is a complex subject (Sauveur and Mongin, 1978) and Washburn (1984) has cautioned against too simple an interpretation of the relationship between sodium and phosphate. Gross disturbances of ionic balance will affect egg production severely, while Harms (1982b) has pointed out that smaller effects seen on shell quality should be interpreted on the basis of the effects of other ions on calcium and phosphorus metabolism.

In studies on individual hens Volynchook (unpublished) measured plasma ions in ten 60-week-old Hubbard hens, fed on a conventional diet. Plasma was sampled about 30 min before the end of the dark period, this time being chosen as reflecting the period when the contribution of the skeleton would be clearest and the contribution of the diet would be absent. Hens were sampled for complete sequences. Multiple regression analysis indicated a significant negative relationship between shell weight and plasma phosphate concentration. When the relationship between plasma phosphate and shell weight of the first egg of the sequence was examined it was much more marked (*Figure 18.3*). These data indicate that, for first eggs in these birds at least, at low plasma phosphate (0.4 mM) shell mass is independent of phosphate concentration, but as plasma phosphate increases a decline in shell mass occurs. It must be stressed that these birds were all receiving the same diet containing about 6.5 g total phosphorus and about 4.5 g available phosphorus with 38 g calcium per kg. These variations therefore represent

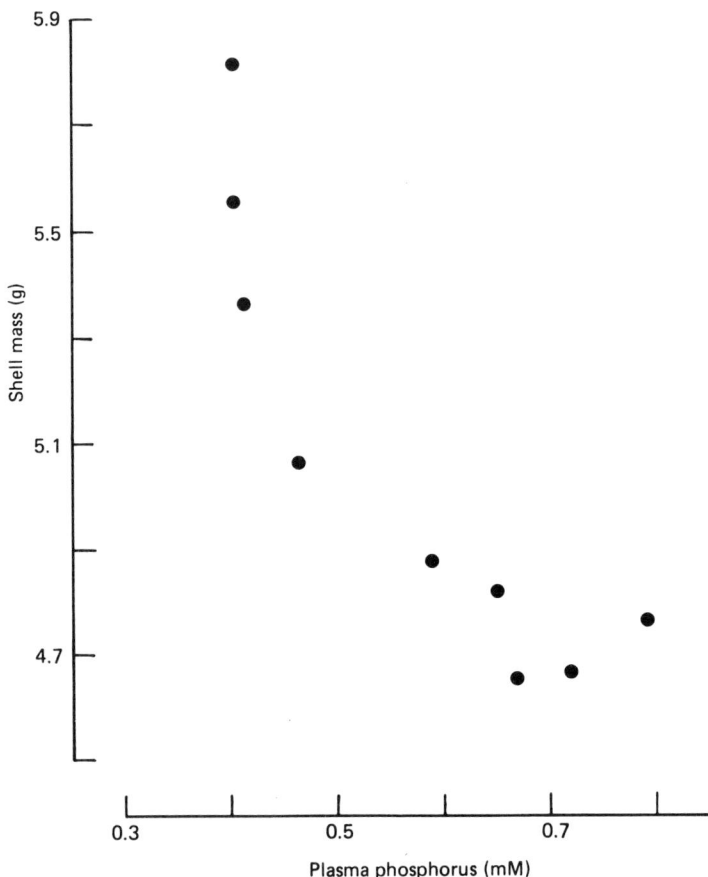

Figure 18.3   Relationship between shell mass of first egg of the sequence and plasma phosphorus concentration at the end of the dark period in Hubbard hens in late lay receiving a diet containing about 4.5 g available P and 38 g Ca/kg

variations between individuals (see above) and the relationship shown indicates that some of the variation among individuals in respect of shell mass is explained by variations in plasma phosphate.

These data may be interpreted in two ways. One interpretation would be based on the assumption that a higher plasma phosphorus 'causes' a weaker shell and that the problem in such individuals is the elimination of excess phosphorus. The other interpretation would be based on the assumption that some individuals require larger amounts of calcium to produce a shell, because, possibly, they are inefficient utilizers of calcium, and the high plasma phosphorus is an indication of the extent of skeletal mobilization necessary in such cases. The former interpretation suggests that shell mass is the dependent variable while plasma phosphorus is the independent, while the latter suggests the converse. Unfortunately, correlative data of this type do not allow distinction of these alternatives.

Finally it must be emphasized that these observations were made on very few hens and data for one were excluded as atypical. There is an evident need to confirm these observations and extend measurement to include younger birds.

Vitamin D

The eluciation of the metabolism of vitamin D and the role of this vitamin in calcium and phosphorus metabolism has been one of the phenomena of nutritional research over the last two decades. The subject is reviewed frequently and Taylor and Dacke (1984) and Soares (1984) have summarized salient features of relevance to poultry. These will not be repeated here. Assuming that ample vitamin D is included in the diet, is there any evidence that its metabolism might sometimes limit its effectiveness? Soares (1984) reported an investigation of the plasma concentration of the active metabolite, 1,25-dihydroxycholecalciferol (1,25 $(OH)_2D_3$), in lines of hens selected for production of thick or thin shells. There was a higher concentration of the metabolite in the plasma of birds producing thick shells. There are also observations which suggest insufficient production of 1,25 $(OH)_2D_3$ in late lay when shell quality is likely to decline (see Taylor and Dacke, 1984). Unfortunately the feeding studies on the effect of vitamin D metabolites on shell quality have been concerned with 25-hydroxycholecalciferol, the substrate for 1,25 $(OH)_2D_3$, which is unlikely to compensate for a deficiency of the latter because it seems that decline in the activity of the enzyme involved (1-hydroxylase) is more likely to be the limiting factor.

This area may be another cause of variation between individuals or may be one linked to those already discussed. The relationship between vitamin D metabolism and individuals in late lay seems worthy of further investigation, although techniques for measurement of vitamin D metabolites are complex.

Implications

VARIATION AND EXPERIMENTATION

Even allowing that the original examination of variation among individuals reported in an earlier section was confounded by structured variation within sequences, there is considerable variation in patterns of change in shell quality among individuals. There are some individuals that produce poor shells for large proportions of the laying year. This raises the question of whether these birds are more likely to respond to remedial treatments such as higher intakes of calcium or separate calcium feeding. This remains to be tested, but it should be noted that eggs from such birds will be a small proportion of those in a random sample, a situation exacerbated, if, as can be the case, the birds involved are also poor producers. If the effects of a remedial treatment, therefore, are tested on means of random samples of eggs, detection of response is likely to be difficult.

The structured variation in shell quality during the sequence imposes further constraints. Ideally, eggs similar in their position in the sequence must be compared in the evaluation of treatments. This means that eggs must be related to the individuals which laid them and those individuals must be sampled for complete sequences or large parts thereof. This applies not only to studies of remedial treatments, but also to correlation studies for the selection of individuals in breeding experiments. Although this seems to demand much more monitoring and effort in experiments, it is likely that understanding this structured variation and using individuals, possibly in cross-over designs, will allow fewer birds to be used in experiments. The approach of Choi *et al.* (1981) could be extended as a compromise. If eggs from individuals were collected for five days, eggs collected during the middle three days of the period could be characterized at least as far as first eggs, last eggs and 'mid-sequence' eggs.

CALCIUM SUPPLY

It is well known that calcium need for maximizing production rate in hens is less than that for maximizing shell quality (see for example, ARC, 1975). An intake of about 3.5 g calcium/day might satisfy the former, while responses in the latter occur up to 5 g/day, although the response above 4 g/day is small and diminishing. With respect to feeding practice, the use of discrete sources of calcium in combination with a low-calcium mash has not recommended itself much in this country. Evidence from empirical trials is equivocal (Belyavin and Boorman, 1981), possibly for reasons of experimental design (*see above*), but the theoretical basis for such practice is sound. In providing a dietary source of calcium during much of shelling this should obviate much of the need for skeletal mobilization. This might improve the situation in birds where such mobilization is a limiting factor, has implications for the unwanted accumulation of phosphate (*see below*), and should reduce the variation within the sequence since this effect is seen as partly the result of the proportion of shelling which is dependent upon such mobilization (*see above*). Limited experience with some birds late in lay does suggest exercise of some caution in this respect. Contrary to general experience it was noted that a significant minority of birds offered low-calcium mash and oyster shell failed 'to eat for calcium' and ceased laying (Volynchook, unpublished). This observation was made with three different groups of Hubbard hens and incidence increased with age. It seems that in late lay (about 60 weeks of age) the 'drive' to lay may not be strong enough to cause a bird to eat an unpalatable source of calcium.

An alternative way of effecting more continuous supply of dietary calcium is to use modified lighting patterns. Sauveur and Mongin (1983) showed substantial improvements in shell quality (7.7 per cent increase in shell weight in conventionally-reared birds) associated with 6 h repeating light–dark cycles. The authors postulated that this effect was due to the more regular supply of dietary calcium allowed by the repeated short light periods. Such patterns do however allow the adoption of ahemeral cycles and the improvement in shell quality is of the same order as that reported in response to 27 or 28 h conventional ahemeral patterns (Melek, Morris and Jennings, 1973; Yannakopoulos and Morris, 1979). It is notable that the effects on egg production rate and egg weight observed by Sauveur and Mongin (1983) using repeating short cycles were typical of those observed in conventional ahemeral patterns of longer than 24 h. Possibly therefore eggshell quality benefits from the longer residence in the shell gland typical of such ahemeral patterns and the supply of calcium is not the main effect. Raine (1984) has commented recently that nutritional treatments can be expected to produce improvements in shell quality of about 1 to 2 per cent, while lighting patterns may produce much larger improvements such as those described above. The interaction between lighting patterns and nutritional treatments has not been much studied and there is a need for the exploration of potential in this direction.

PHOSPHORUS AND ITS INTERACTIONS

Nutritional and economic considerations demand that careful attention be paid to the phosphorus requirement of the laying hen. As with any nutrient there will be a minimum demand for the productive process, but as discussed earlier, there may be deleterious effects on the shell from relatively small excesses. One of the complicating difficulties is the availability of phosphorus from plant sources; total phosphorus has little meaning when comparing diets differing greatly in ingredient composition, while the generality of published values for available phosphorus from ingredients is not known. Current assessments of needs for hens (Harms, 1982b; Raine, 1984; Said *et al.*, 1984) in agreement with earlier assessments (ARC, 1975; Belyavin, 1979) suggest that the requirement for maintaining production and shell quality in caged birds is about 400 mg available phosphorus/day. In experimental work significant decline in shell quality has not usually been observed until intakes of available phosphorus are in excess of 500 mg/day, often considerably. There have been reports to the contrary however; Ousterhout (1980) showed a significant decline in shell quality when increasing dietary available phosphorus from about 3.5 to 4.5 g/kg.

The relationship between plasma phosphorus and shell quality means that nutritional and management practices tending to minimize the former will tend to have a beneficial effect on shell quality. In this context correct ionic balance is important, but it is not clear how large a margin for error there is in the recommendation of 200 mEq $(K^+ + Na^+ - Cl^-)$/kg diet (Sauveur and Mongin, 1978) usually used. Two aspects are of especial note in respect of shell formation. First, strongly acidotic diets (eg. high Cl^-) are likely to cause poor shells and second, sodium bicarbonate can decrease high plasma phosphorus. High concentrations of chloride, without the balancing ion sodium, are unlikely to occur in natural diets, but since such diets always contain excess potassium, additions of sodium bicarbonate without compensatory decrease in sodium from other sources will alter the balance $K^+ + Na^+ - Cl^-$. In light of this complexity it is not surprising that results of experiments on additions of sodium bicarbonate are often equivocal.

Another approach to minimizing plasma phosphorus during shelling is to minimize skeletal mobilization of calcium and phosphate. It is clear from the data of Mongin and Sauveur (1979) that feeding calcareous grit can prevent the large enhancement of plasma phosphorus usually seen during the dark period (*see Figure 18.3*). If skeletal stores are not depleted, their replenishment will not be necessary and the need for phosphorus in this role will largely disappear. Thus, as Mongin and Sauveur pointed out, discrete persistent sources of calcium should decrease the need for phosphorus. If short repeated light–dark cycles also allow more continuous supply of calcium from ˙the gut and thereby prevent skeletal mobilizatiori, they should have the same effect on phosphorus requirement. Possibly, therefore, in such treatments a small excess of phosphorus is being supplied and further improvements in shell quality would accrue from decreasing dietary phosphorus.

Results on individual birds, described above, show that some of the variation among individuals in shell quality is due to individual variation in phosphorus metabolism. Thus even at normal dietary phosphorus content, some birds are showing effects of phosphorus 'excess'. This is assumed to arise because of skeletal mobilization primarily, and it may be predicted that treatments which minimize such mobilization would tend to decrease the variation among individuals in shell quality.

ACKNOWLEDGEMENTS

Our studies reported herein were supported by grants from the British Egg Marketing Board Research and Education Trust and the Agriculture and Food Research Council.

References

AGRICULTURAL RESEARCH COUNCIL (1975). *The Nutrient Requirements of Farm Livestock. No. 1, Poultry*. Agricultural Research Council; London

ARSCOTT, G.H., RACHAPAETAYAKOM, P., BENIER, P.E. and ADAMS, F.W. (1962). *Poultry Science*, **41**, 485–488

ATWOOD, H. (1929). *Poultry Science*, **8**, 137–140

BELYAVIN, C.G. (1979). *Egg-shell quality in the older hen*. PhD thesis. University of Nottingham

BELYAVIN, C.G. and BOORMAN, K.N. (1981). In *Quality of Eggs*, pp. 165–174. Ed. G. Beuving, C.W. Scheele and P.C.M. Simons. Spelderholt Institute for Poultry Research; Beekbergen, The Netherlands

BERG, L.R. (1945). *Poultry Science*, **24**, 555–563

BEUVING, G., SCHEELE, C.W. and SIMON, P.C.M. (1981). *Quality of Eggs*. Spelderholt Institute for Poultry Research; Beekbergen, The Netherlands

BUSS, E.G. and GUYER, R.B. (1981). In *Quality of Eggs*, pp. 239–249. Ed. G. Beuving, C.W. Scheele and P.C.M. Simons. Spelderholt Institute for Poultry Research; Beekbergen, The Netherlands

CHOI, J.H., MILES, R.D., ARAFA, A.S. and HARMS, R.H. (1981). *Poultry Science*, **60**, 824–828

ETCHES, R.J. and SCHOCH, J.P. (1984). *British Poultry Science*, **25**, 65–76

FRAPS, R.M. (1955). In *Progress in the Physiology of Farm Animals*, Vol. II, pp. 671–740. Ed. J. Hammond. Butterworths; London

HARMS, R.H. (1982a). *Feedstuffs*, May 10, pp. 25–26

HARMS, R.H. (1982b). *Feedstuffs*, May 17, pp. 25–28

HOWES, J.R. (1966). *Poultry Science*, **45**, 1092–1093

HURWITZ, S., BAR, A. and COHEN, I. (1973). *American Journal of Physiology*, **225**, 150–154

JUNQUEIRA, O.M., COSTA, P.T., MILES, R.D. and HARMS, R.H. (1984). *Poultry Science*, **63**, 123–130

MELEK, O., MORRIS, T.R. and JENNINGS, R.C. (1973). *British Poultry Science*, **14**, 493–498

MILES, R.D. and HARMS, R.H. (1982). *Poultry Science*, **61**, 175–177

MILLER, E.R., HARMS, R.H. and WILSON, H.R. (1977a). *Poultry Science*, **56**, 586–589

MILLER, E.R., HARMS, R.H. and WILSON, H.R. (1977b). *Poultry Science*, **56**, 1501–1503

MONGIN, P. (1968). *World's Poultry Science Journal*, **24**, 200–230

MONGIN, P. and SAUVEUR, B. (1979). *British Poultry Science*, **20**, 401–412

OUSTERHOUT, L.E. (1980). *Poultry Science*, **59**, 1480–1484

RAINE, H. (1984). *Poultry World*, 29 November, pp. 10–11

ROLAND, D.A., SLOAN, D.R. and HARMS, R.H. (1973). *Poultry Science*, **52**, 506–510

SAID, N.W., SULLIVAN, T.W., SUNDE, M.L. and BIRD, H.R. (1984). *Poultry Science*, **63**, 2007–2019

SAUVEUR, B. and MONGIN, P. (1978). *British Poultry Science*, **19**, 475–485

SAUVEUR, B. and MONGIN, P. (1983). *British Poultry Science*, **24**, 405–416

SCOTT, M.L., HULL, S.J. and MULLENHORF, P.A. (1971). *Poultry Science*, **50**, 1055–1063

SIMKISS, K. and TAYLOR, T.G. (1971). In *Physiology and Biochemistry of the Domestic Fowl*, Vol. 3, pp. 1331–1343. Ed. D.T. Bell and B.M. Freeman. Academic Press; London

SOARES, J.H. (1984). *Poultry Science*, **63**, 2075–2083

TAYLOR, T.G. (1965). *British Poultry Science*, **6**, 79–87

TAYLOR, T.G. and DACKE, C.G. (1984). In *Physiology and Biochemistry of the Domestic Fowl*, Vol. 5, pp. 125–170. Ed. B.M. Freeman. Academic Press; London

VOLYNCHOOK, J.G., BOORMAN, K.N. and BELYAVIN, C.G. (1982). *World's Poultry Science Journal*, **38**, 138

WASHBURN, K.W. (1984). In *Proceedings and Abstracts of 17th World's Poultry Congress and Exhibition, Helsinki (Finland)*, pp. 43–46

WILHELM, L.A. (1940). *Poultry Science*, **19**, 246–253

YANNAKOPOULOS, A.L. and MORRIS, T.R. (1979). *British Poultry Science*, **20**, 337–342

19

INFLUENCE OF NUTRITIONAL FACTORS ON HATCHABILITY

R.A. PEARSON
Agricultural Research Council, Poultry Research Centre, Roslin, Midlothian, Scotland

Introduction

Unlike the mammalian embryo/fetus, which obtains a continuous supply of nutrients from its mother during development, the avian embryo has to rely on a discrete package of nutrients—the egg. If the embryo is to develop into a viable chick, the egg must contain all of the nutrients necessary for growth and development.

Given an adequate supply of nutrients, a disease-free hen in a thermo-neutral environment will produce eggs of a relatively constant composition and physical state. If these eggs are fertile and incubated correctly, then most will hatch after 21 days to produce a viable chick. Any variations in this 'system' will lead to a reduction in the chances of a viable chick being produced. When poultry rations for breeding birds are being formulated it is necessary to understand the extent to which the diet can influence the composition of the eggs produced, as this will directly affect hatchability and subsequent viability of the chicks produced. In this chapter the effects of nutrition on hatchability of fertile eggs will be considered. It is not intended to discuss ways in which nutrition influences fertility, although this will also affect the number of chicks produced by a breeding hen.

The role of nutrition in hatchability is a complex one, affected by many components in the environment which interact to determine the fate of the fertile egg. The effect of nutritional factors on the ability of the hen to produce eggs that will give rise to viable offspring will be considered and current recommendations will be discussed. This account is concerned mainly with the chicken, because it has received most attention by researchers, but many of the principles discussed are also applicable to the turkey and the duck.

The composition and size of hatching eggs

The composition of an unincubated hen's egg weighing 60 g is given in *Figure 19.1* and *Table 19.1*. Egg size does not appear to markedly affect the

First published in *Recent Advances in Animal Nutrition – 1982*

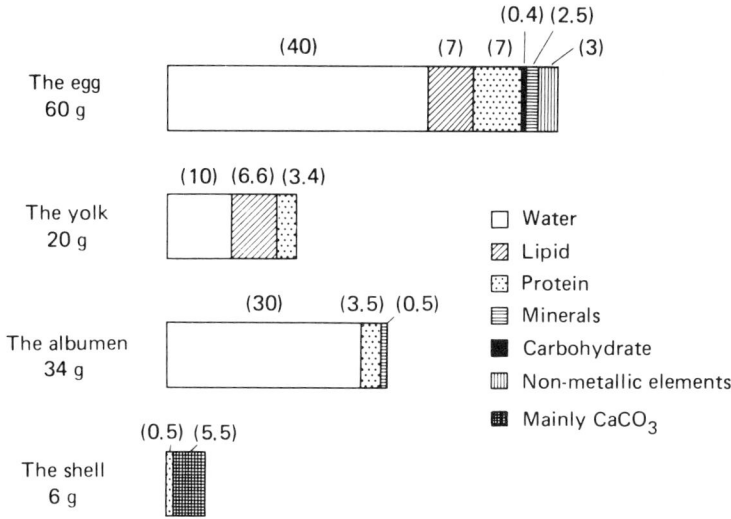

Figure 19.1 The composition of the hen's egg

hatching success of turkey and chicken eggs (Landauer, 1973; Reinhart and Moran, 1979); however, it can influence subsequent chick performance. The size of a chick at hatching is positively correlated with egg weight (e.g. Skoglund and Tomhave, 1949; Gardiner, 1973), and most of any residual variation in chick weight is accounted for by differences in weight loss from eggs during incubation (Tullett and Burton, 1982). Studies of the relationship of egg size and chick weight to subsequent progeny growth and viability have produced varied results. Al-Murrani (1978) and McNaughton *et al.* (1978) found marked effects of chick weight on performance and viability, but other researchers have observed only small changes (e.g. Wiley, 1950; Kosin *et al.*, 1952; Gardiner, 1973; Proudfoot and Hulan, 1981). The manner in which chick weight is achieved, however, has not been generally considered and may account for the differences reported

Table 19.1 VITAMIN AND MINERAL NON-SHELL CONTENTS OF A TYPICAL UNINCUBATED HEN'S EGG WEIGHING 60 g (Data from Bolton, 1961)

Vitamins	Range	Minerals	Range
Vitamin A (i.u.)	0.5–416	Ca (mg)	32
Vitamin D (i.u.)	5.0–880	P (mg)	118
Vitamin E (µg)	297–1400	Mg (mg)	30
Vitamin K (µg)	44	Na (mg)	70
Thiamin (µg)	38–340	K (mg)	76
Riboflavin (µg)	76–300	Cl (mg)	70
Nicotinic acid (µg)	32–38	S (mg)	98
Pantothenic acid (µg)	464–810	Fe (mg)	1.0 ·
Biotin (µg)	8.6–14.0	Cu (mg)	0.05–0.33
Vitamin B$_{12}$ (µg)	130–730	I (mg)	0.004–0.01
Folic acid (µg)	4.2–4.4	Mn (mg)	0.005–0.025
Pyridoxine (µg)	130–150	Zn (mg)	0.7–1.0
Choline (mg)	238–300		

(Tullett and Burton, 1982). Data from O'Neil (1955) indicate that broiler chicks representing a larger percentage of the fresh egg weight (over 69 per cent) are generally heavier at 6 weeks of age and have a lower mortality.

Energy

It is difficult to establish specific effects of maternal dietary energy intake on hatchability. There is some evidence that commercial layers can adjust their voluntary food intake to meet their daily energy requirements (Morris, 1968; Cherry, 1979), although the adjustments in intake which are made as the energy content of a diet changes may not be very precise. Gardner and Young (1972) observed that yolk weight increased relative to other components of the egg when the energy content of a commercial laying hen diet increased. This suggests that energy intake increased as the density of the diet increased, resulting in an increase in the amount of lipid available for production of yolk lipoproteins.

In formulating diets for *ad libitum* consumption, nutrients are generally added in relation to the energy content of the diet. When this is not the case, it is conceivable that voluntary food intake on high-energy rations could be so reduced as to cause a deficiency in intake of a specific nutrient. In breeder diets, where the accumulation of nutrients in the egg is important, this can have drastic effects on hatchability. Broiler breeders, unlike commercial layers, rapidly become obese if allowed food *ad libitum*. However, the excessive energy consumption that occurs under these circumstances has no noticeable effect on hatchability, although it can have striking effects on fertility.

Table 19.2 THE EFFECT OF DAILY ENERGY AND PROTEIN ALLOWANCE ON HATCHABILITY (% FERTILE EGGS) OF EGGS LAID BY BROILER BREEDER HENS FROM 22 TO 36 WEEKS OF AGE (Data from Pearson and Herron, 1981)

Daily allowance	Protein (g/bird)	
	27.0	21.0
Energy (MJ AME)		
1.88	79.8	79.3
1.73	78.1	81.0
1.52	73.1[a]	80.1

[a]Significantly (P<0.01) different from other groups.

Overconsumption of energy by broiler breeders is prevented by regulation of nutrient intake, and quantitative food regulation is now an integral part of most broiler breeder management systems. In general, over-restriction of energy intake during lay will lead to a decrease in egg output before any effects on hatchability are observed. However, on regulated feeding systems the balance between energy and protein intake can have a marked effect on hatchability. When the ratio between protein and energy intake is high, hatchability can be depressed (*Table 19.2*); this is particularly noticeable when egg output is high and energy intake is low.

Protein

The essential amino acids need to be present in sufficient quantity in the egg if the embryo is to develop into a viable chick capable of hatching (Freeman and Vince, 1974). Deficiencies or imbalances of these amino acids in the egg are therefore detrimental to embryonic development.

The laying hen obtains the essential amino acids required in the synthesis of egg protein from the diet. Tissue protein reserves are limited (Harms *et al.*, 1971), and in the long term overall dietary deficiencies of amino acids or protein are associated with reductions in egg size and production. In cases of marginal deficiency only a reduction in egg size is observed, but in a severe deficiency egg production may cease altogether. When protein intake is reduced or a particular amino acid is limiting, the amino acid composition of total egg protein (Evans *et al.*, 1950; Ingram *et al.*, 1951; Lunven *et al.*, 1973), albumen (*Table 19.3*) and egg shells (Butts and Cunningham, 1973) remains unchanged. However, since the total amount of protein synthesized is reduced, an overall decrease in the amount of each amino acid in the egg results. This makes it difficult to assess the direct effects of protein and amino acid intake on hatchability.

Free amino acids account for about 3 per cent of the nitrogenous fraction of the egg and, unlike egg protein, the quantities of these amino acids seem to be a function of amounts present in the diet. Dietary deficiencies of methionine or lysine or both were associated with changes in the contents of all the free amino acids in yolk (Larbier *et al.*, 1972). It was suggested that the balance of amino acids in the diet influenced the renewing of tissue protein and from that the concentration of free amino acids in egg yolk. There is apparently no evidence to suggest that changes in free amino acids in the egg have a detrimental effect on hatchability, although they may influence post-hatching growth rates (Blum *et al.*, 1979).

The response to a particular amino acid or protein intake is often complicated by other factors, and indirect effects on egg formation may be observed. Excess dietary protein seems to increase the requirements for vitamin B_{12} (Patel and McGinnis, 1977) and biotin (Whitehead and Bannister, 1981), and may also influence the requirements for other essential nutrients (Pearson and Herron, 1982), such as calcium (Linkswiler *et al.*, 1981).

Lipid

Yolk lipid is the main source of energy for the developing embryo, accounting for between 84 and 98 per cent of the material oxidized. Synthesis of yolk lipoprotein occurs in the hen's liver, mainly from the breakdown products of dietary carbohydrate. However, some dietary lipid may be incorporated, and, in adverse dietary conditions, lipid-mobilized adipose tissue may also be used. It is generally assumed that the relative composition of the major yolk components, apart from fatty acids, is unlikely to be affected by dietary lipids, particularly as uptake of these substances into yolk seems to require a specific mechanism, the coated vesicles (e.g. Perry and Gilbert, 1981). Therefore, irrespective of liver

Table 19.3 THE EFFECT OF DL-VALINE (VAL), DL-ISOLEUCINE (ISO), DL-THREONINE (THR) AND L-ARGININE (ARG) SUPPLEMENTATION OF A 10% PROTEIN DIET ON THE AMINO ACID CONTENTS OF EGG ALBUMEN (Data from Kashani et al., 1980)

Added amino acids (% of diet)				Plasma total essential AAs (μmol/100 ml plasma)	Total essential AAs	Amino acid content of albumen (g/kg albumen)						
VAL	ISO	THR	ARG			METH	CYS	ISO	VAL	LYS	ARG	THR
–	–	–	–	136	6.10	4.3	2.6	5.9	7.7	7.7	6.6	5.2
0.2	–	–	–	151	5.70	3.9	2.9	5.6	7.2	7.5	6.3	4.8
0.2	0.2	–	–	148	6.32	4.5	2.8	6.0	8.0	8.3	7.0	5.2
0.4	0.2	–	–	157	5.97	4.1	2.7	5.8	7.3	8.3	6.5	5.0
0.4	0.2	0.2	–	139	6.29	4.6	2.8	6.2	7.9	7.8	6.9	5.3
0.4	0.2	0.2	0.3	158	5.42	3.8	2.2	4.9	6.7	7.3	6.0	4.7

function, at most only the total quantity entering the oocyte appears capable of regulation at the ovarian level. Experimental evidence tends to support this view. When Vogtman and Clandinin (1975) fed diets containing rapeseed oil low in erucic acid, refined regular rapeseed oil or tallow to laying hens, they observed that the various dietary treatments did not affect total lipid content of the yolk but did affect fatty acid composition of yolk lipids. Even high-fat diets do not appear to change total yolk lipid, although the proportion of egg-yolk fatty acids can often be related to the relative concentration of fatty acids in dietary triglyceride (Chen *et al.*, 1965; Sell *et al.*, 1968; Naber, 1979).

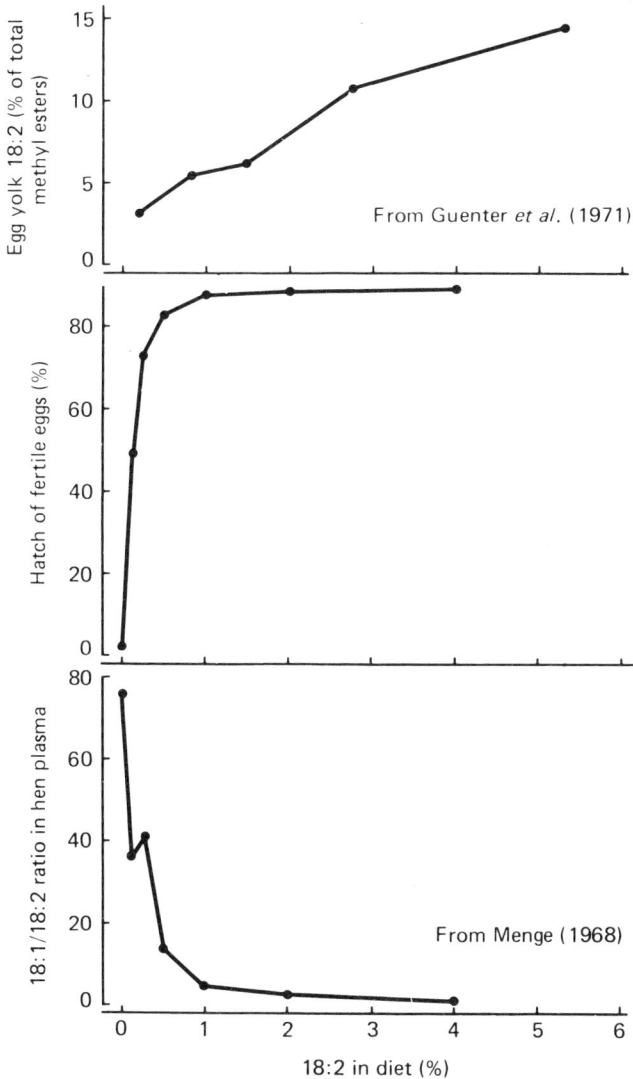

Figure 19.2 Relationship between linoleic acid in the diet and egg yolk and hatchability

Some fatty acids, namely linoleic acid and arachidonic acid, are essential in rations for breeding birds if they are to produce viable chicks. Linoleic acid is stored in the body for some time, so that deficiencies are not always immediately apparent. However, the amounts of these essential fatty acids found in yolk are ultimately related to the intake in the diet (*Figure 19.2*), and dietary deficiencies of linoleic acid are associated with a reduction in egg output and lipid content of the yolk and an increase in early embryo mortality (Menge *et al.*, 1965; Calvert, 1967; Menge, 1968; Guenter *et al.*, 1971). In a severe deficiency linoleic acid and arachidonic acid could not be detected in the yolk and hatchability was reduced to zero (Menge *et al.*, 1965).

In contrast to the main lipoproteins, there are minor components in the yolk (cholesterol, fat-soluble vitamins, carotenoid pigments: Gilbert, 1971) which appear to be directly related to the amounts present in the diet. High dietary levels of cholesterol cause increased plasma and yolk cholesterol concentrations (Chung *et al.*, 1965).

It seems likely, therefore, that some substances such as cholesterol accumulate in the oocyte incidentally with other components because the mechanisms for the uptake of the main components of yolk are not selective.

Vitamins

Jeroch (1971a–c, 1972), in a series of experiments on riboflavin require-ments, was able to show a relationship between dietary intake of a vitamin and the production of viable offspring. Chick growth and development during the first week of age was influenced by the riboflavin obtained from the egg, which, in turn, depended on the parental riboflavin (*Figure 19.3*). Similar relationships between dietary intake of a vitamin, vitamin content of the egg and subsequent hatchability and/or viability of the offspring have been observed for vitamins A (Hill *et al.*, 1961), D (Bethke *et al.*, 1936), E (Bartov *et al.*, 1965) and K (Griminger, 1964); pantothenic acid (Snell *et al.*, 1941; Beer *et al.*, 1963); nicotinic acid (Ringrose *et al.*, 1965); pyridoxine (Fuller *et al.*, 1961; Weiss and Scott, 1979); folic acid (Sunde *et al.*, 1950); and biotin (Brewer and Edwards, 1972). Many of the vitamins cannot be stored; thus, it is not surprising that the amount available for incorporation in the egg is dependent on dietary intake. Specific carrier proteins have been shown for thiamin (Muniyappa and Adiga, 1979), riboflavin (Farrell *et al.*, 1970), vitamin B_{12} (Sonneborn and Hansen, 1970) and biotin (White *et al.*, 1976), and it seems likely that carrier proteins, as yet unidentified, may exist for other vitamins. The carrier proteins appear to be a mechanism for ensuring that a sufficient quantity of each vitamin will enter the egg, given an adequate dietary supply.

Minerals

Many inter-relationships exist among the various inorganic elements, and between these elements and vitamins, amino acids and dietary fat. This

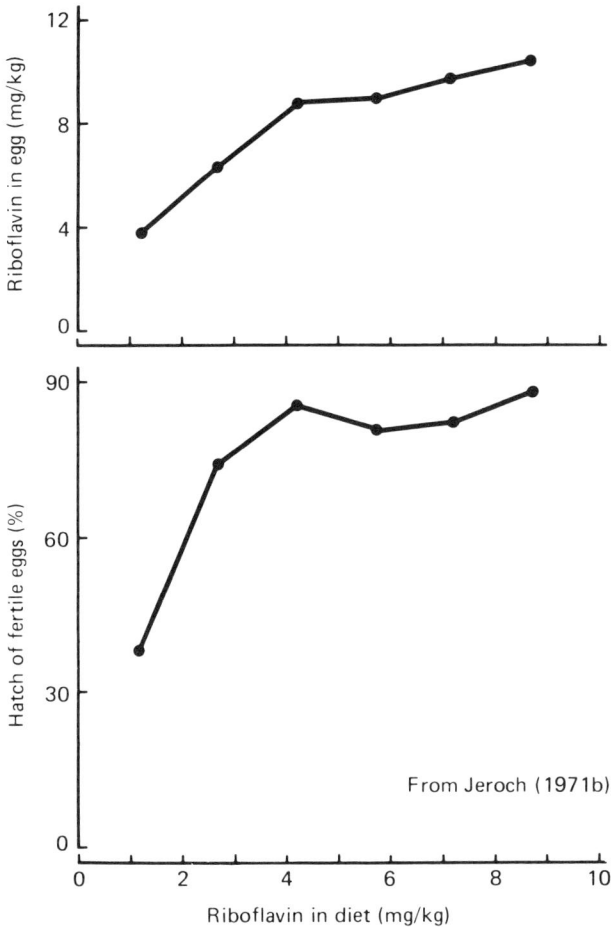

Figure 19.3 Relationship between riboflavin in the diet and egg and hatchability

makes it difficult to establish whether a response to a specific mineral intake is a direct one, a secondary consequence of an effect elsewhere or the result of an interaction which has influenced the availability of the element. Hence, mineral requirements for hatchability can be difficult to establish.

Most ions—in particular, sodium, potassium and chloride—have widespread effects throughout the body, and consequently there are very efficient regulatory mechanisms to maintain plasma ionic balance under normal dietary intakes. However, it appears that when marked dietary deficiencies or excesses of some elements are present, the homoeostatic mechanisms can break down and plasma levels of the mineral decrease when deficient diets are fed, or increase when excessive levels are present in the diet. As the uptake of minerals into the egg depends mainly on the amount in plasma, this influences the content in the egg. With the exception of iron and zinc, which are found in yolk, the minerals are present in both the yolk and the albumen. Minerals are obviously needed for embryonic development. Severe deficiencies in the diets of breeding hens of each of the essential minerals Ca, P, Mg, Mn, Zn, Fe, Cu, Mo, I and Se result in a total failure in hatchability and/or embryonic abnormalities (Beer, 1969; Scott *et al.*, 1976). Excesses of some minerals (in particular, Se, Mg, Mo and Na) can be just as detrimental. If cereals containing high levels of selenium are fed to breeding birds, embryonic development is affected and the eggs usually fail to hatch (Franke and Tully, 1935; Fitzsimmons and Phalaraksh, 1978). In less extreme cases it is possible to see a relationship between dietary intake of some minerals (Se, Mg, Mn, I, Cu, Fe), the amount present in the egg and the mineral status of the chick. A maternal iron deficiency has a marked effect on egg yolk iron, iron status of the embryo and embryonic survival (Morck and Austic, 1981). Similarly, a maternal selenium deficiency will also reduce hatchability and the viability of the offspring (*Figure 19.4*). A direct relationship has

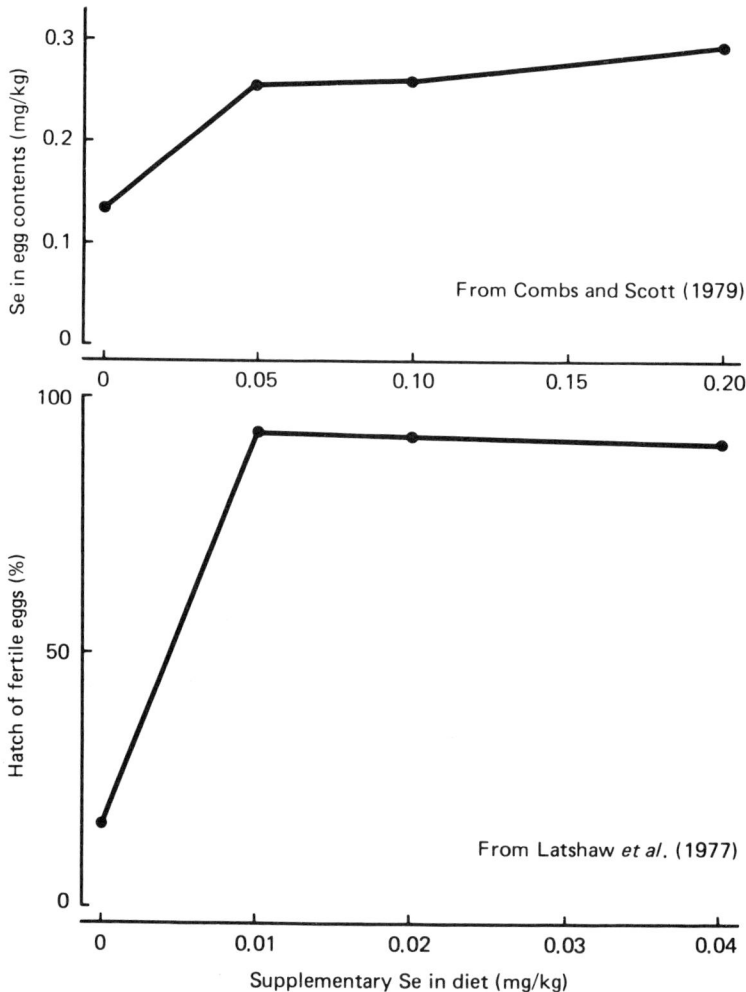

Figure 19.4 Relationship between selenium in the diet and egg and hatchability

been found between the iodine content of the maternal diet and the iodine content of the thyroids of the embryos; when breeding hens were given iodine-deficient diets, hatchability was depressed, hatching time was prolonged and absorption of the yolk sac was retarded (Rogler *et al.*, 1961). Dewar *et al.* (1974) investigated the transfer of some minerals from the egg to the embryo during its development. They observed that only with Fe and Na was there any evidence to suggest that total amount of the element present in the egg was correlated with the amount transferred to the embryo. No such relationship existed for Cu, Mg, Mn, K and Zn, the other minerals studied.

These findings suggest that under normal dietary conditions the mechanisms of uptake into the egg appear to be adequate to meet the requirements of the embryo for development. However, under conditions of deficiency or excess the mechanisms of uptake of many of the minerals cannot be regulated, and hatchability and embryonic development may be affected adversely.

Other nutritional factors, including feed ingredient effects

The isolation of vitamin B_{12} provided an explanation for many of the reports of unidentifiable substances in animal proteins improving hatchability. Subsequent increases in hatchability and viability when some grass meals, fish solubles and yeast products are added to breeder diets (e.g. Kurnick *et al.*, 1956; Touchburn *et al.*, 1957, 1972; Jensen, 1976; Harms *et al.*, 1977) suggest that other factors may also be present. To date, no individual substances have be.,n identified in these feed ingredients which might be related to their ability to improve performance.

While some feed ingredients have a beneficial effect on hatchability, others can have a deleterious effect. Rapeseed meal is a good example: the transfer of iodine to egg yolk is reduced when rapeseed meal is fed in high concentrations (Roos and Clandinin, 1975), and chicks hatched from these eggs have hypertrophied thyroids and reduced body weights (March *et al.*, 1972). The deleterious substances have not been positively identified, but the reduced transfer of iodine to egg yolk is greatest when a high-glucosinolate rapeseed meal is included in the diet (Goh and Clandinin, 1977). Other observations suggest that it may be the thiocyanate rather than the total glucosinolate concentration that is largely responsible (Papas *et al.*, 1979).

Requirements for hatchability

It is possible to draw some general conclusions. Where specific synthetic mechanisms exist (e.g. albumen proteins, yolk lipoproteins) and where uptake mechanisms presumably operate over a wide range, then the bird can apparently maintain the relative composition of the egg irrespective of wide fluctuations in daily intake. For this reason it is difficult to determine an actual requirement for protein, energy or lipid for hatchability, although dietary intake of these nutrients will influence egg number and egg size. Experiments have shown that hatchability of fertile eggs will generally remain high provided that the ratio between protein and energy intake does not increase beyond about 15 g protein to 1 MJ apparent metabolizable energy (Pearson and Herron, 1981, 1982). Therefore it would seem inadvisable to increase the protein content of low-energy breeder diets beyond a certain point.

In contrast, some components of the egg (in particular, some of the minerals and the vitamins) appear to enter during the uptake of other components or their mechanisms of uptake are less able to withstand wide fluctuations in dietary intake. In these cases composition, and therefore the ability of the embryo to develop and hatch, depend more directly on dietary components. Effects of dietary deficiencies of vitamins and minerals on hatchability are well documented (e.g. Beer, 1969), and recommendations of dietary intakes for breeding birds are available (*Tables 19.4, 19.5*) (Agricultural Research Council, 1975; National Research Council,

Table 19.4 VITAMIN REQUIREMENTS OF BREEDING CHICKENS AND TURKEYS PER kg OF DIET

Vitamin	*Chickens*		*Turkeys*	
	ARC	*NRC*	*ARC*	*NRC*
Vitamin A (i.u.)	2700	4000	2700	4000
Vitamin D (i.u.)	600	500	2000	900
Vitamin E (i.u.)	–	10	60	25
Vitamin K (mg)	1.0	0.5	–	1
Thiamin (mg)	–	0.8	–	2
Riboflavin (mg)	4.0	3.8	3.5	4
Nicotinic acid (mg)	10.0	10	–	30
Pantothenic acid (mg)	6.5	10	16	16
Pyridoxine (mg)	4.0	4.5	–	4
Biotin (mg)	0.15	0.15	0.2	0.15
Folic acid (mg)	0.5	0.35	0.7	1.0
Vitamin B_{12} (mg)	0.002	0.003	–	0.003
Choline (mg)	1100	500	1350	1000

Table 19.5 MINERAL REQUIREMENTS OF BREEDING CHICKENS AND TURKEYS PER kg OF DIET

	Chickens		*Turkeys*	
	ARC	*NRC*	*ARC*	*NRC*
Calcium (g)	27–32	27.5	25	22.5
Total phosphorus (g)	5.0	5.0	5.0	7.0
Sodium (g)	1.0	1.5	1.5–2.0	1.5
Potassium (g)	2.5	1.0	4.4	4.0
Chloride (mg)	900	800	–	800
Magnesium (mg)	400	500	450	500
Manganese (mg)	50	33	100	35
Iron (mg)	–	80	–	60
Copper (mg)	–	4.0	–	6.0
Zinc (mg)	60	65	–	65
Iodine (mg)	0.2	0.3	–	0.4
Selenium (mg)	–	0.1	–	0.2

1977). However, the recommendations do not distinguish between heavy or light strains of breeding birds, and are mainly based on results of experiments with the lighter laying strains of breeders. The need for a reassessment of the requirements of these essential nutrients—particularly for heavy strains such as broiler breeders—is obvious, particularly with the changes in genotype and feeding practices that have occurred since much of the research was carried out.

The need for a reappraisal of vitamin requirements for hatchability was recognized by Leeson *et al.* (1979a,b) in Canada. They compared the vitamin requirements of heavy breeders (Rhode Island Red) with those of a lighter strain (White Leghorns) and observed that the heavy breeders required greater quantities of vitamin B_{12}, vitamin E, folic acid, nicotinic acid, pantothenic acid and riboflavin for optimum hatchability. They found that embryo mortality during the 8–14 day period of incubation was a more sensitive indicator of the nutritional inadequacy of a breeder diet than was chick growth *per se* and emphasized the need for more detailed studies, particularly with broiler breeders.

This has been endorsed by recent studies in the UK. Research on the biotin requirements of breeding birds at the Poultry Research Centre (Whitehead, personal communication) has shown that the age of a flock may have an effect on the uptake of a dietary vitamin into the egg. At all levels of supplementary biotin (0–270 µg/kg) eggs from 30-week-old broiler breeders contained lower concentrations of biotin in the yolk than did larger eggs from the same flock at 35 weeks of age. A carry-over effect was seen in the chicks which had lower plasma biotin concentrations at 1 day old. Growth and viability tended to be lower in chicks from 30-week-old breeders on low-biotin diets than from the same birds at 45 weeks of age. Supplemental biotin in the breeders' diet appeared to alleviate this effect. Age could conceivably influence requirements of other vitamins and possibly some of the minerals. The results highlight the importance of vitamin supplementation of breeder rations in the early part of lay.

Beer (1969) suggested that the emphasis had changed from direct to indirect causes of nutritional deficiencies. This is supported by the observations in commercial hatcheries. Many hatcheries now examine dead-in-shell as a routine procedure, although it is by no means standard practice, and clear-cut mineral or vitamin deficiencies are seldom observed. This makes identification of possible dietary inadequacies difficult should hatchability problems be encountered.

For these reasons any detailed study of the requirements of modern breeding birds for hatchability should include an assessment of the more indirect factors likely to influence a hen's ability to lay hatching eggs. Egg storage prior to incubation is now a common practice both on the farm and in the hatchery, and has also been adopted as a means of transporting potential breeding stock to other countries. Storage of eggs is known to affect embryonic development and hatchability (e.g. Mather and Laughlin, 1977), and the longer eggs are stored the less likely they are to hatch. Maternal nutritional factors may well be involved, although, to date, their role in hatchability following storage has received little attention.

Beer (1969) mentioned some of the effects of disease and environment, while Scott (1966) compiled a list of 13 factors likely to influence a laying hen's vitamin requirements. It is impossible within the scope of this review to list all the indirect factors that can affect nutritional requirements for hatchability. However, the importance of such factors as nutrient availability, dietary interactions and ingredient effects cannot be overlooked in formulating breeder rations from the range of feedstuffs currently available.

TURKEYS

Hatchability in turkey breeders is often lower than in chickens and declines with flock age. The decline seems to be associated with changes in egg quality (Nestor *et al.*, 1972, 1977), which may be a consequence of changes in the efficiency of utilization of dietary nutrients with age. Examination of dead-in-shell has revealed an increase in the frequency of nutritional deficiency symptoms in embryos with flock age (Hodgetts and Sweet, 1975; Cherms, 1980). The presence of unidentifiable factors in turkey hen nutrition which affect hatchability and progeny growth has been recorded (e.g. Touchburn *et al.*, 1972), although, to date, few studies on the role of specific nutrients on turkey hatchability have been undertaken.

Feeding practices

The uncertainty regarding, in particular, the vitamin requirements of adult broiler breeders and the turkey tend to support the practice of adding a generous amount of supplement to the breeder ration. Given this, it is worth calculating the difference in hatchability required to pay for marginal increases in vitamin supply.

Using figures produced by the National Farmers' Union (1981): A broiler breeder flock lays an average of 128 eggs per hen, averages 81 per cent hatchability and therefore produces 103.7 hatched chicks per hen over the laying period. With average prices of £12.27 per 100 chicks hatched, the payments on chicks hatched average £12.72 per hen. For every 1 per cent increase in hatchability, at these chick prices the producer will receive an extra 15–16p per hen. If the current prices of breeder feeds and supplements are used, approximately 1.53 per cent of the cost of a ration is taken up by the vitamin supplement. It costs 634 p to feed a broiler breeder over the laying period, so that 9.65 p of this cost can be attributed to the vitamin supplement. If these figures are combined, hatchability only has to increase by 0.64 per cent (or 1 chick per hen) to pay for a 100 per cent increase in the level of vitamin supplement. In the turkey there is an even stronger case economically for feeding generous levels of vitamins in the diet. The cost per hen of the vitamin supplement in a breeder diet is approximately 10 times less than the cost of a hatched turkey poult.

Conclusions

Studies of the effect of maternal nutrition on egg formation have provided an insight into possible ways in which dietary intake can affect hatchability.

An adequate intake of those nutrients whose concentration in the egg depends directly on dietary supply is important for optimum hatchability and chick viability. It is suggested that a detailed re-examination of the recommendations for these nutrients could be undertaken on the modern strains of breeding bird, given the advances in performance that have occurred since many of the requirements were established.

References

AL-MURRANI, W.K. (1978). *Br. Poult. Sci.*, **19**, 277
AGRICULTURAL RESEARCH COUNCIL (1975). *Nutrient Requirements of Farm Livestock No. 1. Poultry.* London; Agricultural Research Council
BARTOV, I., BUDOWSKI, P. and BORNSTEIN, S. (1965). *Poult. Sci.*, **44**, 1489
BEER, A.E. (1969). In *The Fertility and Hatchability of the Hen's Egg*, p.93. Eds T.C. Carter and B.M. Freeman. Edinburgh; Oliver and Boyd
BEER, A.E., SCOTT, M.L. and NESHEIM, M.C. (1963). *Br. Poult. Sci.*, **4**, 243
BETHKE, R.M., RECORD, P.R., WILDER, O.H.M. and KICK, C.H. (1936). *Poult. Sci.*, **15**, 336
BLUM, J.C., SIMON, J. and LARBIER, M. (1979). *Ann. Biol. Anim. Biochem. Biophys.*, **19**, 303
BOLTON, W. (1961). In *Biochemist's Handbook*, p. 766. Ed. C. Long. London; Spon
BREWER, L.E. and EDWARDS, H.M. (1972). *Poult. Sci.*, **51**, 619
BUTTS, J.N. and CUNNINGHAM, F.E. (1973). *Poult. Sci.*, **52**, 1662
CALVERT, C.C. (1967). *Poult. Sci.*, **46**, 967
CHEN, P.H., COMMON, R.H., NIKOLAICZUK, N. and MACRAE, H.F. (1965). *J. Fd Sci.*, **30**, 838
CHERMS, F. (1980). *Feedstuffs*, **52**(12), 45
CHERRY, J.A. (1979). In *Food Intake Regulation in Poultry*, p.77. Eds K.N. Boorman and B.M. Freeman. Edinburgh; British Poultry Science
CHUNG, R.A., ROGLER, J.C. and STADELMAN, W.S. (1965). *Poult. Sci.*, **44**, 221
COMBS, G.F. and SCOTT, M.L. (1979). *Poult. Sci.*, **58**, 871
DEWAR, W.A., TEAGUE, P.W. and DOWNIE, J.N. (1974). *Br. Poult. Sci.*, **15**, 119
EVANS, R.J., DAVIDSON, J.A. and BUTTS, H.A. (1950). *Poult. Sci.*, **29**, 104
FARRELL, H.M., BUSS, E.G. and CLAGETT, C.O. (1970). *Int. J. Biochem.*, **1**, 168
FITZSIMMONS, R.C. and PHALARAKSH, K. (1978). *Can. J. Anim. Sci.*, **58**, 227
FRANKE, K.W. and TULLY, W.C. (1935). *Poult. Sci.*, **14**, 273
FREEMAN, B.M. and VINCE, M.A. (1974). *Development of the Avian Embryo.* London; Chapman and Hall
FULLER, H.L., FIELD, R.C., RONCALLI-AMICI, R., DUNAHOO, W.S. and EDWARDS, H.M. (1961). *Poult. Sci.*, **40**, 249
GARDINER, E.E. (1973). *Can. J. Anim. Sci.*, **53**, 665
GARDNER, F.A. and YOUNG, L.L. (1972). *Poult. Sci.*, **51**, 994
GILBERT, A.B. (1971). In *The Physiology and Biochemistry of the Domestic Fowl*, Vol. 3, p. 1379. Eds D.J. Bell and B.M. Freeman. London; Academic Press

GOH, Y.K. and CLANDININ, D.R. (1977). *Br. Poult. Sci.*, **18**, 705

GRIMINGER, P. (1964). *Poult. Sci.*, **43**, 1289

GUENTER, W., BRAGG, D.B. and KONDRA, P.A. (1971). *Poult. Sci.*, **50**, 845

HARMS, R.H., MANLEY, J.G. and VOITLE, R.A. (1977). *Proc. 32nd Distillers Feed Conf.*, p. 44. Cincinnati, Ohio; Distillers' Feed Research Council

HARMS, R.H., MORENO, R.S. and DAMRON, B.L. (1971). *Poult. Sci.*, **50**, 595

HILL, F.W., SCOTT, M.L., NORRIS, L.C. and HEUSER, G.F. (1961). *Poult. Sci.*, **40**, 1245

HODGETTS, B. and SWEET, W.N. (1975). *Turkeys*, **23**(3), Suppl.

INGRAM, G.R., CRAVENS, W.W., ELVEHJEM, C.A. and HALPIN, J.G. (1951). *Poult. Sci.*, **30**, 431

JENSEN, L.S. (1976). *Proc. 31st Distillers' Feed Conf.*, p. 33. Cincinnati, Ohio; Distillers' Feed Research Council

JEROCH, H. (1971a). *Arch. Tierernähr.*, **21**, 151

JEROCH, H. (1971b). *Arch. Tierernähr.*, **21**, 249

JEROCH, H. (1971c). *Arch. Tierernähr.*, **21**, 713

JEROCH, H. (1972). *Arch. Tierernähr.*, **22**, 97

KASHANI, A.B., CARLSON, C.W. and NELSON, R.A. (1980). *Poult. Sci.*, **59**, 2519

KOSIN, I.L., ABPLANALP, H., GUTIERREZ, J. and CARVER, J.S. (1952). *Poult. Sci.*, **31**, 247

KURNICK, A.A., SVACHA, R.L., REID, B.L. and COUCH, J.R. (1956). *Poult. Sci.*, **35**, 658

LANDAUER, W. (1973). *Storrs agric. Exp. Sta. Mono. 1 (Revised) Suppl.* University of Connecticut

LARBIER, M., BLUM, J.C. and GUILLAUME, J. (1972). *Ann. Biol. Anim. Biochem. Biophys.*, **12**, 125

LATSHAW, J.D., ORT, J.F. and DIESEM, C.D. (1977). *Poult. Sci.*, **56**, 1876

LEESON, S., REINHART, B.S. and SUMMERS, J.D. (1979a). *Can. J. Anim. Sci.*, **59**, 561

LEESON, S., REINHART, B.S. and SUMMERS, J.D. (1979b). *Can. J. Anim. Sci.*, **59**, 569

LINKSWILER, H.M., ZEMEL, M.B., HEGSTED, M. and SCHUETTE, S. (1981). *Fed. Proc.*, **40**, 2429

LUNVEN, P. LE, CLEMENT DE ST MARCQ, C., CARNOVALE, E. and FRATONI, A. (1973). *Br. J. Nutr.*, **30**, 189

McNAUGHTON, J.L., DEATON, J.W., REECE, F.N. and HAYNES, R.L. (1978). *Poult. Sci.*, **57**, 38

MARCH, B.E., BIELY, J. and SOONG, R. (1972). *Poult. Sci.*, **51**, 1589

MATHER, C.M. and LAUGHLIN, K.F. (1977). *Br. Poult. Sci.*, **18**, 597

MENGE, H. (1968). *J. Nutr.*, **95**, 578

MENGE, H., CALVERT, C.C. and DENTON, C.A. (1965). *J. Nutr.*, **86**, 115

MORCK, T.A. and AUSTIC, R.E. (1981). *Poult. Sci.*, **60**, 1497

MORRIS, T.R. (1968). *Br. Poult. Sci.*, **9**, 285

MUNIYAPPA, K. and ADIGA, P.R. (1979). *Biochem. J.*, **177**, 887

NABER, E.C. (1979). *Poult. Sci.*, **58**, 518

NATIONAL FARMERS' UNION (1981). *Annual Broiler Breeder Bulletin, No. 9, Suppl.* Spalding; National Farmers' Union

NATIONAL RESEARCH COUNCIL (1977). *Nutrient Requirements of Poultry.* Washington, D.C.; National Academy of Sciences

NESTOR, K.E., BROWN, K.I. and TOUCHBURN, S.P. (1972). *Poult. Sci.*, **51**, 104

NESTOR, K.E., TOUCHBURN, S.P. and MUSSER, M.A. (1977). *Poult. Sci.*, **56**, 8

O'NEIL, J.B. (1955). *Poult. Sci.*, **34**, 761

PAPAS, A., CAMPBELL, L.D., CANSFIELD, P.E. and INGALLS, J.R. (1979). *Can. J. Anim. Sci.*, **59**, 119

PATEL, M.B. and McGINNIS, J. (1977). *Poult. Sci.*, **56**, 45

PEARSON, R.A. and HERRON, K.M. (1981). *Br. Poult. Sci.*, **22**, 227

PEARSON, R.A. and HERRON, K.M. (1982). *Br. Poult. Sci.*, **23**, 71

PERRY, M.M. and GILBERT, A.B. (1981). *Trans. Biochem. Soc.*, **9**, 177P

PROUDFOOT, F.G. and HULAN, H.W. (1981). *Poult Sci.*, **60**, 2167

REINHART, B.S. and MORAN, E.T. (1979). *Poult. Sci.*, **58**, 1599

RINGROSE, R.C., MANOUKAS, A.G., HINKSON, R. and TEERI, A.E. (1965). *Poult. Sci.*, **44**, 1053

ROGLER, J.C., PARKER, H.E., ANDREWS, F.N. and CARRICK, C.W. (1961). *Poult. Sci.*, **40**, 1554

ROOS, A.J. and CLANDININ, D.R. (1975). *Br. Poult. Sci.*, **16**, 413

SCOTT, M.L. (1966). *Vitam. Horm. Lpz.*, **24**, 633

SCOTT, M.L., NESHEIM, M.C. and YOUNG, R.J. (1976). *Nutrition of the Chicken*, 2nd edn. New York; Scott

SELL, J.L., CHOO, S.H. and KONDRA, P.A. (1968). *Poult. Sci.*, **47**, 1296

SKOGLUND, W.C. and TOMHAVE, A.E. (1949). *Delaware agric. exp. Sta. Bull. 278*

SNELL, E.E., ALINE, E., COUCH, J.R. and PEARSON, P.B. (1941). *J. Nutr.*, **21**, 201

SONNEBORN, D.W. and HANSEN, H.J. (1970). *Science, N.Y.*, **158**, 591

SUNDE, M.L., CRAVENS, W.W., BRUINS, H.W., ELVEHJEM, C.A. and HALPIN, J.G. (1950). *Poult. Sci.*, **29**, 220

TOUCHBURN, S.P., BIELY, J. and MARCH, B. (1957). *Poult. Sci.*, **36**, 591

TOUCHBURN, S.P., CHAMBERLIN, V.D. and NABER, E.C. (1972). *Poult. Sci.*, **51**, 96

TULLETT, S.G. and BURTON, F.G. (1982). *Br. Poult. Sci.*, **23**, 361

VOGTMAN, H. and CLANDININ, D.R. (1975). *Br. Poult. Sci.*, **16**, 55

WEISS, F.G. and SCOTT, M.L. (1979). *J. Nutr.*, **109**, 1010

WHITE, H.B., DENNISON, B.A., DELLA FERA, M.A., WHITNEY, C.J., McGUIRE, J.C., MESLAR, H.W. and SAMMELWITZ, P.H. (1976). *Biochem. J.*, **157**, 395

WHITEHEAD, C.C. and BANNISTER, D.W. (1981). *Br. Poult. Sci.*, **22**, 467

WILEY, W.H. (1950). *Poult. Sci.*, **29**, 595

20

RECENT DEVELOPMENTS IN THE FIELD OF ANTICOCCIDIAL AGENTS FOR POULTRY

A.C. VOETEN
Gezondheidsdienst voor Dieren in Noord-Brabant, Boxtel, The Netherlands

Introduction

Before discussing the use of anticoccidial agents it is appropriate to consider recent developments in our knowledge of the field of coccidiosis.

Coccidiosis occurs in mammals and avian species. Almost all avian species are susceptible; in domestic poultry the disease is caused by the protozoan, *Eimeria* and is characterized by intestinal lesions. These intestinal lesions are the result of different developmental stages of *Eimeria* in the gut wall, which reproduce there both sexually and asexually. The disease can vary from a mild, subclinical form to a very severe clinical form with a high mortality rate. Coccidiosis is more prevalent in younger chicks and current husbandry practices with broiler chicks promotes the occurrence of the disease. Coccidiosis in broiler chicks can be prevented by the use of anticoccidial products incorporated in the feed. Furthermore both clinical and subclinical coccidiosis outbreaks can be either acute or chronic.

Although broiler chicks can be affected by all types of coccidiosis, under most practical conditions one usually finds *Eimeria acervulina*, *E. maxima* and *E. tenella*. The species *E. acervulina* and *E. maxima* are generally responsible for subclinical cases which can occur as both acute and chronic outbreaks. *E. tenella* is usually associated with clinical cases; it is not known whether this species can cause subclinical forms of the disease, and if so what level of damage results.

Broiler chicks with coccidiosis shed large numbers of oocysts in the faeces. These are the forms of *Eimeria* that live outside the body, and can remain viable for several years. On the farm they may be found both inside and outside the building with a considerable cross infection between the two. The carry-over effects of oocysts from one batch of birds to the next therefore poses a recurring danger, with that species of *Eimeria* which was allowed to propagate during the previous batch posing the greatest danger to incoming birds. However, since oocysts can be carried by man and materials from farm to farm, other *Eimeria* spp. can also occur.

The manner of coccidiosis occurrence in broiler chicks

As mentioned earlier, small intestinal coccidiosis, caused by both *E. acervulina* and *E. maxima* infections, generally occurs subclinically, in both the acute and chronic

First published in *Recent Advances in Animal Nutrition – 1985*

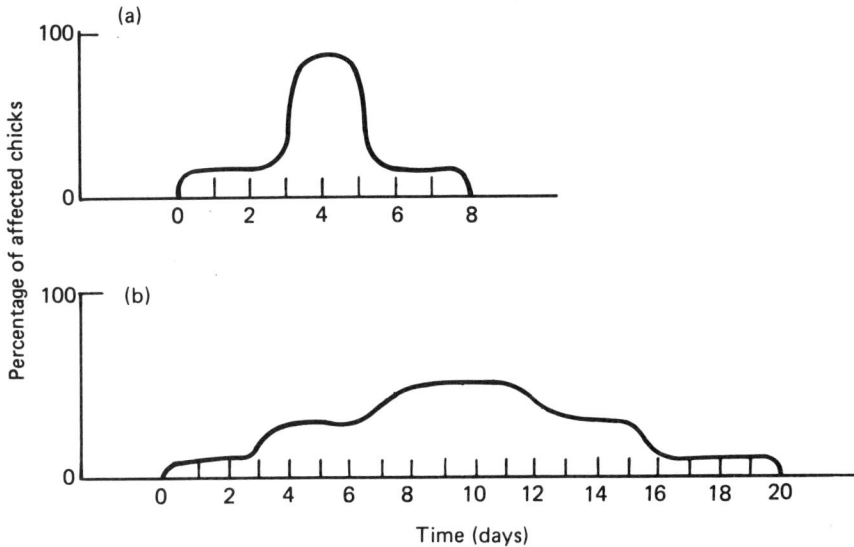

Figure 20.1 Pattern of infection with *E. acervulina* or *E. maxima* in broiler chicks with acute (a) or chronic (b) forms of the disease

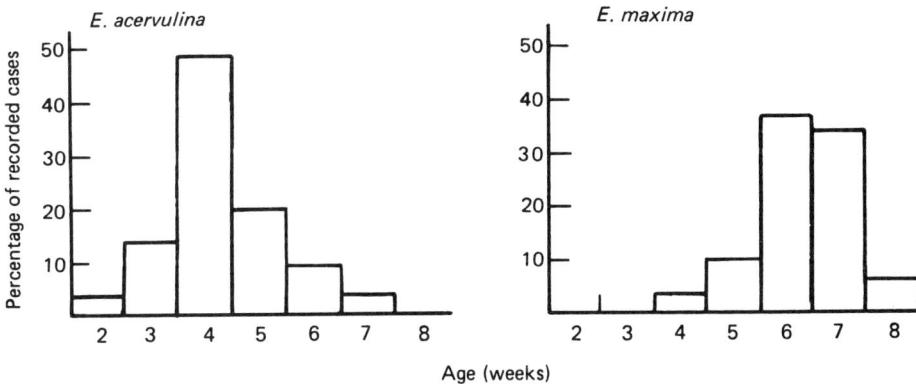

Figure 20.2 Effect of age of chick on the incidence of acute infections of subclinical coccidiosis, caused by either *E. acervulina* or *E. maxima*

form. In the acute subclinical form, almost all birds can be affected within a few days, but with the chronic form the rate of build-up and the proportion of infected birds are much lower (*Figure 20.1*).

If the coccidiosis can occur unhindered, i.e. in the absence of any effective treatment, then *E. acervulina* infections generally occur in the third, fourth or fifth week of life, while those associated with *E. maxima* generally occur in the fifth, sixth or seventh week of life (*Figure 20.2*). Chronic coccidiosis usually occurs in the same weeks of life, but takes a longer course.

Effect of coccidiosis on growth rate and feed conversion efficiency

Although *E. acervulina* infections predominantly occur during the fourth week of life, chicks can be artificially infected at other ages. The reductions in growth rate and feed conversion efficiency were assessed at the end of the growing phase (six weeks age) in chicks subjected to acute subclinical coccidiosis at different ages. In the case of *E. acervulina*, the greatest influence on feed conversion efficiency occurred when the birds were affected during the third week of life, but this had only a small detrimental effect on the growth rate. The reduction in growth rate was most severe when the birds were affected towards the end of their growing phase (*Figure 20.3*). Similar results were obtained with subclinical infections of *E. maxima* (*Figure 20.3*).

The actual extent and time course of damage from chronic forms of subclinical coccidiosis are not known, although it is known that the damage is appreciable. As

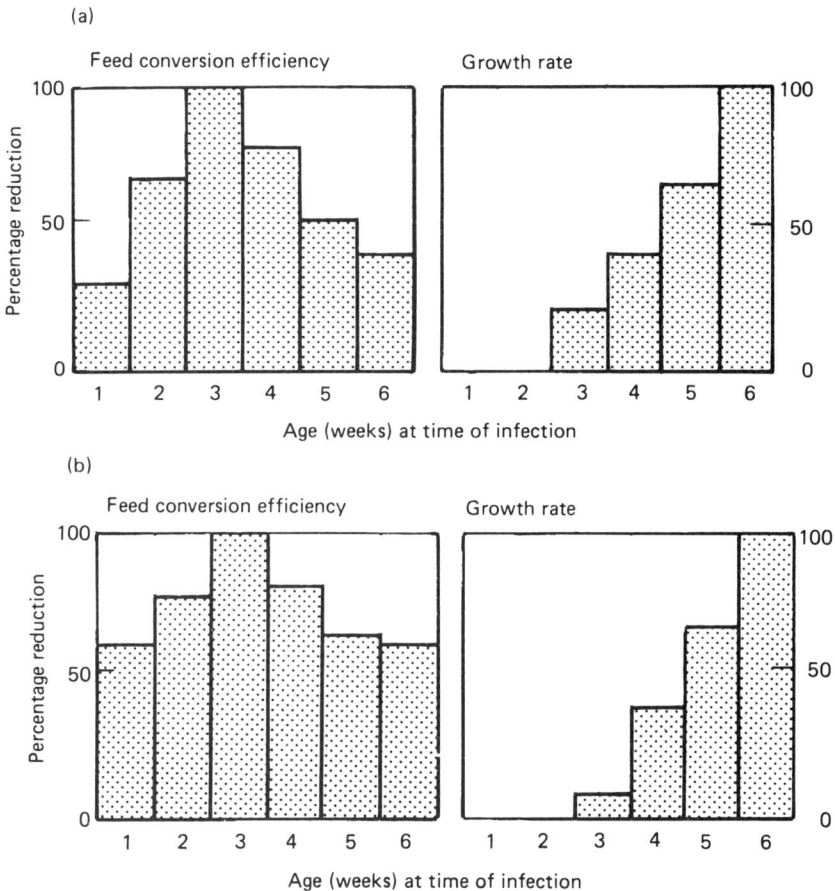

Figure 20.3 Effect of age at infection with *E. acervulina* (a) or *E. maxima* (b) on the reduction in feed conversion efficiency and growth rate up to the end of six weeks of age in broiler chicks. The vertical axis represents the relative reduction in performance expressed as a percentage of the maximum levels recorded

indicated above, the damage from subclinical coccidiosis manifests itself in an increased feed conversion ratio, which can amount to an extra 0.1 kg feed/kg liveweight gain by the end of the growing period. In such case the broilers can be 100 g lighter at slaughter.

For preventive measures, isolation and disinfection can be considered, but it would be difficult to be optimistic about this at present. Furthermore, preventive vaccinations for broiler chicks are not yet available, although the weapons to combat coccidiosis totally may well be obtained in time from such areas of molecular biology. Consequently, the only means of combatting the problem at the moment is the recourse to pharmaceutical materials.

Pharmaceuticals for the prevention of coccidiosis

Pharmaceuticals which are used for the prevention of coccidiosis can be either coccidiostatic or coccidiocidal in their mode of action. Those products which are coccidiostatic in effect prevent the multiplication of *Eimeria* spp. in the gut. On the other hand, coccidiocidal products kill the development stage within the gut. This classification is somewhat artificial as it is not definitely known for many products whether their effect is coccidiocidal or coccidiostatic. Furthermore, experience has indicated that products which were initially coccidiocidal have a static effect after long use, and with time a reduced activity results. Consequently, within this discussion we shall limit ourselves to the term anticoccidial products which encompasses both modes of action.

Numerous anticoccidial products have been developed in the past 25 years, and these are listed in *Table 20.1*. Some are single products while others are combinations of products.

Table 20.1 LIST OF AVAILABLE ANTICOCCIDIAL AGENTS AUTHORIZED FOR USE WITH POULTRY

	Trade name	*Chemical name*
Single products chemical products	Amprolium	amprolium
	Arpocox	arprinocid
	Coyden	metichlorpindol
	Cycostat	robenidine
	Deccox	quinoline
	Nicrazin	nicarbazine
	Stenerol	halofuginone
	Zoalene	3,5 dinitro-*o*-toluamide (DOT)
ionophore preparation	Avatec	lasalocid sodium
	Elancoban	monensin sodium
	Monteban	narasin
	Saccox	salinomycin sodium
	Cygro	ammonium prinicin
Combination products		
	Aprol plus	amprolium and ethopabate
	Lerbek	metichlorpindol and methylbenzoquate
	Pancoxin plus	amprolium, ethopabate, sulphaquinoxaline and pyrimethamine

The development of resistance to anticoccidiosis products

In the event that an anticoccidial product is used continuously on the farm for any given period of time, then there is a great chance that the *Eimeria* spp. will become partially or totally resistant to it. This does not have to occur simultaneously with all *Eimeria* spp. For example, a resistance may develop against an *E. acervulina* strain, but not against an *E. maxima* strain. In practice, it appears that the development of resistance to *E. necatrix* and *E. brunetti* usually does not pose a problem. However, a great number of *E. acervulina* and *E. maxima* strains have developed resistance to many of the available products, while the development of resistance by *E. tenella* strains varies from product to product.

When an *Eimeria* spp. develops resistance to an anticoccidial product this can occur either in a 'one step' fashion or in a 'step by step' manner (*Figure 20.4*). An

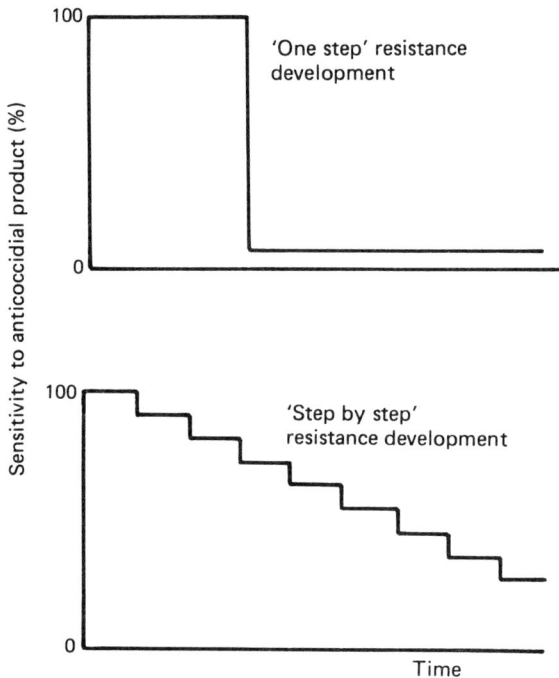

Figure 20.4 Different patterns of change with time in the sensitivity of *Eimeria* spp. to anticoccidial products

example of an anticoccidial product against which 'one step' resistance has developed is Deccox. When an *Eimeria* strain on a farm has developed resistance, then the presence of the anticoccidial product in question is of little or no use. An *E. acervulina* or E. maxima infection in this case would behave as if the product was not present, with the result that subclinical acute coccidiosis would occur.

In the case where an *Eimeria* strain develops 'step by step' resistance the situation is different. Examples of products against which such a pattern of resistance has developed are the ionophores and nicarbazine. In this instance the *Eimeria* spp. will suddenly be in the state to start multiplying, with the result that subclinical, chronic coccidiosis may occur. This can be of a mild or more serious nature, resulting in varying degrees of impairment in performance. Such changes in resistance are typical of those observed with ionophore anticoccidial products.

When the occurrence of coccidiosis in a large number of closed integrated broiler farms was analysed it was possible to draw several conclusions. The continual use of monensin over several years was initially associated with hardly any coccidiosis. After one or two years' use, coccidiosis did occur fairly regularly, but the impairment in performance was minimal. After more extensive use, coccidiosis occurred on almost all farms, and on numerous farms an effect on performance could be determined. At an even later stage the damage became more severe with the frequent occurrence of subclinical coccidiosis, which tended to be chronic in its course. Although the time course of the development of resistance varied from farm to farm, the same series of stages listed below occurred in all cases:

Stage 1 almost no cases of subclinical coccidiosis.
Stage 2 regular cases of subclinical, chronic coccidiosis, but the influence on feed conversion efficiency and growth rate was minimal.
Stage 3 many cases of subclinical, chronic coccidiosis; the influence on feed conversion efficiency and growth rate was quite large.
Stage 4 almost no efficacy attained with the anticoccidial preparation/product and any cases of coccidiosis remained chronic and were therefore difficult to treat; the effect on performance was severe.

The problems of continual use of ionophore anticoccidial products

The various stages of damage listed above occur when the same ionophore is used over several years on the same farm. If a second ionophore is used its behaviour and effect is different from the previous one. Although we can be confident that a 100 per cent cross-resistance does not exist, a certain degree of cross-tolerance is

Table 20.2 THE EFFECT OF CHANGING FROM MONENSIN TO SALINOMYCIN ON THE INCIDENCE OF COCCIDIOSIS IN BROILERS. THE INCIDENCE OF SUBCLINICAL COCCIDIOSIS WAS ASSESSED THREE TIMES THROUGHOUT THE GROWING PHASE IN THE LAST TWO CROPS OF BIRDS FED MONENSIN AND THE FIRST TWO CROPS FED SALINOMYCIN

| Unit | Monensin | | | | | | Salinomycin | | | | | |
	Previous crop 2			Previous crop 1			Trial crop 1			Trial crop 2		
1	—	M	—a	—	M	A	—	M	A	—	A	M
2	—	—	—	—	—	—	—	A	—	—	—	—
3	—	T	—	—	T+M	—	—	M	—	—	M	—
4	—	—	—	—	—	—	—	—	—	—	—	M
5	—	T	—	—	—	—	—	—	—	—	—	—
6				—	M	—	—	M	—	—	—	—
7	—	M	—	—	—	—	—	M+A	—	—	M	—
8	—	—	—	—	—	—	—	—	—	—	A	—
9	—	—	—	—	T	—	—	A	—	—	M	—
10	—	—	—	—	—	—	—	—	—	—	—	—
11	—	—	—	—	M	—	—	M	—	—	—	—
12	—	—	—	—	—	—	—	A	A	—	—	—
13	—	A	—	—	M	—	—	—	—	—	T	—
14	—	M	—	—	M+A	—	—	M	M	—	M	—

aA = *Eimeria acervulina*; M = *Eimeria maxima*; T = *Eimeria tenella*; — None recorded

Table 20.3 THE EFFECT OF CHANGING FROM MONENSIN TO SALINOMYCIN ON THE PERFORMANCE OF BROILERS. ALL VALUES RELATE TO THE EUROPEAN BROILER INDEX (SEE TEXT) AND ARE GIVEN FOR THE FINAL TWO CROPS FED MONENSIN AND THE FIRST TWO CROPS FED SALINOMYCIN

Unit	Monensin		Salinomycin	
	Previous crop 2	*Previous crop 1*	*Trial crop 1*	*Trial crop 2*
1	139	155	156	149
2	166	168	174	162
3	184	179	188	164
4	173	151	166	155
5	183	188	193	195
6	—	144	196	164
7	172	157	194	176
8	125	135	163	160
9	153	138	178	176
10	163	157	176	158
11	167	167	165	164
12	157	140	188	184
13	133	139	178	166
14	129	155	166	149
Mean ± s.e.	156 ± 17.9		172 ± 13.3	

Table 20.4 THE EFFECT OF CHANGING FROM MONENSIN TO LASALOCID ON THE INCIDENCE OF COCCIDIOSIS IN BROILERS. THE INCIDENCE OF SUBCLINICAL COCCIDIOSIS WAS ASSESSED THREE TIMES THROUGHOUT THE GROWING PHASE IN THE LAST TWO CROPS OF BIRDS FED MONENSIN AND THE FIRST TWO CROPS FED LASALOCID

Unit	Monensin						Lasalocid					
	Previous crop 2			*Previous crop 1*			*Trial crop 1*			*Trial crop 2*		
1	—[a]	AT	—	—	A	—	—	—	—	—	T	—
2	—	—	—	—	—	—	—	—	—	—	—	—
3	—	—	—	—	TM	—	—	A	—	—	M	—
4	—	A	T	—	T	—	—	A	—	—	A	—
5	—	—	—	—	—	—	—	A	—	—	A	—
6	—	—	M	—	—	M	—	—	M	—	—	A
7	—	—	M	—	—	M	—	A	—	—	—	—
8	—	T	M	—	AT	M	—	A	—	—	M	A
9	—	A	—	—	—	—	—	A	—	—	—	A
10	—	T	—	—	—	—	—	M	—	—	A	T
11	—	—	—	—	—	—	—	—	—	—	—	—

[a]A = *Eimeria acervulina*; M = *Eimeria maxima*; T = *Eimeria tenella*; — None recorded

bound to have developed. The experiments presented below were designed to investigate the extent to which this cross-tolerance existed.

On two separate integrated broiler operations monensin, which had been used on all farms for some four to five years, was replaced by another ionophore in the feed. With one operation this was salinomycin and with the other it was lasalocid. Coccidiosis examinations were carried out three times during the growing phase on the last two crops of birds fed monensin, and this was repeated with the first two crops which received the new ionophore. Data of the performance for all four crops of birds were also collected. *Tables 20.2* and *20.3* show the results for the changeover from monensin to salinomycin, and *Tables 20.4* and *20.5* give the results for the changeover from monensin to lasalocid.

Table 20.5 THE EFFECT OF CHANGING FROM MONENSIN TO LASALOCID ON THE
PERFORMANCE OF BROILERS. ALL VALUES RELATE TO THE EUROPEAN BROILER
INDEX (SEE TEXT) AND ARE GIVEN FOR THE FINAL TWO CROPS FED MONENSIN AND
THE FIRST TWO CROPS FED LASALOCID

Unit	Monensin		Lasalocid	
	Previous crop 2	*Previous crop 1*	*Trial crop 1*	*Trial crop 2*
1	200	137	184	172
2	—	173	193	186
3	—	176	171	177
4	163	142	171	171
5	180	170	151	173
6	175	163	163	171
7	153	154	166	155
8	154	146	—	175
9	153	197	166	161
10	136	172	185	178
11	—	—	187	164
Mean ± s.e.	163.6 ± 18.5		172.5 ± 10.8	

From the data of both field trials it is apparent that the occurrence of coccidiosis
was hardly reduced by changing to the new ionophore, although it was noted that
the impairment in performance was markedly reduced. This was manifest in a
higher European Broiler Index (EBI = growth per animal per day × per cent
survival:feed conversion factor × 10), together with a decrease in the variation of
performance between farms. This improvement in EMI could be traced back to a
better feed conversion efficiency and an improved growth rate. Since the number of
coccidiosis cases was hardly reduced it is apparent that the extent of the disease was
reduced from stage 3 to stage 2. Since then the damage from coccidiosis has
increased again and returned to the stage 3 level.

The interrupted use of anticoccidial products

Since the continuous use of any one anticoccidial product is not possible, it is
necessary to interrupt their use. The number of farms contaminated with
coccidiosis is of fundamental importance and this can be determined in several
ways. For a good coccidiosis management programme it is essential to be up to date
with such information because it is important that the use of any anticoccidial
product be terminated prior to it losing all of its efficacy.

Furthermore, it is more important that some anticoccidial products are used
during certain times of the year. For example, an anticoccidial product with a
water-sparing effect, such as monensin, would preferably be used during those
months when there is the greatest chance of wet litter problems (summer months).

The efficacy of several anticoccidial products increases when they have not been
used for a period of time. For example, the activity of the ionophores, after not
being used for 12 months or more, improves and allows the incidence of subclinical
coccidiosis to be reduced from stage 3 to stage 2. Other products, such as Coyden,
Nicarbazine or Cycostat, also regain some of their activity if they have not been
used for a period of time (up to several years). However, this restoration of activity
does not always happen; practical experience has indicated that the re-use of
biquinolates after many years of non-use still gives disappointing results.

Unfortunately precise data relating to this phenomenon for all anticoccidial products are not available.

The influence of anticoccidial products on the growth rate is also an important factor for consideration in any coccidiosis management programme. The use of some materials will actually reduce growth rate and this is especially true with nicarbazine, and less so with monensin. Once these products are withdrawn there is a compensatory effect on growth rate. Indeed in cases where nicarbazine is used for three to four weeks the growth rate may be decreased by as much as 10 per cent. However, experience has shown that this initial loss in growth rate is totally regained by compensatory growth after its withdrawal, so much so in fact that the compensatory growth achieved after the product has in several cases been greater than the initial growth depression during the period when the material was included in the diet.

When choosing an anticoccidial product one will also have to take price into consideration. A great variation in price exists between the various products available. In the Netherlands there is as much as a tenfold difference in price.

Finally, there may be technical problems of incorporating anticoccidial products into the feed. These can be related to the facilities available to the feed compounder, or the fear of accidental cross-contamination of other feeds produced in the mill.

An action plan for the use of anticoccidial products

On the basis of experience from the interrupted use of anticoccidial products it is possible to develop an appropriate strategy for coccidiosis management. In principle, such a programme will be based on rotation, in which, after a period of use, one anticoccidial product will be exchanged for another. In addition a shuttle programme can be employed in which two different anticoccidial products are used at different growth stages with the same crop of birds.

There is only one reason for using a shuttle programme, and that is when nicarbazine is used. In such cases nicarbazine is used for three to four weeks, and thereafter another material is used. Furthermore it is recommended that the use of ionophorous anticoccidial products are alternated with chemical products and that within each category the actual materials used are also rotated.

There are particular reasons for using ionophorous products in the summer. The ionophore, monensin, has a water-sparing effect, which makes it particularly useful for use in the summer. Secondly, despite the use of ionophores, subclinical chronic coccidiosis still occurs. This is of less concern in the summer as the higher external house temperature allows an increase in ventilation rate, which in turn results in a reduction in the potential risk from viral infection. The combination of coccidiosis, a high NH_4 content in the house and virus infection frequently results in an *E. coli* infection.

A quick change of chemical products offers most benefits during the winter months. There are a number of chemical anticoccidial products which, when not having been used for some one to two years, are very effective for two to three months. A rotation involving two or three such materials has many advantages in the winter months. To afford a higher degree of protection to the birds, these can even be incorporated into a shuttle programme with nicarbazine. Experience has shown that nicarbazine can be used in this manner for several years without

Table 20.6 A SUGGESTED MANAGEMENT PROGRAMME FOR THE CONTROL OF COCCIDIOSIS INVOLVING IONOPHORES, TOGETHER WITH EITHER A SHUTTLE PROGRAMME OR A ROTATIONAL PROGRAMME INVOLVING CHEMICAL PRODUCTS

Summer period (6–8 months)	*Winter period* (4–6 months)
Annual changeover of ionophore:	Chemical products used[a] in either (a) a shuttle programme with nicarbazine or (b) a rotational programme
Salinomycine	Coyden (lerbek)
Avatec	Cycostat
Monensin	Arpocox
	Stenerol

[a]Chemical products used for two months at a time in either programme

problems. A satisfactory suggested management programme for the use of anticoccidial products which incorporates many of these points is given in *Table 20.6*.

It must be emphasized that if one does have a management schedule, it is unwise to leave it too readily. Feed compounders are extremely sensitive to complaints, and, in the cases where there is an accumulation in the incidence of complaints among customers, compounders frequently take panic-orientated decisions to change the anticoccidial product being used. All too often a sudden deterioration in food conversion efficiency, bad litter conditions or an increase in the variation of time to slaughter are blamed on loss of effectiveness of the anticoccidial agent. Generally, these problems cannot be correlated with an increased resistance to the products being used, and are likely to be the result of some other nutritional, environmental or management problem.

In order to keep abreast of coccidiosis prevention it is useful to use routine coccidiosis examinations. This can be done by using indicator animals, that is by checking five birds per 10 000 for the presence of *Eimeria*. On those farms where coccidiosis is not a problem this examination can be conducted once, during the fourth week of life. On those farms that have a history of coccidiosis problems the examination should be conducted twice, during the third and fifth week of life. Severely affected farms may need to conduct a weekly examination. Instead of using indicator animals one can conduct an examination of the litter. This has the disadvantage that it requires a large amount of laboratory work. Furthermore, the oocysts in the litter frequently do not survive for long, with the result that the findings are not always easy to interpret. A reduction in the number of oocysts in the litter by a factor of 50 to 100 per week is quite usual.

An example of the results from an operation which has been conducting coccidiosis examinations over the last ten years on its farms are given in *Figure 20.5*. Also shown in this figure are the anticoccidial products which were used over different portions of the overall period, and in many instances there was a reduction in the incidence of infected birds when there was a change in the anticoccidial product employed.

If, despite all preventive measures, there is a coccidiosis outbreak the animals can be treated with 60 mg sulphadimidine sodium per kg bodyweight per day. The medication must be given in the drinking water for a period of 5 h/day for three days. Two comments must be made here. Such treatment will have an immediate and beneficial effect on the feed conversion efficiency and growth rate in cases of

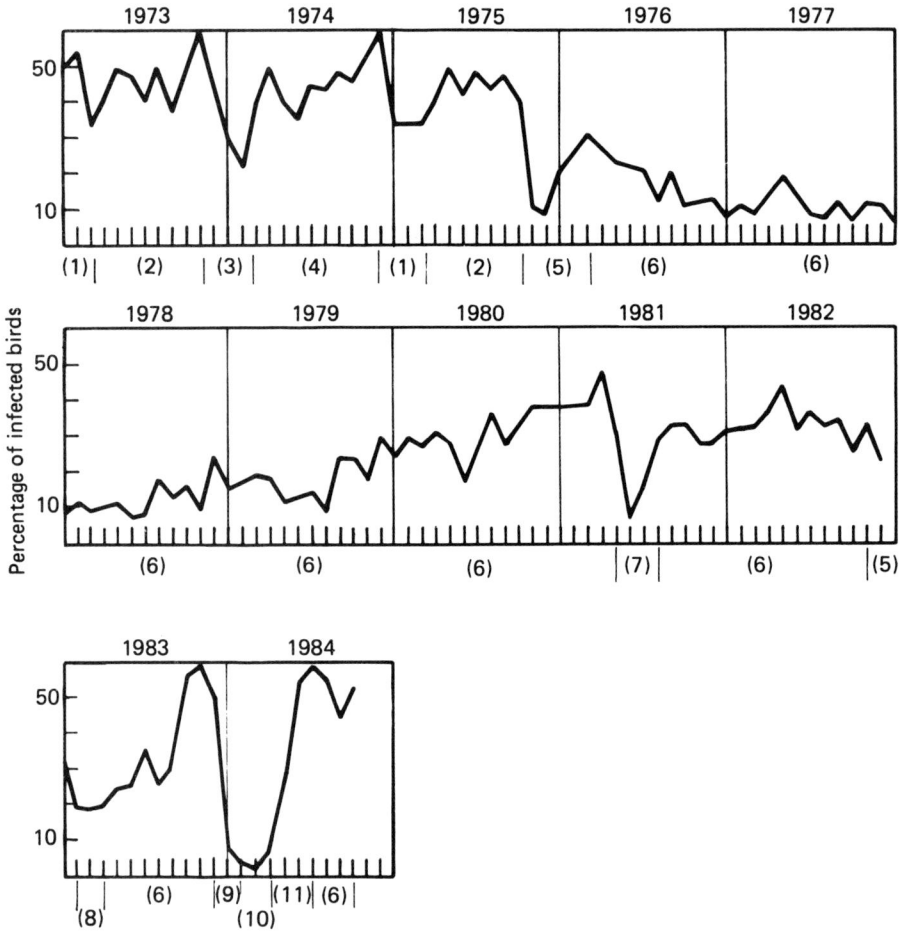

Figure 20.5 Incidence of coccidiosis in indicator broiler chicks from one operation over a ten-year period. Figures in parenthesis beneath the horizontal axis indicate the anticoccidial product being used as follows: (1), Coyden; (2), Zoalene; (3), Deccox; (4), Amprol plus; (5), Cycostat; (6), Monensin; (7), Arpocox; (8), Lerbek; (9), Shuttle programme of Nicarbazine and Stenerol; (10), Shuttle programme of nicarbazine and Narasin; (11), Narasin

subclinical acute coccidiosis, at least when treatment is initiated during the peak of the infection and for some days thereafter. However, it will have little or no effect when ionophorous anticoccidial products are being administered and the chronic forms of subclinical coccidiosis is present.

21

NUTRITION–DISEASE INTERACTIONS OF LEG WEAKNESS
IN POULTRY

D.R. WISE
Department of Clinical Veterinary Medicine, University of Cambridge

Reluctance to walk and/or impaired gait may be the main presenting signs in a variety of disorders of the nervous, muscular and skeletal systems. Some of these are now well defined while the aetiologies of others remain obscure. Reference to past literature is frequently confusing owing to several distinct entities being given the same name: perosis, for example. Equally, one abnormality may be given different names by different authors, as in the case of tibial dyschondroplasia and focal osteodystrophy, which describe the same pathological entity.

The general subject of leg weakness in poultry has previously been reviewed by Nairn and Watson (1972) while specifically skeletal abnormalities are the subjects of reviews by Riddell (1975a, b) and by Wise (1975). It is therefore the intention of this chapter to dwell only on those conditions which are currently of economic importance to the poultry industry and to make but passing reference to other abnormalities.

Twisted, bent and bowed legs and rotated tibias

These abnormalities, all associated with gross deformities of the leg skeleton, have been grouped under one heading although, in future, when more is known about them, there may be justification for their separation. Ferguson *et al.* (1974), however, considered that they were all manifestations of the same syndrome and in this chapter they will be treated as such.

The syndrome is seen in all broiler flocks, with an incidence of 5% being far from unusual. In broiler types grown to older ages, severe losses due to culling and downgrading may ensue. One or both legs may be involved and birds may become 'cow hocked' or 'bow legged'. The distal end of the tibia and the proximal end of the tarsometatarsus become bent outwards (*Figure 21.1*) or inwards. Alternatively, the whole tibia may rotate so that the distal condyle faces laterally instead of forward. On occasions, the gastrocnemius tendons slip off the condyles. 'Perosis' is a term often used to describe the syndrome but for reasons which will become apparent later, this is totally misleading. In fact Nairn and Watson (1972) made a plea that the term should be dropped

First published in *Recent Advances in Animal Nutrition – 1978*

from the vocabulary of the avian pathologist. Bowed legs and, less frequently, rotated tibias are also seen in turkey flocks, being more frequent in stags than hens (Buffington, Kleven and Jordan, 1975). Bergmann and Pietsch (1976) have also described a 4% incidence of tibial torsion in a flock of guinea-fowl.

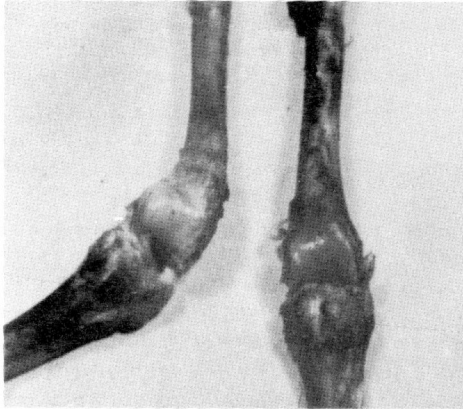

Figure 21.1 Hock joints in a case of unilateral twisted leg

It has frequently been observed that environmental factors can influence the incidence of these types of skeletal defects. Rodenhoff and Dämmrich (1971) found a lower incidence of leg deformities in broilers grown on free range than in those reared conventionally. Incidence in caged broilers tends to be higher than in those grown on litter (Reece *et al.*, 1971; Seth and Clandinin, 1973; Andrews *et al.*, 1974). Andrews *et al.* (1974) also found that incidence varied with type of cage flooring.

Different strains of broilers have been reported to show different incidences of the syndrome under identical conditions (Dämmrich and Rodenhoff, 1970; Dingle, 1977). Somes (1969) identified an autosomal recessive gene in one strain of chickens which was responsible for twisted tibiotarsal and bent tarso-metatarsal bones. All birds carrying a recessive pair of genes were affected but to varying degrees in either one or both legs.

Certain additives to poultry rations or feed ingredients have been associated with increased incidences of skeletal deformities. Briggs and Lillie (1946) found that high levels of the thyroid-blocking agent, thiouracil, caused leg deformities in chicks but D.R. Wise (unpublished) has recently been unable to reproduce similar abnormalities in turkey poults with this drug. Bowen and Waldroup (1969) produced crooked toes and hock abnormalities in chickens by feeding them a ration containing 5% propylene glycol. Holmes and Roberts (1963) and Seth and Clandinin (1973) found that high levels of rape-seed meal increased the incidence of leg deformities. This appeared to be unrelated to the goitro-genic activity of the meal but may have been associated with high tannin levels, as probably was a similar leg problem associated with feeding certain types of sorghum (Rostagno, Featherston and Rogler, 1973). Possibly of more importance to British compounders is the work of Sharby *et al.* (1972). These workers incorporated experimentally moulded maize into practical rations and fed them to broiler chicks. They used several isolates of *Fusarium* and *Aspergillus* in separate trials. All isolates resulted in high incidences of bent and twisted legs. Experimental birds also had reduced growth rates and produced sticky droppings. At autopsy, such birds had distended caecae, containing dark-brown material.

Wagstaff *et al.* (1968) infected chicks with *Candida albicans*. Those that were thus infected but fed a practical ration remained normal while those that received a semi-purified diet containing all known nutrients at the correct levels developed a high incidence of skeletal defects. Uninfected chicks fed the synthetic ration remained unaffected.

The pathology and aetiology of this type of skeletal disorder has been studied by several workers. The syndrome is not associated with poor mineralization (Dämmrich and Rodenhoff, 1970; Sharby *et al.*, 1972; Ferguson *et al.*, 1974). Seth and Clandinin (1973) found that extra manganese supplementation failed to affect incidence of bone abnormalities induced by rape-seed meal. Ferguson *et al.* (1974) omitted manganese, choline and niacin supplementation from practical-type broiler rations and failed to increase the incidence of bone problems. They even found that their experimental subjects converted food more efficiently on the ration unsupplemented with manganese. Dingle (1977) was unable to influence the incidence of twisted legs in broilers with extra supplements of manganese, choline, niacin, folic acid or biotin although he increased the incidence with low-protein and with high-zinc diets. The only way that Dingle was able to effect a highly significant reduction in incidence of the condition was by massive tryptophan supplementation (0.5–2.0%). Supplementation with 0.3% tryptophan, however, was without effect.

In summary, therefore, it may be stated that the cause of twisted, bowed and bent legs and rotated tibias is obscure or complex. Genetics, environment and food composition all appear to have an influence. Simple nutrient deficiencies, however, cannot be implicated. It should not be assumed that the primary problem necessarily resides in the skeleton. Rapidly growing bones are subject to deformity when abnormal forces are applied to them. Such forces may be the consequence of poor conformation or of nerve or muscle damage or disease. It has been demonstrated experimentally, for example, that surgical transection of the gastrocnemius tendon of one leg will cause the tibia of that leg to twist (Riddell, 1975c). Osbaldiston and Wise (1967) suggested that leg deformities could be the consequence of spinal cord damage resulting from spondylolisthesis. This hypothesis, however, is probably not valid. Riddell (1973) bred a line of chickens with a high incidence of spondylolisthesis but could find no association between this condition and twisted legs.

Chondrodystrophy, enlarged hock disorder, turkey syndrome '65

This group of disorders may superficially resemble those previously described. However, the aetiology is quite distinct. There is a generalized disorder of the growth plates of long bones such that linear growth is impaired while mineralization and appositional growth remain normal. Grossly, therefore, the most rapidly growing bones which are those most severely affected appear shortened and thickened. Furthermore, they frequently become secondarily bowed. The most obvious signs are at the hock joints (*Figure 21.2*), which are apparently enlarged and knobbly. Affected birds frequently become 'cow hocked' or 'bow legged' and sometimes there are slipped gastrocnemius tendons. The description may appear similar to that given for the twisted-leg condition. However, in twisted leg there is no reduction in skeletal length and there are no characteristic histopathological changes in growth plates as there are in chondrodystrophy. It is for the reason that both entities have been described, in the past, under the title of perosis that this term ought no longer to be used.

It is well established that many primary nutritional deficiencies can cause chondrodystrophy. These include manganese (Wilgus, Norris and Heuser, 1936), choline (Jukes, 1940), biotin (Jukes and Bird, 1942), niacin (Briggs *et al.*, 1943),

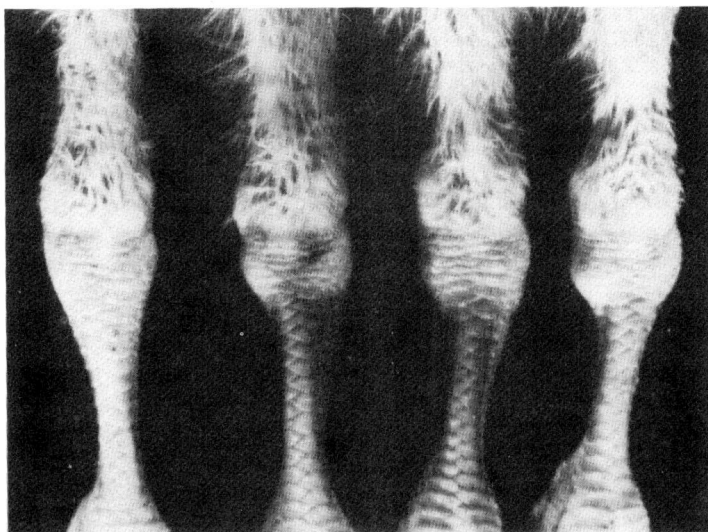

Figure 21.2 Legs of 3-week-old turkeys. That on the left is normal; the other legs are chondrodystrophic

folic acid (Daniel, Farmer and Norris, 1946), zinc (O'Dell and Savage, 1957) and pyridoxine (Gries and Scott, 1972a). However, these simple deficiency states are extremely rare, so much so that D.R. Wise (unpublished) has never seen evidence of any of them in the field.

The only chondrodystrophy of economic importance in the modern poultry industry is that seen in turkeys. Although it is identical in all its pathological manifestations to that which can be produced by, for example, a deficiency of niacin, it does not stem from a simple lack of nutrients although its flock incidence may apparently be affected by the quality of the starter rations fed (Wise, Boldero and Thornton, 1973). The condition develops in young turkeys at 2–4 weeks of age. Severe cases have to be culled but poults with only mild deformities may make apparent recoveries by 8 weeks of age. The development of the syndrome appears usually to be associated with early mycoplasma infection, normally *Mycoplasma meleagridis* infection (Peterson, 1968; Wannop, Butler and Pearson, 1971; Wise, Boldero and Thornton, 1973; Wise, Fuller and Thornton, 1974). Only a proportion of poults hatching with congenital *M. meleagridis* actually develop chondrodystrophy and the mechanism by which this infective agent apparently causes a block to the proper nutrition of the growth plates of long bones remains obscure. Chondrodystrophy is now of declining importance as vertical transmission of *M. meleagridis* is being strictly controlled by eradication or by antibiotic dipping of eggs. Chondrodystrophy in turkeys may not always be the consequence of mycoplasma infection. The condition can be reproduced experimentally in mycoplasma-free poults either with very high brooding temperatures or with implantations of androgenic and of oestrogenic hormones at 1 day of age. The classical condition of turkey syndrome '65, which occurred in many flocks in the UK at high incidences between 1964 and 1966, is characterized by stunting, poor feathering and chondrodystrophy in a proportion of affected birds (Report of Working Party, 1965). The syndrome is now very rarely seen but its cause is probably not solely due to mycoplasmosis and thus remains obscure.

Rickets and osteomalacia

Bones consist of water, organic matter and ash. With increasing age, water is replaced by mineralized organic matter. The degree of mineralization of organic matter also tends to increase slowly with age. In poultry, most of this bone maturation occurs within the first fortnight of life (Wise, 1970a). If bones are deficient in organic matter but are well mineralized, they will be thin and brittle and will break easily. If, however, there is defective mineralization, they will be more rubbery than normal. In growing birds, such a defect is called rickets whereas in adults it is known as osteomalacia. Rickets is a condition which is not uncommonly encountered in turkey flocks but which is rarely diagnosed in other classes of poultry. The majority of birds in affected flocks are normally involved and the commonest age to encounter the problem is between 2 and 3 weeks. If rickets is diagnosed early, treatment with water-soluble vitamin D_3 preparations and a change of feed are normally so effective that there may often be no check in growth or development of gross deformities. Cage layer fatigue is a manifestation of osteomalacia but there is also accompanying osteoporosis (organic matrix deficiency). This condition is now unimportant and will receive no further attention.

Groth (1962) and Groth and Frey (1966) produced experimental rickets in fowl with deficiencies of calcium, phosphorus and vitamin D_3 and described the various pathological changes produced. Calcium deficiency was associated with disorganization of the growth plates of long bones and with replacement of marrow in the metaphyses with fibrous tissue. Phosphorus deficiency, however, caused no disruption of growth plate architecture but merely a delay in calcification of the hypertrophic cartilage. Vitamin D_3 deficiency was reported to resemble calcium deficiency in its pathological effects on bones. This may not be surprising since the main established role of the vitamin is to stimulate calcium-binding protein and hence calcium absorption from the gut. However, unpublished results of D.R. Wise are not in agreement with those of Groth and Frey. Trials with vitamin D_3 deficiency in turkeys suggested that the pathological changes in bones (*Figure 21.3*) were essentially similar to those described for phosphorus deficiency and also resembled those encountered in field cases of rickets.

Field experience suggests that rickets is hardly ever associated with calcium deficiency. Should such a problem arise, flock owners would observe drastic stunting of birds well before leg weakness or skeletal problems became apparent. Undoubtedly, most cases of rickets are due to compounding errors and can be ascribed to vitamin D_3 deficiency or to a lack of available phosphorus. There are, however, occasional cases where problems cannot be explained in this way. Hurwitz, Bar and Meshorer (1973) described severe rickets in turkey poults in Israel in which no food deficiencies could be discovered and which was unresponsive to parenteral vitamin D_3 therapy. There was found to be a lower than normal level of calcium-binding protein in the gut. Huff *et al.* (1977) and Huff, Doerr and Hamilton (1977) studied the effects of aflatoxins and ochratoxins on avian bone, commenting that both fragile and rubbery bones are two problems which some observers have associated empirically with mycotoxicoses. They found that these toxins decrease bone strength and also provided evidence to

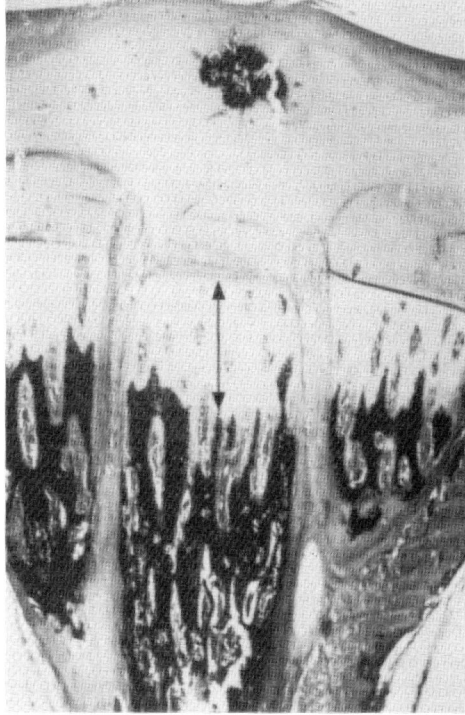

Figure 21.3 Rickets caused by vitamin D_3 deficiency in a 3-week-old turkey. This is a photomicrograph of a section of the proximal end of the tarso-metatarsus stained for calcium with a von Kossa stain. The area between the arrows is abnormal and should be stained black

suggest that they could exacerbate vitamin D_3 deficiency by interfering with mineral absorption.

It is interesting to speculate on the possibility that subclinical or undiagnosed rickets may be quite common and may be associated with later leg weakness. The typical rachitic deformity of birds is an anteroposterior curvature of the tibia. In very severe cases there may also be dorsoventral flattening of the rib cage and kyphosis of the vertebral column in the lumbar region. Enlarged costo-chondral junctions are not a prominent feature of avian rickets although the capitula and tubercula of the ribs may be enlarged. If one examines modern broiler flocks, a high proportion of birds will be found to have severely or slightly curved tibias which resemble those that can be produced experimentally with a mild vitamin D_3 deficiency. Such experimental deficiencies are often self-correcting without dietary change or treatment because of the declining vitamin D_3 requirements as specific growth rate falls. It is possible, therefore, to hypothesize that rickets is common but is normally mild and undiagnosed in broilers. Nevertheless, the curved tibias, which rapidly become properly mineralized but which never straighten could account for locomotor difficulties in later life. However, it should be stated that such deformities, which are now apparently common in broilers, but which were far less frequent a decade ago, may have nothing to do with subclinical rickets and may even be the inevitable consequence of improved growth. Riddell (1975c) found that, when broilers were obliged to carry all their weight on one of their legs, severe curvatures of the tibias of those legs developed.

Tibial dyschondroplasia (focal osteodystrophy)

This condition is extremely common in all meat types of poultry but is rarely the cause of leg weakness. It is characterized by the presence of uncalcified 'plugs' of avascular, hypertrophic cartilage in the proximal metaphyses of the tibiotarsi and, less frequently, the tarsometatarsi (*Figure 21.4*).

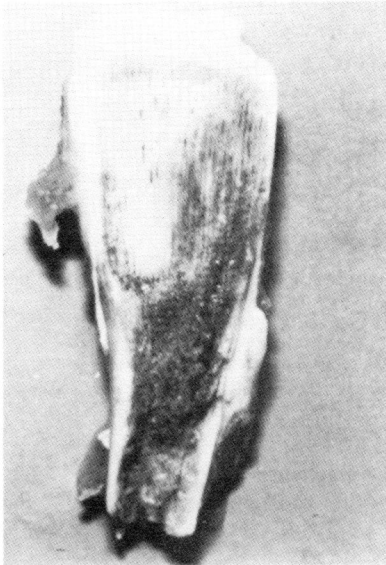

Figure 21.4 Tibial dyschondroplasia in a 6-week-old duck. The affected bird showed no evidence of leg weakness

The aetiology of the deformity is complicated and not fully understood. Leach and Nesheim (1972) and Riddell (1975d) have shown that its development is dependent upon genotype and diet in broiler chickens. A high chloride level in the diet appears to exacerbate the problem. Wise and Jennings (1972) were of the opinion that tibial dyschondroplasia was due to excessive pressure causing failure of the metaphyseal blood supply to the upper tibial growth plate. This excessive pressure was said to result from either a primary skeletal or postural abnormality or from a very rapid early growth rate. Riddell (1975e) was able to reproduce the lesion by surgical interference to the metaphyseal blood supply. Restriction of early growth rate has been shown to reduce the severity and incidence of tibial dyschondroplasia in broilers (Riddell, 1975f) and in ducks (Wise and Knott, 1975). However, Riddell (1975f) was unable to correlate growth rate with the development of tibial dyschondroplasia in individual chickens and considered that rapid growth was no more than a contributing factor. The probable explanation for the localization of dyschondroplastic lesions in the proximal growth plates of the tibiotarsus and tarsometatarsus is that these are the two fastest-growing plates in the skeleton (Church and Johnson, 1964; Riddell, 1975f).

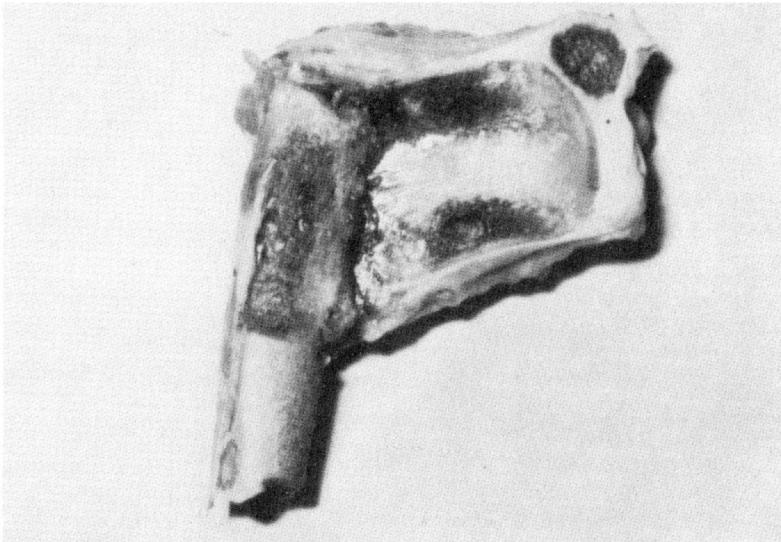

Figure 21.5 Lateral view of proximal end of tibiotarsal bone. There has been a pathological fracture of the top end of the bone resulting from a severe lesion of dyschondroplasia. The detached end has subsequently reattached to the shaft of the bone at near right angles to its original alignment

The only occasions when tibial dyschondroplasia gives rise to locomotor difficulty are when the lesions in the tibia are very severe and the 'plugs' of cartilage bulge outwards into the bone cortex. This may lead to severe bowing or even fracture of the proximal end of the tibia (*Figure 21.5*). The older broiler or roaster flocks of chickens are those most likely to display such problems. Not all such flocks are afflicted, but the clinical incidence may reach 2–3% in a few.

Spondylolisthesis ('kinky back')

This condition occurs only in broiler types of chicken. Spondylolisthesis always occurs at the level of the sixth thoracic vertebra. The cranial end of the body of this vertebra becomes twisted downwards, pressing on the body of the seventh thoracic vertebra and causing it to become wedge-shaped. A degree of deformity may be present in up to 20% of a flock of broiler chickens of certain strains but it does not necessarily cause clinical signs. It is only when the support given to the sixth by the seventh thoracic vertebral facets is lost that spinal cord damage occurs (*Figure 21.6*), leading to severe leg weakness or hind-leg paralysis (*Figure 21.7*). Although a few clinical cases of spondylolisthesis occur between 3 and 6 weeks of age in most broiler flocks, a flock incidence of 1% would be regarded as very high. The pathology of 'kinky back' has been described in detail by Wise (1970b).

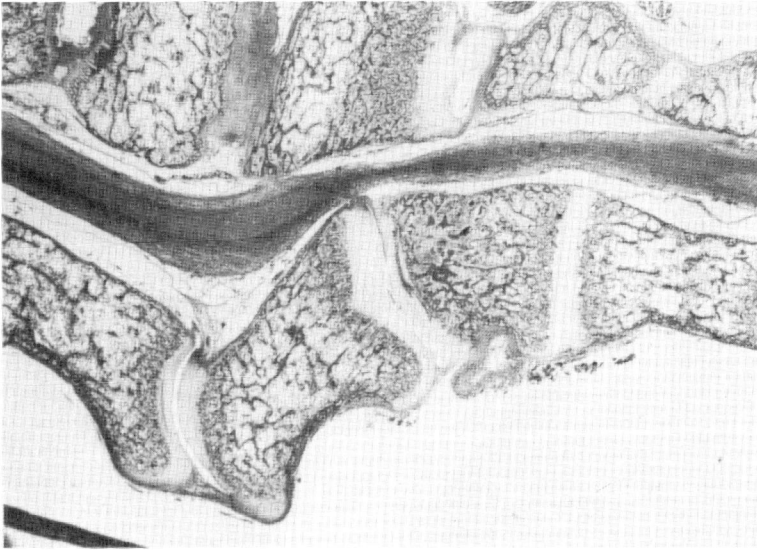

Figure 21.6 Spondylolisthesis in a broiler chicken. Photomicrograph of thoracic 5–7 region of vertebral column. Note damaged spinal cord underlying the joint between the sixth and seventh thoracic vertebral bodies

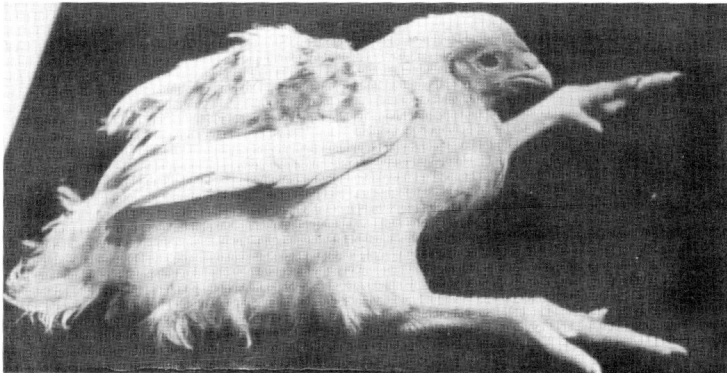

Figure 21.7 Paraplegia caused by spondylolisthesis

Wise (1973), Riddell (1973) and Khan (1976) have all emphasized the importance of genotype in the aetiology of this abnormality. Wise (1973) also
demonstrated that severe restriction in growth rate for the first seven days after
hatching completely prevented the development of spondylolisthesis regardless
of the subsequent rate of growth.

'Kinky back' should be distinguished from the condition of scoliosis in which
there is a lateral, as opposed to vertical, displacement of the vertebral column.
This abnormality, which is also heritable, does not lead to spinal cord damage
and is not a cause of leg weakness.

Aseptic necrosis of the femoral head

Lameness due to degeneration of the head and neck of the femur has been
diagnosed in the USA and Australia and is apparently responsive to dietary
inclusions of between 0.5 and 2.5 ppm of molybdenum as sodium molybdate
(Payne and Bains, 1975). The condition has not been described in the UK but,
from the evidence presented, it would appear sensible to include molybdenum
in broiler and, possibly, turkey supplements as routine.

Laing (1976) reported severe degenerative lesions in the hip joints of turkeys
which became lame between the ages of 8 and 14 weeks. The abnormalities
were most severe in stags and it was suggested that turkeys were anatomically
predisposed to the problem. D.R. Wise and K.N.P. Ranaweer (unpublished)
found that small lesions in the necks of femurs were almost invariably found in
turkeys of both sexes from the age of 4 weeks. They were usually bilateral,
always situated medially and histological examination revealed that there was
localized destruction of bone which was replaced by fibrous tissue. The lesions
generally appeared to become progressively less active between 4 and 17 weeks
of age, and were not associated with leg weakness. On occasions, Wise and
Ranaweera encountered more severe hip changes which in some way resembled
those described by Laing (1976). These were found in stag turkeys of between
17 and 19 weeks of age which were displaying extremely poor posture and
reluctance to walk. However, other turkeys in the same groups and suffering
similar clinical signs did not display severe hip lesions. It was therefore concluded
that these advanced degenerative changes in the hip were more likely to be the
result of rather than the cause of poor deportment.

Osteomyelitis

This condition, alone or coexisting with synovitis, is not uncommon in poultry
whose bones are still active. Sporadic cases or minor flock outbreaks may occur.
The condition is usually of bacterial origin (Nairn, 1973) but Riddell and Topp
(1972) discussed mycotic spondylitis of the first thoracic vertebra of chickens.

There are two distinct sets of clinical signs depending upon whether the leg
bones or vertebral columns are involved. If the former, affected birds become
acutely lame but rarely show evidence of gross skeletal deformity. The signs are
due to one or more small abscesses in the leg bones in the regions of the growth

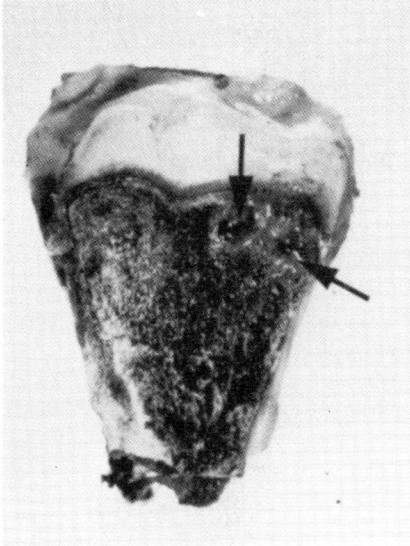

Figure 21.8 Osteomyelitis. Abscesses beneath proximal tibiotarsal growth plate

plates (*Figure 21.8*). Sometimes, but not always, contiguous joints contain pus. In some flocks, minor outbreaks of osteomyelitis confined to the vertebral column occur (Carnaghan, 1966; Wise, 1971). The thoracic 5–7 region of the column is the usual area involved and the consequence is the collapse of one or more vertebral bodies with subsequent spinal cord damage and paraplegia (*Figure 21.9*). The signs are thus identical to those of spondylolisthesis.

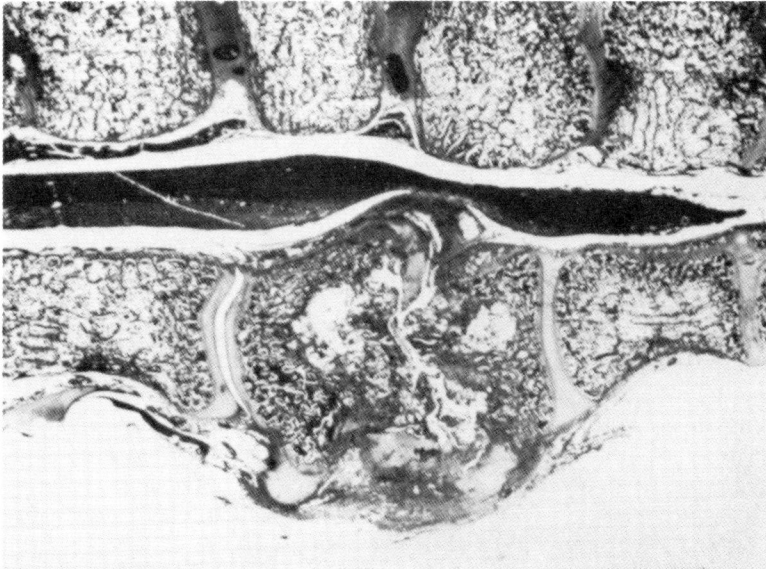

Figure 21.9 Osteomyelitis. Collapse of sixth and seventh thoracic vertebral bodies and spinal cord damage

Synovitis and viral tenosynovitis

Synovitis is characterized by pus in the joints, which causes them initially to be hot, swollen and painful and results in lameness. Sporadic cases or minor flock outbreaks are encountered in all classes of poultry. Synovitis is normally caused by bacteria but mycoplasmas are sometimes involved. If the infection occurs in the foot, the whole foot pad may become swollen giving rise to the condition known as 'bumble foot'.

Viral tenosynovitis (viral arthritis) appears to be confined to meat types of chicken and has emerged as a distinct entity only in the past decade. The condition occurs in the UK but is not common. The causative agent is considered to be a reovirus and up to 50% of birds in affected flocks may show clinical signs of lameness and poor growth. Mortality, however, is usually low. The disease, normally seen from approximately 5 weeks of age, is accompanied by a swelling of the flexor tendon sheaths running along the posterior aspects of the shank bones, giving them a plump, rounded appearance. The sheaths of the gastro-cnemius tendons just above the points of the hocks, are also swollen. At autopsy, the sheaths can be seen to be oedematous and gelatinous. No pus and only small quantities of free fluid can be expressed from the inflamed tissues. There are no gross abnormalities visible in other organs.

Ruptured gastrocnemius tendons (green leg)

Rupture of gastrocnemius tendons used to be seen in 4- to 9-month-old laying and breeding chickens of both sexes. Its maximum incidence appeared to be at the time of full sexual maturity and there appeared to be a hereditary predisposition to the problem (Carnaghan and Hanson, 1958). The problem

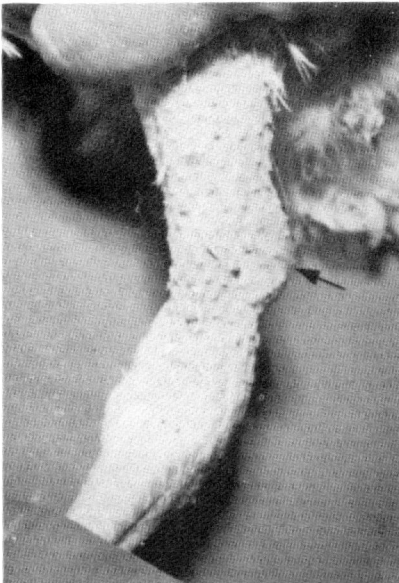

Figure 21.10 Ruptured gastrocnemius tendon

is now seen only in broiler chickens grown on to heavy weights. Incidence increases with age from 9 to 15 weeks and may approach 5% in certain flocks.

The tendon may be completely or partly ruptured immediately above the point of the hock (*Figure 21.10*). The condition may be bilateral or unilateral and leads to lameness. At autopsy, depending upon the severity of the rupture and the length of time that it has been present, one may see accompanying haemorrhage, nodular, fibrous swelling or green discoloration of the affected area. The swelling above the point of the hock should be clearly differentiated from the acquired bursitis (*Figure 21.11*) which is found on the back of the hock

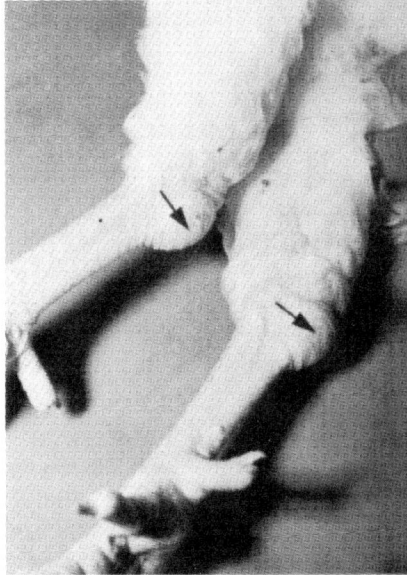

Figure 21.11 Acquired bursae at back of hocks on the pressure points. These are indicative of poor litter conditions or leg weakness

and develops, just as do breast blisters, on the pressure points of the birds during sitting. The condition is most pronounced in heavy birds when litter conditions are bad or when there is a leg weakness problem.

The cause of ruptured tendons is unclear. It has been shown experimentally that one end result of infection with the viral tenosynovitis agent may be the rupture of these tendons. However, it is by no means clear that this is the cause of the field condition, which usually occurs in the absence of the more classical signs of viral tenosynovitis. There is some evidence from the field that the problem can be minimized, as can so many other leg weakness conditions, by lowering the growth rate during the first fortnight of life.

Shaky-leg syndrome in turkeys

It is very common in turkey flocks to encounter a high proportion of birds between 8 and 12 weeks of age which are reluctant to rise and, having risen,

which tremble and quiver on their hocks. Such birds, when driven, appear to lack exercise tolerance and collapse after a short distance. The problem appears much more severe in stags than in hens. The condition may not progress and certain birds or flocks even appear to get better. Others, however, particularly stags of heavy strains, deteriorate and develop a 'tipped forward' posture.
A proportion of such cases subsequently become 'bow legged'. The advanced stages are associated with a deteriorating bird performance. The syndrome is possibly the most serious disease entity faced by the producer of heavy turkeys both in Europe and the USA.

Little is known of the aetiology of the condition and there are few references to it in the literature. Sanger *et al.* (1974) investigated the syndrome and concluded that they had eliminated genetic, nutritional and bacterial causes. They described pathological bone changes which they unconvincingly claimed were indicative of viral infection. K.N.P. Ranaweera and D.R. Wise (unpublished) were unable to find any characteristic and consistent abnormalities in the bones, muscles or nerves of affected turkeys although they were able to mimic the field condition in experimental turkeys by implanting them at 6 weeks of age with androgenic steroids. Often, the field syndrome appears to be immediately preceded by a period of scouring. The cause of the scouring may be viral but, on certain occasions, there is strong circumstantial evidence that it is feed-associated.

J.S. McDougall (personal communication) infected turkeys at 1 day of age with reticuloendotheliosis (RE) virus. Infected birds scoured at approximately 6 weeks of age and subsequently developed severe leg weakness before a proportion died with tumours some six weeks later. It seems clear, however, that RE virus is not a major cause of the shaky-leg syndrome.

Plantar pododermatitis

The condition plantar pododermatitis is characterized by lesions on the ventral foot pads of meat chickens and turkeys. Nairn and Watson (1972) considered that they were the result of some corrosive substance in the litter, possibly arising from the droppings. They stated that the lesions start with necrotizing degeneration of the epithelial cells followed by ulceration and inflammation. A proportion of affected birds show evidence of leg weakness. Schmidt and Luders (1976), however, described an investigation of a similar condition in turkeys. They found that all the stags in a flock had ulcerated feet but none of the hens on the same litter, which was dry, was similarly affected. They suggested that the extra weight of the stags was responsible for a hyperkeratosis which, in turn, led to ulceration of the foot pads.

The condition undoubtedly occurs commonly in the UK, particularly in turkeys and in broilers grown to greater than normal weights. Its contribution to leg weakness, however, is not clear. It is of interest to note that similar ulceration may be found on the pressure points of older broilers – over the keels and at the backs of the hocks – and may lead to severe downgrading. The conformation of the turkey, however, does not predispose it to this problem. While biotin and pantothenic acid deficiencies may lead to dermatitis, it should be noted that it is highly unlikely that such deficiencies can ever be implicated in the type of pododermatitis described above. They may, however, play a role in the aetiology of scabby foot pads, which are very occasionally encountered in turkey poults in the first fortnight of life.

Feed-associated disorders of the nervous system

Vitamin A deficiency results in poor bone remodelling owing to inhibition of osteoclast production. Affected birds become ataxic because of compression of the central nervous system. Wolbach and Hegsted (1952) explained the compression on the basis of retarded bone growth without retardation of soft tissue growth. However, Howell and Thompson (1967) suggested that increased growth of subperiosteal bone was the important fact in central nervous system compression and were able to produce ataxia in vitamin A-deficient fowls in which endochondral bone growth had already ceased. Vitamin A deficiency is an extremely rare cause of leg weakness in the modern poultry industry and it is much more likely that poultry suffer from hypervitaminosis A. Dorr and Balloun (1976) found that a level of 16×10^6 iu of vitamin A per tonne of feed caused marked growth depression and decreased bone strength in turkeys. This level is only some four times the recommended requirements and almost within the range that is included in standard, commercial rations.

Vitamin E deficiency causes encephalomalacia with gross damage to the cerebellum, particularly in the presence of dietary unsaturated fats. Crazy chick disease is still encountered occasionally in young broiler chickens and turkeys. In certain cases, however, there is doubt that a deficiency of vitamin E is the actual cause. It is interesting to note, therefore, that Yoshida and Hoshii (1976)

Figure 21.12 Curled toes in young turkey poisoned with 3-nitro-4-hydroxyphenylarsonic acid (100 ppm). Riboflavin deficiency might well produce identical signs

were able to reproduce all the clinical signs of the disease by feeding dilauryl succinate. These signs could not be prevented or reversed with either vitamin E or selenium and were not associated with increased levels of peroxides or decreased levels of tocopherols in the plasma. However, dietary inclusions of vitamin C or ethoxyquin minimized the dilauryl succinate-induced condition.

Riboflavin deficiency is associated with demyelination of peripheral nerves and consequent locomotor difficulty. The classical disease associated with this deficiency is known as curled-toe paralysis. Gries and Scott (1972b) were able to demonstrate curled toes in experimental chickens with the deficiency. However, Wyatt *et al.* (1973) and Johnson (1976) were unable to produce a high incidence of curled toes in their experimental deficiencies although the former report describes abnormality of posture and gait while the latter mentions complete or partial hind-limb paralysis. Wise, Hartley and Fowler (1974) described a condition similar to curled-toe paralysis, which occurred in turkeys fed 3-nitro-4-hydroxyphenylarsonic acid (*Figure 21.12*). It was found that dietary levels of this substance greater than 50 ppm caused demyelination of peripheral nerves.

Very rarely, birds hatch with nervous disorders which respond to thiamine therapy. Charles, Roland and Edwards (1972) described this condition in turkeys and quail. The problem was prevented by supplementing breeder rations with extra thiamine.

Disorders of the nervous system unassociated with feed

Any infection of the central or peripheral nervous systems of poultry may obviously lead to leg weakness. Such infections include Newcastle disease, Marek's disease, avian encephalomyelitis, turkey meningo-encephalitis and botulism. These diseases are mentioned because they must be considered in the differential diagnosis of leg weakness conditions but their further discussion is not warranted.

References

ANDREWS, L.D., SEAY, R.L., HARRIS, G.C. and NELSON, G.S. (1974). *Poult. Sci.,* **53**, 1141

BERGMANN, V. and PIETSCH, M. (1976). *Mh. VetMed.,* **31**, 581

BOWEN, T.E. and WALDROUP, P.W. (1969). *Poult. Sci.,* **48**, 608

BRIGGS, G.M. and LILLIE, R.J. (1946). *Proc. Soc. exp. Biol. Med.,* **61**, 430

BRIGGS, G.M., LUCKEY, T.D., TEPLEY, L.J., ELVEHJEM, C.A. and HART, E.B. (1943). *J. biol. Chem.,* **148**, 517

BUFFINGTON, D.E., KLEVEN, S.H. and JORDAN, K.A. (1975). *Poult. Sci.,* **54**, 457

CARNAGHAN, R.B.A. (1966). *J. comp. Path. Ther.,* **76**, 9

CARNAGHAN, R.B.A. and HANSON, B.S. (1958). *Br. vet. J.,* **114**, 360

CHARLES, O.W., ROLAND, D.A. and EDWARDS, H.M. (1972). *Poult. Sci.,* **51**, 419

CHURCH, L.E. and JOHNSON, L.C. (1964). *Am. J. Anat.,* **114**, 521

DÄMMRICH, K. and RODENHOFF, G. (1970). *Zentbl. VetMed.,* **17B**, 131

DANIEL, L.J., FARMER, F.A. and NORRIS, L.C. (1946). *J. biol. Chem.,* **163**, 349

DINGLE, J.G. (1977). *Abstracts 6th International Congress of the World Poultry Veterinary Association, Atlanta, Georgia*

DORR, P. and BALLOUN, S.L. (1976). *Br. Poult. Sci.,* **17**, 581

FERGUSON, A.E., SUMMERS, J.D., LESLIE, A.J. and CARLSON, H.C. (1974). *Can. vet. J.,* **15**, 185

GRIES, C.L. and SCOTT, M.L. (1972a). *J. Nutr.,* **102**, 1259

GRIES, C.L. and SCOTT, M.L. (1972b). *J. Nutr.,* **102**, 1269

GROTH, W. (1962). *Zentbl. VetMed.,* **9**, 1009

GROTH, W. and FREY, H. (1966). *Zentbl. VetMed.,* **13A**, 302

HOLMES, W.B. and ROBERTS, R. (1963). *Poult. Sci.,* **42**, 803

HOWELL, J.M. and THOMPSON, J.N. (1967). *Br. J. Nutr.,* **21**, 741

HUFF, W.E., CHAO-FU CHANG, J.D., GARLICH, J.D. and HAMILTON, P.B. (1977). *Poult. Sci.,* **56**, 1724

HUFF, W.E., DOERR, J.A. and HAMILTON, P.B. (1977). *Poult. Sci.,* **56**, 1724

HURWITZ, S., BAR, A. and MESHORER, A. (1973). *Poult. Sci.,* **52**, 1370

JOHNSON, W.D. (1976). *Dissert. Abstr. Int.,* **37B**, 1136

JUKES, T.H. (1940). *J. Nutr.,* **20**, 445

JUKES, T.H. and BIRD, F.H. (1942). *Proc. Soc. exp. Biol. Med.,* **49**, 231

KHAN, M.A. (1976). *Dissert. Abstr. Int.,* **36B**, 5371

LAING, P.W. (1976). *Vet. Rec.,* **99**, 391

LEACH, R.M. and NESHEIM, M.C. (1972). *J. Nutr.,* **102**, 1673

NAIRN, M.E. (1973). *Avian Dis.,* **17**, 504

NAIRN, M.E. and WATSON, A.R.A. (1972). *Aust. vet. J.,* **48**, 645

O'DELL, B.L. and SAVAGE, J.E. (1957). *Fedn Proc. Fedn Am. Socs exp. Biol.,* **16**, 394

OSBALDISTON, G.W. and WISE, D.R. (1967). *Vet. Rec.,* **80**, 320

PAYNE, C.G. and BAINS, B.S. (1975). *Vet. Rec.,* **97**, 436

PETERSON, I.L. (1968). *Poult. Sci.,* **47**, 1708

REECE, F.N., DEATON, J.W., MAY, J.D. and MAY, K.N. (1971). *Poult. Sci.,* **50**, 1786

REPORT OF WORKING PARTY (1965). *Vet. Rec.,* **77**, 1292

RIDDELL, C. (1973). *Avian Path.,* **2**, 295

RIDDELL, C. (1975a). *Vet. Bull., Weybridge,* **45**, 629

RIDDELL, C. (1975b). *Vet. Bull., Weybridge,* **45**, 705

RIDDELL, C. (1975c). *Avian Dis.,* **19**, 497

RIDDELL, C. (1975d). *Avian Dis.,* **19**, 443

RIDDELL, C. (1975e). *Avian Dis.,* **19**, 483

RIDDELL, C. (1975f). *Avian Dis.,* **19**, 490

RIDDELL, C. and TOPP, R. (1972). *Avian Dis.,* **16**, 1118

RODENHOFF, G. and DÄMMRICH, K. (1971). *Zentbl. VetMed.,* **18A**, 297

ROSTAGNO, H.S., FEATHERSTON, W.R. and ROGLER, J.C. (1973). *Poult. Sci.,* **52**, 765

SANGER, V.L., DAHLGREN, B.R., COVER, M.S. and LANGHAM, R.F. (1974). *Avian Dis.,* **18**, 378

SCHMIDT, V. and LUDERS, H. (1976). *Berl. Münch. tierärztl. Wschr.,* **89**, 47

SETH, P.C.C. and CLANDININ, D.R. (1973). *Poult. Sci.,* **52**, 1158

SHARBY, T.F., TEMPLETON, G.E., BEASLEY, J.N. and STEPHENSON, E.L. (1973). *Poult. Sci.,* **52**, 1007

SOMES, R.G. (1969). *J. Hered.,* **60**, 163

WAGSTAFF, R.K., JENSEN, L.S., TRIPATHY, S.B. and KENZY, S.G. (1968). *Avian Dis.,* **12**, 185

WANNOP, C.C., BUTLER, E.J. and PEARSON, A.W. (1971). *Vet. Rec.,* **88**, 30

WILGUS, H.S., NORRIS, L.C. and HEUSER, G.F. (1936). *Science, NY,* **84**, 252

WISE, D.R. (1970a). *Br. Poult. Sci.,* **11**, 333

WISE, D.R. (1970b). *Res. vet. Sci.,* **11**, 447

WISE, D.R. (1971). *Res. vet. Sci.,* **12**, 169

WISE, D.R. (1973). *Res. vet. Sci.,* **14**, 1

WISE, D.R. (1975). *Avian Path.,* **4**, 1

WISE, D.R., BOLDERO, M.K. and THORNTON, G.A. (1973). *Res. vet. Sci.,* **14**, 194

WISE, D.R., FULLER, M.K. and THORNTON, G.A. (1974). *Res. vet. Sci.,* **17**, 236

WISE, D.R., HARTLEY, W.J. and FOWLER, N.G. (1974). *Res. vet. Sci.,* **16**, 336

WISE, D.R. and JENNINGS, A.R. (1972). *Vet. Rec.,* **91**, 285

WISE, D.R. and KNOTT, H. (1975). *Res. vet. Sci.,* **18**, 193

WOLBACH, S.B. and HEGSTED, D.M. (1952). *Archs Path.,* **54**, 13

WYATT, R.D., TUNG, H.T., DONALDSON, W.E. and HAMILTON, P.B. (1973). *Poult. Sci.,* **52**, 237

YOSHIDA, M. and HOSHII, H. (1976). *J. Nutr.,* **106**, 1184

22

PROBLEMS OF DIARRHOEA AND WET LITTER IN MEAT POULTRY

M. PATTISON
Sun Valley Poultry Ltd, Hereford, UK

Normal faeces

Faeces from the digestive tract which mix with urine in the cloaca constitute droppings. Their appearance varies considerably but, typically an individual dropping appears as a rounded brown mass with a characteristic white cap of uric acid (Hill, 1971).

The contents of the caecal tubes are discharged from time to time and appear as discrete masses of dark brown glutinous material. The ratio of caecal to normal droppings varies depending on the composition of the diet. In the healthy bird, enteric faeces are evacuated 12 to 16 times daily and caecal faeces once or twice daily. Although the factors provoking caecal discharge are not fully understood they presumably relate to the quantity of dead bacteria and indigestible material that accumulates in these organs which are essentially 'waste collectors'.

The faecal components of the droppings are conveniently divided into those of dietary origin, i.e. the indigestible residues, and those of metabolic origin. The latter comprise the desquammated cells from the digestive tract mucosa, products of bacterial fermentation, residual digestive secretions and endogenous proteins. Dietary change will affect the output of these various components: for example, bile acid output varies with intake of saturated and unsaturated lipids and endogenous faecal nitrogen is elevated when the protein level of the diet is increased. Short-chain volatile fatty acids are formed by bacterial fermentation and are found in particularly high concentration in caecal faeces.

Faecal components of dietary origin will also vary in concentration. The quantity of fibre depends on the digestibility of crude fibre and similarly starch grains appear in varying concentration in faeces, depending on digestibility.

The consistency of droppings will vary considerably depending on urine output. The urine of the fowl can be collected separately only by surgical procedures. It can be defined as a semifluid product of the kidneys which, on standing, separates into a white precipitate and a supernatant fluid. Eighty-five per cent of the waste nitrogen it contains is in the form of uric acid and the rest in the form of urea, ammonia, amino acids and creatinine. Uric acid is synthesized in the liver and after absorption into the blood stream is excreted by tubular filtration and secretion. In certain situations broilers drink excessively; this increases urine output and as a result the

First published in *Recent Advances in Animal Nutrition – 1987*

droppings have a more fluid consistency. Water usage in different drinker systems will be referred to later.

Diarrhoea

This is the term used for more frequent evacuation of watery faeces. In general, the causes are either of nutritional or infectious origin. Both these causes are influenced by environmental factors, such as ventilation, and design of equipment in the poultry house. The end result is wet litter, which may result in downgrading of the birds in the factory.

NUTRITIONAL CAUSES

Minerals

High sodium, high chloride and high potassium levels all encourage increased water intake. It is important to maintain the correct balance between mineral levels not only to prevent diarrhoea, but also to achieve maximum growth rate. The normal range of mineral levels in broiler diets should be maintained within the following levels:
    Sodium 0.15 – 0.20%
    Potassium < 0.8%
    Chloride 0.12 – 0.15%
If the salt levels rise above 0.4%, increased water intake can be expected.

Fat

Any part of dietary fat which is not digested or absorbed must be excreted in the form of faecal fat. Excess fat tends to make the faeces sticky and the droppings adhere to the foot pads and can form a solid crust on the litter surface. Young chicks are not able to digest fats with a high proportion of saturated fatty acids, especially stearic acid which is present in high proportions in animal fats such as tallow. Consequently starter diets for chicks would normally contain fats of vegetable origin. As the chick gets older, tallow can be introduced in progressively greater quantities. The digestibility of fat is affected by the level of oxidized fatty acids and by the presence of unsaponifiable matter. Bray (1985) described a good fat for poultry as containing 1.9% unsaponifiable matter and 1.2% oxidized fatty acids, compared with a poor fat which contained 14.1% and 9.7% of these constituents, respectively. The overall level of fat is also important and would not normally be more than 9%. A reduction in fat digestibility may affect the digestibility of other nutrients in the diet.

Protein

Soyabean meal is the major source of vegetable protein in broiler and turkey diets. When it is fed in excessive quantities, diarrhoea and litter problems often occur. In general, it is wise to limit its inclusion to no more than 20% of the ration. Soyabean meal is naturally high in potassium, which increases thirst. Urease levels are important to show the nutritive value of the protein and that heat processing has been correct. The presence of complex carbohydrates, which are poorly digested may also cause a problem.

By the nature of their contents meat and bone meal and poultry offal meal may contain protein of poor quality, for example, feathers and collagen. The storage time is important as rancidity develops quickly, making the product less acceptable to the digestive system. Even with poultry offal meal containing antioxidant it should be used within a week of production. This material tends to be very indigestible for the young chick, so it is not usually found in starter diets. Total inclusion should be limited in finisher to no more than 5%. Skim milk powder may also have an adverse effect on droppings.

Carbohydrate

Tapioca can cause droppings problems in poultry if included at high levels (30%), possibly due to the presence of indigestible starches or the presence of toxins.

Barley is not usually present in high levels in broiler diets, but it does have certain carbohydrates which have a low coefficient of digestibility and its use may increase water consumption. Wheat flour in badly formed pellets may encourage birds to drink more water.

Mycotoxins

Finally, it has been shown that the mycotoxins ochratoxin and citrinin cause marked pathological changes in the kidneys, but their effect on water consumption is less clear. These toxins are commonly found in home produced grains, but it seems unlikely that they are often present at high enough levels to cause a problem.

All these ingredients can have an individual effect, but often in commercial diets there is an additive effect between several undesirable ingredients producing wet litter.

INFECTIOUS AGENTS

There are a number of infectious agents, which can cause enteritis in poultry and may lead to diarrhoea.

Bacteria

Escherichia coli infection is very common and generally causes septicaemia with pericarditis and perihepatitis. Diarrhoea may occur as a result of the sick birds drinking more rather than the effect of the *E. coli per se* on the gut mucosa. There is no evidence of the presence of enterotoxigenic *E. coli* which is one of the principal causes of diarrhoea in calves (Snodgrass *et al.*, 1986) and pigs (Taylor,

1986). If drinking water is heavily contaminated with bacteria, water consumption seems to increase.

Campylobacter jejuni is a very common organism in the intestinal tract of broilers aged two weeks or more (Neill, Campbell and Greene, 1984). These authors found that on some farms, isolation of *Campylobacter* appeared to coincide with the sudden appearance of wet litter. However, *Campylobacter* was also isolated on farms with normal litter, so as with other enteropathogenic organisms perhaps the strain of organism is important. This study showed that there were at least five distinct serotypes of *C. jejuni*.

Salmonella isolation in general is not related to the occurrence of wet or damp litter as it does not cause enteritis in poultry. *S. typhimurium* infection can cause disease in young chicks, but not generally in older broilers and turkeys.

There is a recent report (Davelaar *et al.*, 1986) of infectious typhlitis in chickens caused by spirochaetes. This organism was isolated in laying chickens on a farm which also kept pigs. The pathology of the condition resembled a mild form of swine dysentery caused by *Treponema hyodysenteriae*. It is not clear whether this organism could be a problem to broilers or turkeys, but certainly it is not unusual to see broilers at slaughter with enlarged caecae containing very fluid contents. This phenomenon requires further investigation.

Protozoa

Coccidia species *Eimeria* especially *E. necatrix*, *E. brunetti*, *E. acervulina* and *E. tenella* may also cause enteritis and diarrhoea. Generally the control of coccidiosis is good, with the widespread use of ionophor coccidiostats, but sometimes subclinical coccidiosis can be complicated by secondary clostridial infection causing necrotic enteritis.

Viruses

It is possible that several viruses may be implicated as causes of diarrhoea. McFerran *et al.* (1983) carried out longitudinal surveys on 11 farms in Northern Ireland and found many different viruses in the intestinal tract of broilers (*Tables 22.1* and *22.2*).

Rotavirus infection usually appeared after three weeks and four electropherotypes were recognized. Type one showed a group antigen with mammalian isolates. Although rotavirus is an important cause of diarrhoea in calves, it was impossible to say whether it had the same effect in chickens. There was a general background of adeno- and reovirus infection. Also parvo-like and entero-like viruses were found by direct electron microscopy.

The entero-like viruses have now been implicated as a cause of runting in broilers (McNulty *et al.*, 1984). They were found in gut contents of one-week old broilers, which subsequently showed runting.

They have a diameter of 31 nm and grow in the cytoplasm of epithelial cells lining the intestinal tract, especially in the middle region. A crude inoculum of intestine, which also contained reovirus, produced faecal changes, depressed weight gains and caused poor feathering. The faeces became mushy, like worm casts, and the birds congregated around drinkers and became bedraggled. Faeces

Table 22.1 VIRUS ISOLATION AND CLINICAL SIGNS ON ONE FARM IN NORTHERN IRELAND

Age (days)	Direct electron microscopy of faeces	Cell culture	Litter and clinical condition
1	2/6 ELP[a]	None	Excellent
6	7/8 ELP[a]	2/6 Reovirus 2/6 Adenovirus	Birds uneven Litter very wet
13	6/8 Rotavirus	7/7 Reovirus	Some runts Litter drying
20	5/7 Rotavirus	7/7 Reovirus	Many runts Litter very wet
27	2/6 ELP[a] 1/6 Adenovirus	2/5 Reovirus 2/5 Adenovirus	Uneven birds (Runts been culled) Litter drying
34	2/6 ELP[a]	1/6 Reovirus 5/6 Adenovirus	Litter very wet again
41	NVO[b]	4/4 Adenovirus	
48	1/3 Adenovirus	3/4 Reovirus 1/4 Adenovirus	Birds uneven Litter drying
56	NVO[b]	—	Litter good

After McFerran *et al.* (1983)
[a] Enterovirus-like particles. Numerator = number of specimens positive; denominator = number examined
[b] No virus isolated

stuck to the claws and by one week had formed balls on the feet. In these experiments, reovirus did not produce this condition, so it was felt that the enterovirus was the important organism.

Stunting syndrome has been associated with other viruses. Farmer and Taylor (1985) reported isolation of an agent called FEW, which was a virus particle 45–55 nm causing 20–60% mortality in experimental infection and 30% reduction in growth rate. Calicivirus has been isolated by Wyeth, Chettle and Labram (1981) and a Togavirus-like agent by Frazier, Farmer and Martland (1986). It is likely that stunting syndrome is caused by several agents and is not one syndrome, as the lesions can vary considerably. Griffiths and Williams (1985) have described the runting as due to a temporary maldigestion of food over the first four weeks of life. The intestinal tract cannot absorb complex fats or carbohydrates, such as starch, but it can absorb simple sugars such as glucose and xylose. Thus, the amount of starch and fat in the faeces of runted birds is much greater and gives the droppings a yellowish colour and makes them bulky and sticky. Frazier, Farmer and Martland, (1986) found that the pancreatic duct was obstructed in stunted birds. This led to pancreatic atrophy and malabsorption.

Virological causes of diarrhoea in turkeys have shown a similar cross-section of reo-, adeno-, picorna- and rotaviruses (Andral *et al.*, 1985). Rotavirus certainly appears to be significant. Haemorrhagic enteritis virus infection is probably the most important cause of diarrhoea in turkeys and is seen commonly at 6–8 weeks of age. Astrovirus has been shown to cause pale, frothy accumulations in small intestine and caeca (Saif *et al.*, 1985). A similar condition of malabsorption is seen in turkeys as in chickens, and probably mixed virus infections are involved.

Table 22.2 RESULTS OF DIRECT ELECTRON MICROSCOPIC EXAMINATION OF
FAECES FROM A BROILER FLOCK NC 2

Age (days)	*Viruses seen on direct electron microscopic examination of faeces*						*Litter and birds*
	ELPs	*Rota*	*Reo*	*PLPs*	*Calici virus*	*Adenovirus*	
9	10/12[a]	3/12	2/12	—	—	—	Good
16	10/12	7/12	—	2/12	—	—	Damp in patches
22	2/12	1/12	—	3/12	2/12		Wet patches bigger
26	3/12	3/12	—	1/12	—	—	First burn lesions wet
27	2/12	2/12	—	—	—	—	Wet
29	2/12	3/12	—	—	—	—	Bad
33	1/12	—	1/12	3/12	—	—	Bad
34	1/12	—	—	—	—	—	Bad
36	3/12	1/12	—	7/12	—	—	Birds inactive
41	1/12	—	—	3/12	—	3/12	Bad
43	2/12	4/12	—	—	1/12	3/12	Improving
47	5/12	5/12	—	2/12	—	—	Wet
50	1/12	3/12	—	2/12	—	1/12	Wet
54	1/8	—	—	—		1/6	Wet
58	4/6	—	—	—	—	—	Wet
62	2/6	—	—	—	—	—	Moderate
65	3/6	—	—	—	—	—	Moderate
69	—	—	—	—	—	1/6	Quite good
75	—	—	—	—	—	—	Good

After McFerran *et al.* (1983)
[a] Numerator = number of viruses isolated; denominator = number of specimens examined
ELPs = enterovirus-like particles
PLPs = parvovirus-like particles

Water

Water consumption has a critical effect on faecal composition. An increase in consumption leads to increasingly liquid faeces. This in turn raises the moisture content of the litter and reduces its friability.

Water usage is also important, because wasted, spilt water from drinkers has an adverse effect on litter quality.

Water wastage depends on the drinker design and the management of those drinkers. Recent studies (Lynn, 1986) at the Ministry of Agriculture's Experimental Husbandry Farm at Gleadthorpe has shown that water usage with nipple and cup drinkers may be as much as 0.8 litres per bird less then with bell-type drinkers especially those without ballast. A 2.4 kg bird uses approximately 8–9 litres of water to 49 days of age.

Environment

Poultry litter is usually soft wood shavings, but may be a mixture of shavings and straw or even chopped straw on its own. Removal of moisture is obviously more difficult in cold weather. Supplementary heat is often required to keep up a good standard of ventilation. Condensation can occur if moist cold air is directed onto the litter surface. Also drips from cold pipes or uninsulated tanks may amount to a considerable quantity of water. There may also be a capillary effect on earth floors or if concrete floors are without a damp-proof membrane.

Litter often loses its friability suddenly. This may be due to it reaching a saturation threshold beyond which it is unable to absorb moisture. Bray (1985) reported that a moisture level greater than 46% led to a wet cap forming on the litter. The pH of litter increases from 5.2 to 8.2 and the temperature rises, creating an ideal environment for uric acid splitting bacteria, which form ammonia and this is trapped under the cap of the litter.

Hock scabs and breast burns

The modern broiler tends to spend much of its time sitting on the litter. It has been shown (Martland, 1984 and 1985) that wet litter can induce severe ulceration of the skin over the plantar aspect of the feet, the caudal aspect of the hock joints and over the sternum in both chickens and turkeys. Greene, McCracken and Evans (1985) described these as unsightly brown-black erosions and suggested that the condition is really a contact dermatitis (*Figures 22.1* and *22.2*). They are known by producers as hock scabs and breast burns. Histological examination shows the lesions to consist of acute inflammation with necrosis of the epidermis and in more severe cases, the upper dermis. Generally, hock lesions in turkeys are less frequent because the turkey is a more active bird.

These unsightly lesions can result in downgrading of the carcass and are a cause of serious economic loss to the poultry industry. The lesion on the breast may be single or appear as a series of dots associated with the sternal feather tracts. The lesions on the hocks may be seen as early as 20 days, but if litter conditions improve they may disappear, particularly if the birds are thinned out to give them more room.

Processing

Normally during processing, the legs of birds are cut below the hock to ensure that the bird can be properly trussed and to increase carcass yield. The presence and size of hock lesions can have a significant effect on the grading of the carcass. A typical specification for fresh chickens is shown in *Table 22.3*.

Small hock lesions can be removed with a sharp knife or rasp and still qualify for premium grade. The presence of larger lesions may mean the loss of premium grade status and the difference would be 6p/kg. Further downgrading to B grade would entail another loss of 3p/kg.

If a flock has large numbers of birds with hock scabs or breast burns, the upset to factory routine is considerable and the inability to make an order for premium or 'A' grade whole birds may have long-term effects on customer relations, which are difficult to quantify.

The general effect of diarrhoea in a flock is to make the birds' feathers dirty and if this happens near to killing time, the birds will be dirty when they come into the factory. This is undesirable as it will inevitably increase the overall contamination in the plant.

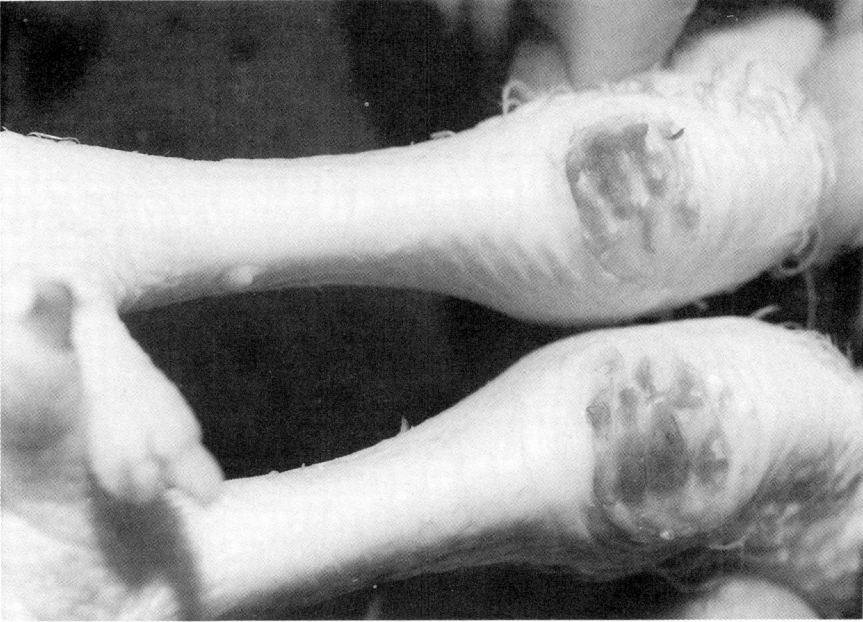

Figure 22.2 Large hock scabs which cannot easily be removed

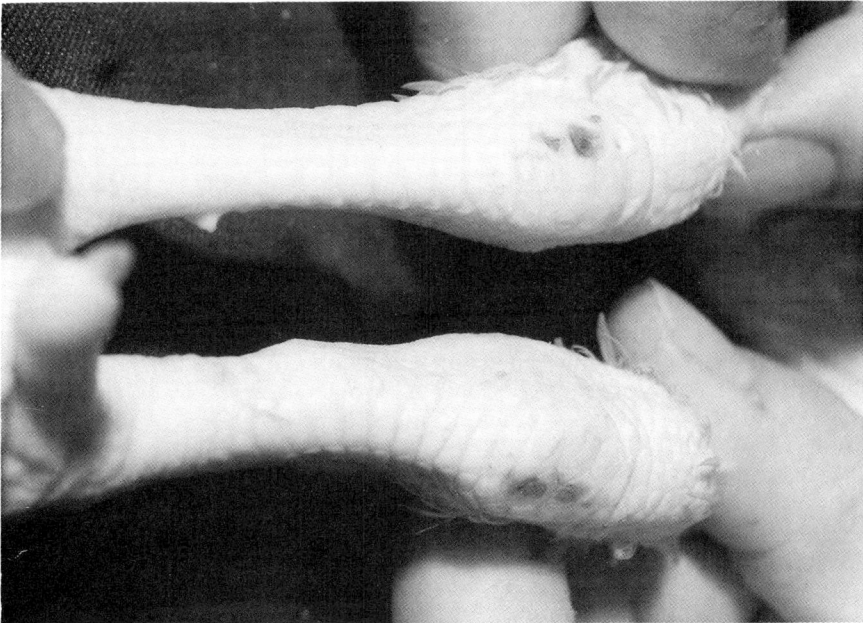

Figure 22.1 Small hock scabs which can easily be removed

Table 22.3 QUALITY CONTROL CRITERIA IN GRADING HOCK SCABS

Premium	'A' Grade	'B' Grade	'C' Grade
Small pinhead size. No more than three per hock — pale red scar where small hock scabs have been removed.	Hock scabs up to $\frac{1}{2}$ cm in size not containing pus. Scars where larger hock scabs have been removed.	Black hock scabs not containing pus. Birds cut to hock thus removing hock scab.	Hock scabs with pus to be removed. Not capable of being cut to hock and removing scab.

Farm

If wet litter occurs through diarrhoea or bad management, fresh shavings will be required to take up the excess moisture and to provide a clean bed for the birds to lie on. Shavings are becoming a more expensive commodity and now cost approximately £2.20 for a 32 kg bale. One bale gives a 7 cm depth of shavings over 5.5 m², so as many as 200 bales could be required to relitter a 1350 m² poultry house. This would cost £440. Also extra labour is required to spread the shavings. If this happens, towards the end of the crop's life, emptying large numbers of bags of shavings can be very stressful to the birds as they have to keep moving about in relatively little space. Great care has to be taken to stop the birds scratching each other as it is easy to cause a panic.

Conclusion

In most cases where litter deteriorates suddenly it is likely that several factors are operating together. There may be an initial dietary or infectious cause for the diarrhoea, which increases the moisture content of the litter. Environmental control of temperature and ventilation is then important in determining whether the litter deteriorates suddenly or gradually becomes friable again.

Continuous monitoring of the presence of hock scabs through a large poultry organization over a two-year period has shown an average incidence of approximately 20%. Males and females appear to be affected equally. The incidence of the condition is not necessarily higher during winter when conditions of environment would be expected to be at their worst. Bad episodes can occur in the summer also. This is in contrast to the findings of McIlroy, Goodall and McMurray (1987) that most cases of acute litter deterioration occurred in the winter months when relative humidity was highest and that males had a higher incidence of lesions than females.

Our own experience shows that good and bad spells for hock scabs go in cycles. During a good spell, the overall average will be about 15%, but a small number of farms will have a much higher level. During a bad spell, the overall average will be about 25% but a greater number of farms will have a figure much higher than this. All these farms are being fed the same ration from the same mill: good and bad results can occur at the same time on the same feed. There may even be considerable differences between houses on the same farm. However, the general cyclical nature of the incidence of hock scabs makes it likely that diet has an influence depending on which raw materials are available at the time. For example, the sudden introduction of new crop wheat may set off outbreaks of scouring. Also, changes in the composition of poultry offal meal or the source of fish or meat and bone meal can have similar effects. However, in most cases it is impossible to carry out laboratory tests in advance of feeding these materials to eliminate the likelihood of causing diarrhoea. McIlroy, Goodall and McMurray (1987) recorded that of three feed manufacturers involved in their survey, one produced significantly fewer cases of litter deterioration than the other two, even though the diets were made to the same nutritional specification. They also noted that the incidence of hock scabs was greater if birds failed to achieve target weights and feed conversions.

Experience shows that it is possible to keep litter dry and friable at all times, but extra gas must be used for brooders to provide heat, if ventilation is to be maintained in cold weather. The cost of this must be set against the losses through downgrading that might occur in the factory. However, birds will perform better on dry, friable litter, so the economic equation is usually in favour of using extra heat, where necessary, to improve ventilation.

It would be expected that increasing stocking density would increase the incidence of poor litter conditions. However, within our own company, stocking density is maintained at a constant level for all farms and the variability in incidence of hock scabs is still enormous.

In summary it can be said that the causes of diarrhoea in poultry are usually multifactorial. If excess moisture cannot be removed from the litter by environmental control, deterioration occurs rapidly. This is likely to lead to lesions of contact dermatitis on the hocks and breasts of birds, with subsequent downgrading in the factory.

References

ANDRAL, B., TOQUIN, D., L'HARIDON, R., JESTIN, A., METZ, M.H. and ROSE, R. (1985). *Avian Pathology*, **14**, 147–162

BRAY, T. (1985). *Poultry World*, 7th March, 9

DAVELAAR, F.G., SMIT, H.F., HOVIND-HOUGEN, K., DWARS, R.M. and VAN DER VALK, P.C. (1986). *Avian Pathology*, **15**, 247–258

FARMER, A.M. and TAYLOR, J. (1985). *Veterinary Record*, **116**, 111

FRAZIER, J.A., FARMER, H. and MARTLAND, M.F. (1986). *Veterinary Record*, **119**, 208–209

GREENE, J.A., MCCRACKEN, R.M. and EVANS, R.T. (1985). *Avian Pathology*, **14**, 23–38

GRIFFITHS, G.L. and WILLIAMS, W. (1985). *Veterinary Record*, **116**, 160–161

HILL, K.J. (1971). In *Physiology and Biochemistry of the Domestic Fowl*, pp. 46–47. Ed. Bell, D.J. and Freeman, B.M. Academic Press, London

LYNN, N.J. (1986). *Poultry International*, April, 20–22

MARTLAND, M.F. (1984). *Avian Pathology*, **13**, 241–252

MARTLAND, M.F. (1985). *Avian Pathology*, **14**, 353–364

MCFERRAN, J.B., MCNULTY, M.S., MCCRACKEN, R.M. and GREENE, J.A. (1983). In *The International Union of Immunological Societies Proceedings No. 66. Disease Prevention and Control in Poultry Production*, pp. 129–138. Ed. Hungerford, T.G. University of Sydney, Sydney

MCILROY, S.G., GOODALL, E.A. and MCMURRAY, C.H. (1987). *Avian Pathology*, **16**, 93–105

MCNULTY, M.S., ALLAN, G.M., CONNOR, T.J., MCFERRAN, J.B. and MCCRACKEN, R.M. (1984). *Avian Pathology*, **13**, 429–439

NEILL, S.D., CAMPBELL, J.N. and GREENE, J.A. (1984). *Avian Pathology*, **13**, 777–785

SAIF, Y.M., REYNOLDS, D.L., SAIF, L.J. and THELL, K.W. (1985). *Turkeys*, July/Aug, 24

SNODGRASS, D.R., TERZOLO, H.R., SHERWOOD, D., CAMPBELL, I., MENZIES, J.D. and SYNGE, B.A. (1986). *Veterinary Record*, **119**, 31–34

TAYLOR, D. (1986). In *Practice Veterinary Record*, March, 40–48

WYETH, P.J., CHETTLE, N.J. and LABRAM, J. (1981). *Veterinary Record*, **110**, 477

23

NUTRIENT REQUIREMENTS OF GAMEBIRDS

J. V. BEER
Game Conservancy Ltd, Fordingbridge, Hampshire, UK

Introduction

The gamebird industry in Britain is substantial and an estimate of the total number of birds hand-reared each summer is 24 million birds (Tapper, personal communication). Four galliform species are involved, the ring-neck or common pheasant, *Phasianus colchicus* which occurs as a variety of races, the red-legged partridge, *Alectoris rufa*, the chukar partridge, *A. chukar* and hybrids, and the grey partridge, *Perdix perdix*. It is estimated that about 86% of those reared are *Phasianus*, 11% *Alectoris* and 3% *Perdix* and the total numbers reared and released has been increasing since the early 1960s (Tapper and Bond, 1987). Indeed there is likely to be an increasing interest in rearing with the current emphasis to take land out of milk or cereal production.

Gamebirds are reared and released by estates into the wild to contribute to winter stocks but rarely direct for the table. During a designated open season, estates and syndicates arrange shooting days, usually on a weekly basis, throughout much of lowland Britain, especially the south east. The harvested birds therefore derive from stock either reared in the wild or hand-reared and progressively adapted to the wild from six to seven weeks of age onwards. Some estates rely entirely on wild-reared birds. A great deal of time, effort and money is spent on establishing a suitable habitat for birds and the character of the British countryside owes much to the production and maintenance of game. Other wildlife and plants also benefit.

Many of the hand-reared birds are produced in quantity by specialized game farms, some of which belong to the Game Farmers Association. Often estates will sell part of their production to offset costs of the shoot. Trading is carried out in a variety of ways involving eggs, day-old chicks, six–eight week old poults, as well as breeders and post-laying-season birds. Custom hatching is fairly common particularly being used by estates who do not want to establish a hatchery, yet who wish to keep control of rearing.

Rearing programmes

Details of many of the possible rearing programmes are covered in a series of Game Conservancy advisory guides (Anon, 1983a, 1983b, 1983c, 1986a, 1986b). A typical

First published in *Recent Advances in Animal Nutrition – 1988*

pheasant rearing programme covering all stages includes collecting free-living adults (mostly first year birds) from December to February followed by penning for egg production at the end of February or in early March. A breeders diet is fed from this time onwards. The first eggs are laid at the end of March or early April and maximum production is reached at the end of the month or in early May. Peak laying continues for about six weeks declining to zero by the end of July. Few estates, however, retain birds this long and most will return them back to the wild in June where they may lay a further clutch to hatch and rear themselves. A hen can produce about five–six eggs/week at maximum and 30–50 eggs, each weighing about 33 g, in a season. Artificial incubation involves anything from 6 to 12 or 13 weekly settings or two to three monthly settings depending on the production required. Incubation requires 24/25 days.

Day-old chicks are reared in various units, often on grass, ranging in size from 100 chicks to many thousands, the latter usually in 500 chick flocks sectioned off in buildings perhaps adapted from unused barns, etc. Heat is required up to about four weeks when the poults are hardened off. At six to eight weeks of age they are ready to be adapted to the wild in a release pen.

During rearing a starter crumb is fed followed by a rearer crumb or small pellet. Poults on grass supplement their diet to a significant degree with natural food and a small amount of kibbled wheat might be added after a few weeks. In the release pen poults further supplement their diet with other greenfood and seeds. Here they will be changed onto a growers diet plus an increasing proportion of cereals especially wheat, and after a few weeks a poult pellet is often introduced. A short time after release the birds learn to fly in and out of the pen, eventually ranging in the neighbouring fields and spinneys without returning to the pen. By the time they are well feathered with an adult plumage (four months) they will be eating natural foods supplemented with cereals. However, it is now commonplace to provide a small amount of a maintenance pellet during the winter, partly to boost trace nutrients as well as to discourage the birds from wandering off the estate. Partridges are reared and released in a similar manner but there are differences in detail. For instance, release is done in much smaller groups.

Game-farm stock is reared in a similar manner but is more likely to be overwintered in large fixed pens; none are released except where a game farmer may be running a shoot for a client.

Gamebird nutrition

Gamebirds are omnivorous and in the wild take a wide range of vegetable and animal foods depending on availability (Dalke, 1937). Animal foods, generally insects, are very important during the first two weeks of life (Hill, 1985). Early attempts at hand-rearing often involved a chicken feed supplemented with such extras as boiled eggs, rabbit, ant eggs, fish meal or milk powder and even custard in Victorian days (Walsingham and Payne Gallwey, 1889). Post-war complete commercial game feeds differed enormously, the protein content varying from 12 to 25% (Anon, 1954).

A survey of available information shows that while some research has been published in detail, other work has been published only as dietary ingredients. In America, Ewing (1963) reviewed the nutrition of gamebirds and Scott (1978) covered 25 years of research into gamebird nutrition carried out by his group. Streib, Streib and Fletcher (1973) and Summers and Leeson (1985) in Canada listed diets and feeding regimes while Leclercq *et al.* (1987) listed the components of diets based on work in France. Apart from the feed components listed by Burdett, Woodward and Wenham (1978) there are few British publications which include gamebirds. The Agricultural Research Council (1975) published data for poultry and gamebirds and a conference on 'Nutrient Requirements of Poultry and Nutritional Research' produced just one comment, on gamebird vitamin requirements (Whitehead, 1986). A list of vitamin requirements is published in the commercial literature (Anon, 1986c).

PROTEIN

It is generally thought that gamebirds are nearer to the turkey than the chicken in their overall nutritional requirements. *Table 23.1* gives a range of crude protein content and dietary energy levels of pheasant and partridge diets. Starter crumbs can be as low as 22–23% crude protein when metabolizable energy (ME) is also low but the consensus is to use 28–30% protein with energy levels near to 12.55 MJ ME/kg. This is fed for two to four weeks followed by a rearer diet with less protein (20–25%) and a lower energy level (about 11.72–12.13 MJ ME/kg) until about five to ten weeks. A growers diet is next fed for several weeks with a gradually increasing proportion of wheat; the protein is normally below 20%. This can be fed as early as six weeks until feathering is virtually complete at four months. A poult pellet with around 15% protein is often fed once the birds are well established in the release pen. Maintenance or winter pellets with 10–15% protein are often fed periodically during the winter. A breeder diet does not need a high protein level and usually has about 20% although some workers suggest diets with crude protein as low as 15.1–12.5%.

AMINO ACIDS

The requirements of gamebirds for amino acids tend to be high (*Table 23.2*) but Leclercq *et al.* (1987) quoted lower figures for partridge than for pheasant. Since reared gamebirds released to the wild must be able to fly and survive, their plumage must be well developed at an early stage for good thermal insulation, water repellency and flying (Scott, 1978; Deschutter and Leeson, 1986). In addition, feather pecking is often a problem and sufficient protein/amino acids must be available for both adequate feather production and replacement—in particular cystine and methionine must be high. Scott, Holm and Reynolds (1963) reported that 26.5% protein diets should be supplemented with 0.1% methionine.

FATS AND FIBRE

The amount of fat present in feeds ranges from 2 to 3.9% during rearing (Summers and Leeson, 1985). Essential fatty acids range from 0.6 to 1.2%. Crude fibre ranges from 3.0% (Summers and Leeson, 1985) to as high as 9% in winter feeds (Wöhlbier, 1974).

Table 23.1 CRUDE PROTEIN (%) AND METABOLIZABLE ENERGY (MJ/kg) IN PHEASANT AND PARTRIDGE DIETS

Diet	Pheasant											Partridge
	1	2	3	4	5	6	7	8	9	10	11	11
Starter												
% Protein	21–30	27.25	27	22–25	23.4	30	26.5–30.0	23.4	30	29.1–29.4	23.1–28.7	17.6–20.4
ME				9.62–10.46	11.38	11.72	12.55	11.38	11.72	11.53–11.89	10.46–12.97	10.88–12.55
Rearer												
% Protein				18–22						23.3–25.5		14.0–16.0
ME				9.62–10.46						12.00–13.36		10.46–12.13
Grower												
% Protein			26				24			18.5–18.7	14.8–17.2	
ME							12.55			12.43–12.66	10.46–12.12	
Poult												
% Protein					19.8	24	20	19.8	24		13.0–15.0	13.0–15.0
ME					11.13	11.97		11.13	12.17		10.46–12.13	10.46–12.13
Winter												
% Protein				13–15			12					
ME				12.13–13.39								
Breeder												
% Protein			20		15.1	19	15.5–20.0	15.1	19	16.9–17.0	12.5–14.5	14.7–17.0
ME					10.75	11.51		10.75		11.59–11.84	10.46–12.13	10.88–12.55

Sources:

1 Norris *et al.* (1936) 2 Scoglund (1940) 3 Stanz (1952) 4 Wöhlbier (1974)
5 Woodard *et al.* (1977) 6 Burdett *et al.* (1978) 7 Scott (1978) 8 Woodard *et al.* (1978)
9 Jee and Wilson (1981) 10 Summers and Leeson (1985) 11 Leclercq *et al.* (1987)

Table 23.2 AMINO ACID CONTENT OF PHEASANT DIETS (%)

Amino acid	Starter			Rearer		Grower		Poult	Breeder		
	1	2	3	2	3	1	2	1	1	2	3
Lysine	1.4	1.75	1.75	1.3	0.8	0.98	0.9	0.7	0.72	0.9	0.8
Methionine	0.5	0.8	0.8	0.5	0.6		0.4	0.27	0.31	0.4	0.4
Methionine + cystine	1.0	1.1		0.5		0.4	0.6	0.54	0.55	0.68	
Tryptophan	0.22	0.42		0.32		0.26	0.15	0.13	0.15	0.27	
Threonine	0.85	1.1		0.9		0.75	0.5	0.41	0.48	0.7	

Sources:
1 Leclercq et al. (1987)
2 Summers and Leeson (1985)
3 Burdett et al. (1978)

ENERGY

The range of energy levels quoted in *Table 23.1* covers 10.46–12.97 MJ ME/kg but the upper part of the range is preferred since high protein diets may be used (Leclercq *et al.*, 1987).

MINERALS

The calcium level needed by the growing bird is 1% and available phosphorus 0.49% (Scott, 1978; Reynnells and Flegal, 1979). The breeder requires 2.8 and 0.34% respectively. The requirements for manganese and zinc are important and the totals should be 95 ppm and 62 ppm respectively (Scott, 1978). Salt should be a maximum of 0.35–0.40% and sodium a minimum of 0.12% (Burdett, Woodward and Wenham, 1978).

VITAMINS

Vitamin requirements for gamebirds tend to be higher than for poultry and are listed by Scott (1978), Summers and Leeson (1985), Burdett, Woodward and Wenham (1978), Leclercq *et al.* (1987) and Anon. (1986c). Where mixes are not specifically available for gamebirds turkey mixes have often been suggested. Because gamebirds are reared and maintained under a variety of conditions it is prudent to use the high levels. For example, the absolute requirement for Vitamin A under ideal conditions for pheasant chicks is much lower than the normal amounts provided in commercial feeds (Scott, 1978).

Scott (1978) discusses the importance of vitamins A, D, K, B_2, niacin and choline for gamebird growth and hatchability. There are indications that high levels of biotin should be used where wheat is present in large amounts (Anon, 1986c). Vitamin C levels may be one factor associated with the fracture of long bones in two and a half week old grey partridge chicks reared indoors on shavings (Beer and Jenkinson, 1982). This condition has not been found in grey partridge reared out-of-doors nor in the pheasant or redleg, however reared.

Growth rates, feed consumption and feed efficiency

It is not possible to give precise body weights because of the varied genetic makeup of gamebirds and conditions of rearing, but a guide for young mixed sexes and adult cocks and hens is given in *Table 23.3* (Beer, 1988). The cock bird is significantly heavier than the hen.

The quantity of crumbs, pellets and grain consumed by hand-reared pheasants is given in *Table 23.4*. The amounts consumed by the red-legged partridge is about one-half to one-third, and the grey partridge one-third to one-quarter of the pheasant. Summers and Leeson (1985) quote feed conversion ratios for pheasants reared in Canada which ranged from 1.71 at two weeks to 4.14 at 18 weeks for cocks, and 1.71 to 5.40 for hens. No figures are available for partridges.

Table 23.3   GUIDE TO MEAN WEIGHTS OF THREE GAMEBIRD SPECIES

Age	Pheasant	Weight (g) Red-legged partridge	Grey partridge
Day old	21	13	9
1 week	45	25	16
2 weeks	85	40	30
3 weeks	150	65	45
4 weeks	200	100	65
5 weeks	280	140	90
6 weeks	380	180	115
7 weeks	450	220	140
8 weeks	550	270	170
9 weeks	650	310	200
10 weeks	720	360	230
12 weeks	870	420	270
17 weeks	1000	490	320
Adult male	1300	540	400
Adult female	1050	460	370

Beer (1988)

Table 23.4   CONSUMPTION OF CRUMBS, PELLETS AND GRAIN BY 100 PHEASANTS

Age (weeks)	1	3	5	7	9	11	13	15	17	19	Breeders
Weekly consumption (kg)	7	18	28	34	41	47	48	45	44	42	56
Total consumption (kg)	7	38	88	153	232	323	422	515	604	690	975 (March-June)

Discussion

The amount of published work on gamebird nutrition is much less than for poultry and since the genetic variation is so large the values for dietary components cannot be exact.

There is increasing evidence that hand-reared pheasants do not survive in the wild as well as wild-reared birds (Hill and Robertson, 1986). Scott (1978) noted that the ability of young pheasants to withstand the stress of cold, drenching rain and resistance to general stress improved by increasing protein from 28 to 34%. Good feathering is needed to combat heat loss but a poorly developed or damaged plumage in hand-reared gamebirds is not uncommon. Deschutter and Leeson (1986) in their review of growth and development of feathers in poultry consider that the levels of methionine and cystine must relate to the growth and rate of production of feathers and their replacement and not just meat. Woodard, Vohra and Snyder (1977) found that feathering was better when the protein level was at least 25% for the first five weeks. Zinc is important in feathering and Scott (1978) indicated that cereal diets for pheasants need to be supplemented to avoid poor development. If the mycotoxin T-2 is present in chick diets it depresses the production of feathers (Wyatt, Hamilton and Burmeister, 1975). The reduction in costs when using lower protein feeds is offset by higher mortality (Woodward, Vohra and Snyder, 1977).

Since released birds undergo various environmental stresses and may not feed adequately despite provision of compound feeds and cereals alongside the natural foodstuffs growing in the pen, losses from these causes and predation may be significant. Thomas (1986) suggested that shortly before release, the energy content of the feeds should be increased to ensure extra internal energy supplies while the birds are changing to the new regime in the wild. Hand-reared red grouse (*Lagopus lagopus scoticus*) are not suitable for release because they show a less well developed gut than a truly wild bird, a feature associated with the feeding of compounded easily digestible feeds, rather than the high fibre, lower digestible heather shoots taken by the wild birds (Moss and Hanssen, 1980). Thomas (1986) fed pheasant chicks an experimental diet containing inert fibrous filler. They grew more slowly than the controls but did eventually reach a normal size and again had larger digestive systems. However, in one small experiment involving the release of more than 300 experimental and control pheasant poults, mortality within two weeks from predation and starvation was high in both groups. He considers that foraging and predator avoidance behaviour should be studied as well as early nutrition and gut development.

The modern tendency to rear intensively various gamebirds as rapidly and to as large a size as possible for release is not likely to contribute to optimum survival. What is needed is a vigorous, lean, hardy bird with a well developed plumage, whose gut is able quickly to make maximum use of natural foods when released to allow the bird to survive stressful situations and to fly well. Feeds should be compounded to help with this aim and also to take account of species differences where relevant.

Acknowledgements

I am most grateful for the help provided by Criddle Peters Feeds Ltd, Heygate and Sons Ltd, Pauls Agriculture Ltd, Roche Products Ltd, Sportsman Game Feeds and Spratt's Game Foods. Also I thank Drs S. Tapper and P. Robertson who commented on the drafts.

References

AGRICULTURAL RESEARCH COUNCIL (1975). *Nutrient Requirement of Farm Livestock, No. 1. Poultry.* Agricultural Research Council, London
ANON (1954). *Annual Report ICI Game Research Station.* Fordingbridge, Hampshire
ANON (1983a). *Pheasant Rearing and Releasing.* Game Conservancy, Fordingbridge, Hampshire
ANON (1983b). *Egg Production and Incubation.* Game Conservancy, Fordingbridge, Hampshire
ANON (1983c). *Red-legged Partridges.* Game Conservancy, Fordingbridge, Hampshire
ANON (1986a). *The Grey Partridge.* Game Conservancy, Fordingbridge, Hampshire
ANON (1986b). *Game in Winter, Feeding and Management.* Game Conservancy, Fordingbridge, Hampshire
ANON (1986c). Roche Vitec 2. Roche, Welwyn Garden City, Hertfordshire
BEER, J.V. (1986). *Annual Review The Game Conservancy,* **17,** 140–143
BEER, J.V. (1988). *Diseases of Gamebirds and Wildfowl.* Game Conservancy, Fordingbridge, Hampshire (in press)

BEER, J.V. and JENKINSON, G. (1982). *Annual Review The Game Conservancy*, **13**, 112–115

BURDETT, B., WOODWARD, P. and WENHAM, T. (1978). *Poultry World*, **131** (43), 17–38

DALKE, P.L. (1937). *Ecology*, **18** (2), 199–213

DESCHUTTER, A. and LEESON, S. (1986). *World Poultry Science Journal*, **42**, 259–267

EWING, W.R. (1963). *Poultry Nutrition*, 5th Edition. The Pay Ewing Company, Pasadena, California

HANSSEN, I., GRAV, H.J., STEEN, J.B. and LYSNES, H. (1979). *Journal of Nutrition*, **109**, 2260–2276

HILL, D. (1985). *Annual Review The Game Conservancy*, **16**, 41–46

HILL, D. and ROBERTSON, P. (1986). *Annual Review The Game Conservancy*, **17**, 76–84

JEE, D. and WILSON, S. (1981). *Poultry World*, **133** (47), 11–30

LECLERCQ, B., BLUM, J.C., SAUVEUR, B. and STEVENS, P. (1987). In *Feeding of Non-Ruminant Livestock*, pp.116–119. Ed. Wiseman, J. Butterworths, London

MOSS, R. and HANSSEN, I. (1980). *Nutrition Abstracts and Reviews—Series B*, **50**, 555–567

NORRIS, L.C., ELMORE, L.J., RINGROSE, R.C. and BUMP, G. (1936). *Poultry Science*, **15**, 454–459

REYNNELLS, R.D. and FLEGAL, C.J. (1979). *Poultry Science*, **58**, 1097–1098

SCOTT, M.L. (1978). *World Pheasant Association Journal*, **3**, 31–45

SCOTT, M.L., HOLM, E.R. and REYNOLDS, R.E. (1963). *Poultry Science*, **42**, 676

SKOGLUND, W.C. (1940). *Bulletin 389*. Pennsylvania State College, School of Agriculture and Experimental Station, Pennsylvania

STANZ, H.E. (1952). *Technical Wildlife Bulletin No. 3*. Wisconsin Conservation Department, Madison

STREIB, A., STREIB, D. and FLETCHER, D.A. (1973). *Pheasants*, Pub. 1514, Canada Department of Agriculture, Ottawa

SUMMERS, J.D. and LEESON, S. (1985). *Poultry Nutrition Handbook*. University of Guelph, Ontario

TAPPER, S. and BOND, P. (1987). *Annual Review The Game Conservancy*, **18**, 167–173

THOMAS, V.G. (1986). *World Pheasant Association Journal*, **11**, 67–75

WALSINGHAM and PAYNE GALLWEY, R. (1889). *Shooting*, 3rd Edition, pp. 246–250. Longmans Green, London

WHITEHEAD, C.C. (1986). In *Nutrient Requirements of Poultry and Nutritional Research*, pp. 173–189. Eds. Fisher, C. and Boorman, K.N. Butterworths, London

WÖHLBIER, W. (1974). *The Supplementary Feeding of Game Animals and Fowl*. Roche, Basle

WOODARD, A.E., ERNST, R.A., VOHRA, P., NELSON, L. and PRICE, F.C. (1978). *Raising Game Birds*, Leaflet 21046. University of California, Berkeley

WOODARD, A.E., VOHRA, P. and SNYDER, R.L. (1977). *Poultry Science*, **56**, 1492–1500

WYATT, R.D., HAMILTON, P.B. and BURMEISTER, H.R. (1975). *Poultry Science*, **54**, 1042–1045

INDEX